Jander/Jahr
Volumetric Analysis

Also of interest

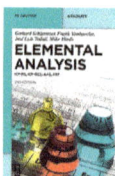

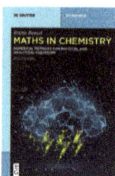

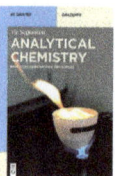

Jander/Jahr

Volumetric Analysis

Titrations with Chemical and Physical Indications

Based on the Translation of the 20th German Edition

Founded by Gerhart Jander and Karl-Friedrich Jahr
Completely Revised by Gerhard Schulze and Jürgen Simon

Continued, Translated and Revised by
Ralf Martens-Menzel, Lena Harwardt
and Hanns-Jürgen Krauss

DE GRUYTER

Authors
Prof. Dr. Ralf Martens-Menzel
FB II, Department of Chemistry
Berlin University of Applied Sciences
Luxemburger Str. 10
13353 Berlin
Germany

Dr. Lena Harwardt
Department of Chemistry
University of Ulm
Albert-Einstein-Allee 11
89081 Ulm
Germany

Dr. Hanns-Jürgen Krauss
Department of Pharmacy
Ludwig-Maximilians-Universität München
House C / Butenandtstr. 5
81377 Munich
Germany

ISBN 978-3-11-134977-0
e-ISBN (PDF) 978-3-11-135012-7
e-ISBN (EPUB) 978-3-11-135029-5

Library of Congress Control Number: 2024948509

Bibliographic information published by the Deutsche Nationalbibliothek
The Deutsche Nationalbibliothek lists this publication in the Deutsche Nationalbibliografie;
detailed bibliographic data are available on the Internet at http://dnb.dnb.de.

© 2025 Walter de Gruyter GmbH, Berlin/Boston, Genthiner Straße 13, 10785 Berlin
Cover image: Matt Turner/iStock/Getty Images Plus
Typesetting: Integra Software Services Pvt. Ltd.

www.degruyter.com
Questions about General Product Safety Regulation:
productsafety@degruyterbrill.com

Foreword to the 20th edition

For 87 years, "Maßanalyse" by Gerhart Jander and Karl Friedrich Jahr has been one of the proven standard works for use in teaching laboratories at universities, colleges and technical colleges. After acquiring an initial basic knowledge of chemistry, this book provides students with an introduction to quantitative analysis and demonstrates ways to determine the content of substances in chemical mixtures precisely and quickly using titrations. In the numerous new editions published since then, the need to systematically present the content in line with current requirements has been taken into account. The practical instructions have always been placed in a theoretical context, so that it has always been clear where the application possibilities of a titration lie and where its limits may be. With the 14th edition in 1986, the text was extensively redesigned and adapted to developments in metrology and instrument technology so that it corresponds to the progress of scientific knowledge and methodology. The terms introduced on the basis of the SI system of units, the German law on units in metrology and the DIN standards, such as amount of substance (Stoffmenge), molar mass (molare Masse), and equivalent (Äquivalent), were adopted in "Maßanalyse". In the 15th edition, a chapter on instrumental volumetric analysis was added, as automation and computer applications have also taken on an important role in quantitative analysis.

The chapter on devices for volume measurement had to be revised, as there were many innovations, but also changed principles for calibration had to be taken into account. Automatic analyzers, as they are used today in industrial laboratories for a fast throughput of serial analyses, are not discussed in detail, as their applications are so focused on specific problems that there is no practical benefit for the students. Nevertheless, the "old" methods of volumetric analysis are by no means losing their importance in the present; on the contrary, they are needed, for example, to produce standard solutions for calibrating automatic analyzers. In the later professional practice of an analyst, for example in an industrial laboratory, the methods used must be validated for quality assurance purposes. Since the 18th edition, several regulations on pharmaceutical and environmental analysis as well as additions to coulometric titrations have been added.

We would like to take this opportunity to thankfully remember Prof. Dr.-Ing. Gerhard Schulze († 2020), who made a considerable contribution to the development of this book. Together with Prof. Dr. Jürgen Simon, he presented the work in its 14th edition in 1986 in a completely revised form and actively contributed to the "Maßanalyse" until the 19th edition in 2017. This work created a very good starting point for the current additions.

We also would like to take this opportunity to thank Mr. Simon, who left the team of authors at his own request, for the valuable work he invested in the editions from 1986 to 2017. His advice on the preparation of the 20th edition is also acknowledged here.

https://doi.org/10.1515/9783111350127-202

In the 20th edition – in addition to further procedures of chemical and pharmaceutical analysis – statistical aspects for the indication of the analysis result, as well as didactic approaches to the topic of titration, have been included.

The book is aimed at students of chemistry as a major and minor subject, biochemistry, food chemistry, pharmacy and environmental technology at universities and colleges, students at technical colleges, pupils at vocational schools, chemical technicians, environmental technicians, chemical-technical and pharmaceutical-technical assistants, and chemical laboratory technicians.

We are grateful for comments on the 19th edition. We would like to thank Ms. Katja Hoffmann for the experimental verification of analytical specifications at the Berlin University of Applied Sciences (Berliner Hochschule für Technik). We would also like to thank the publisher Walter de Gruyter for their constant obligingness and understanding cooperation. We continue to ask the readers and users of our book for suggestions and critical comments.

Berlin, May 2022

Ralf Martens-Menzel
Lena Harwardt
Jürgen Krauss

Contents

1 Introduction and basic concepts

The aim of a quantitative chemical analysis is to answer the question of how much of a substance is contained in a sample. The methods used to achieve this goal can be divided into classical and physical analysis methods.

Classical methods are those in which a mass or volume is determined after a chemical reaction. The quantity of the substance sought is calculated from the measured value using stoichiometric conversion factors, which can be easily derived from the molar masses of the substances involved in the reaction.

With the **physical methods**, on the other hand, a concentration-dependent physical property of the substance in question is measured, and its concentration, mass or amount of the substance is calculated from the value of the measured variable. The respective conversion factor depends on the chemical, physical and instrumental test conditions. It is not derived theoretically, but is determined experimentally by measurements on solutions of known concentration or on substance samples with known contents. As the measurements require special equipment, these analytical methods are also known as **instrumental methods**.

The classical determination methods are essentially based on two analytical principles:
- **gravimetry** or **weight analysis** and
- **titrimetry** or **volumetric analysis**.

1.1 Gravimetry and titrimetry

The analytical principle of gravimetry is based on determining the mass of the product of a precipitation reaction. All gravimetric determination methods are basically characterized by the following procedure: By adding a suitable standard solution to the sample solution, the substance to be determined is converted into a sparingly soluble compound under specified working conditions. The precipitate is separated and weighed after suitable treatment. The following **requirements** must be fulfilled for the successful use of gravimetry:
- The precipitation must be quantitative, i.e. the residue of the substance to be determined remaining in the solution due to the low solubility of the reaction product must be so small that its mass is below the readability of the analytical balance used.
- The precipitate must have a constant and known stoichiometric composition or be convertible by subsequent operations into a compound that fulfills this condition.
- The precipitate must allow an accurate mass determination, i.e. it must not change on the balance.

https://doi.org/10.1515/9783111350127-001

Example: The iron(III) contained in an iron(III) salt solution is to be determined gravimetrically. To do this, the acidic sample solution is heated to the boiling point, and ammonia solution is added drop by drop while stirring until there is an excess. A brown precipitate of iron(III) oxide hydrate, $Fe_2O_3 \cdot x\, H_2O$, is yielded. It is separated from the solution by filtration and freed as far as possible from adhering accompanying substances by washing with hot water containing ammonium nitrate. The separated compound is the **precipitate form**. It fulfills the first of the requirements listed above. However, due to the fluctuating water content and the possible presence of basic salts, the composition is not constant. The precipitate is converted into the **weighing form** by annealing in a porcelain crucible. This weighing form also satisfies the other two requirements. Annealing produces iron(III) oxide, Fe_2O_3, which has a stoichiometric composition and attracts neither water nor carbon dioxide from the air. It can be weighed easily and accurately. From the mass of iron(III) oxide, $m(Fe_2O_3)$, the stoichiometric factor F is used to calculate the mass of the required iron component, $m(Fe)$:

$$m(Fe) = F \cdot m(Fe_2O_3),$$

$$F = \frac{2\, M(Fe)}{M(Fe_2O_3)}.$$

The factor F is obtained from the molar masses of iron, $M(Fe) = 55.845$ g/mol, and iron(III) oxide, $M(Fe_2O_3) = 159.688$ g/mol, to give $F = 0.6994$.

In order to achieve the quantitative separation of the substance to be determined in a gravimetric determination, the respective reagent must be used in excess. This is a characteristic of gravimetric methods.

The analytical principle of titrimetry (volumetric analysis) is based on measuring the volume of a solution of known concentration, which is called the **standard solution**. During the volumetric analysis, just this volume of the standard solution is added to the sample solution that is required for the chemical conversion of the substance to be determined, i.e. the equivalent amount of substance. From the volume of the standard solution required to reach this point (**equivalence point**, theoretical or stoichiometric end point) and its concentration, which must be known exactly, the mass of the substance sought can be calculated if the reaction process is known. The entire process is called **titration**.

Example: The iron(II) contained in a sulfuric acid solution of iron(II) sulfate is to be determined by titration. For this purpose, a potassium permanganate solution is added to the analysis solution. The iron(II) ions are oxidized by permanganate ions to iron(III) ions, while at the same time the violet permanganate ions are reduced to almost colorless manganese(II) ions:

$$5\, Fe^{2+} + MnO_4^{-} + 8\, H^{+} \rightarrow 5\, Fe^{3+} + Mn^{2+} + 4\, H_2O.$$

As soon as the permanganate solution slowly dripping in toward the end of the titration has oxidized practically all the iron(II) ions, the next drop of standard solution is no longer decolored so that the titrated solution appears pink. This means that the end point of the titration has been reached, and the reaction has been quantitatively completed. The potassium permanganate solution is added from a suitable device, a burette, from which the volume used up to the end point of the titration can be read. Based on the reaction equation given above, the desired mass of iron in mg, $m(Fe)$, can be calculated from the volume of the standard solution in ml, $V(M)$, and its concentration according to

$$m(\text{Fe}) = F \cdot V(\text{M}).$$

is calculated. The factor F (conventionally called **analytical equivalent**) in mg/ml results from the molar mass of the iron and the concentration of the permanganate solution, both in terms of equivalents. Both values can be easily calculated from the molar masses of the reacted substances using the equivalent numbers, which can be found in the reaction equation (see p. 41).

Volumetric analysis determinations are linked to the following three conditions:
– The chemical reaction underlying the titration must take place quickly, quantitatively and just as indicated by the reaction equation.
– It must be possible to prepare a standard solution of a defined concentration or to determine the concentration of the solution exactly by suitable means.
– The end point of the titration must be clearly recognizable. It should coincide with the equivalence point at which the reagent quantity is equivalent to the quantity of the substance being sought, or at least come very close to it.

The third requirement usually calls for additional measures, because in only a few cases, the titration end point is recognizable as easily as in the example described, where the standard solution has a strong inherent color that disappears during the reaction. The end point is often made recognizable by the addition of an **indicator**. This is an auxiliary substance that is added to the titration solution and indicates the end of the titration by means of a conspicuous visual change (color change, appearance or disappearance of turbidity or fluorescence).

The point at which the change is recognizable represents the experimental or practical end point of the titration. Ideally, this end point and the equivalence point should be identical. In practice, however, a deviation often occurs, which is called the **titration error** (see p. 98). The error should be kept as small as possible by selecting a suitable indicator.

Another way of finding the titration end point is to use **physical measurement methods**. Their importance has steadily increased, especially in the second half of the twentieth century (see p. 247). As a result, the application possibilities of volumetric analysis have been considerably expanded in many respects.

The use of scales in gravimetry is obvious; however, even volumetric analysis is based on weighing – despite the use of volume measurements. Apart from the fact that the sample is often not available as a solution from the outset, so that a certain portion of the substance has to be weighed and the analysis solution prepared from it, the balance is needed to prepare the standard solutions.

Friedrich Mohr (1806–1879), who rendered outstanding services to the development of volumetric analysis, characterized this peculiarity of titrimetry in the introduction to his classic *Lehrbuch der chemisch-analytischen Titrirmethode* (*sic*) (1855) (i.e. textbook on the chemical-analytical titration method) in the following meaning:

Titration is actually weighing without a balance, and yet all results are understandable in terms of the expression of the balance. In the last instance, everything refers to weighing. However, you only do one weighing where you would otherwise have to do many. The accuracy of one standard weighing is repeated in each test made with the liquid prepared in this way. Several hundred analyses can be made with one liter of sample solution. The preparation of two or more liters of standard solution requires no more time and no more weighing than that of one liter. So, if you have time and leisure, you weigh in advance and use the weighings when you examine.

It is clear from the quotation that the titrimetric determination procedures have the great advantage over gravimetric methods of saving time, which is particularly important when similar analyses have to be carried out frequently.

1.2 Classification of titrations

The titrimetric procedures can be categorized based on various aspects.

Reaction type. A distinction is made between four groups according to the type of chemical reaction that takes place during titrimetric determination:
– **Acid-base titrations**, where protons are transferred from the acid to the base
– **Precipitation titrations**, where a sparingly soluble compound is formed which precipitates
– **Complexation titrations**, where ions combine with ions or molecules to form soluble, less dissociated complexes
– **Redox titrations**, where electrons are transferred from the reducing agent to the oxidizing agent

The processes belonging to the first three groups are based on the combination of ions or ions and molecules; the oxidation numbers of the reactants do not change. The processes in the latter group, on the other hand, are characterized by a change in the oxidation numbers of substances involved.

Endpoint detection. A distinction can be made between
– **chemical indication** of the end point (visual indication) and
– **physical indication** of the end point (instrumental indication).

In the first case, a color indicator is usually added, and the color of which changes at the titration end point. If the standard solution is colored, the color indicator can be omitted (**self-indicating titration**). In a few titrations, the end point can be identified by the formation of a precipitate, such as in the Liebig cyanide determination (see Section 3.2.4).

In the second case, the end point is inferred from the change in a physical variable as measurand. This type of indication has become increasingly important with the development of physical analysis methods. Depending on the type of measurand used to indicate the end point, it can be further subdivided into optical, electrical, thermal and radiometric indication methods.

Titration type. The following classification can be made according to the titration procedure:
1. **Direct titration:** The sample and standard solution are reacted directly with each other.
    1.1 **Direct titration in the narrower sense:** The sample solution is prepared and titrated with the standard solution.
    1.2 **Inverse titration:** A measured volume of the standard solution is prepared and titrated with the sample solution.
2. **Indirect titration:** The substance to be determined is chemically reacted before titration.
    2.1 **Back titration:** A defined volume of standard solution is added in excess to the sample solution. After the reaction has taken place, the unused portion of the standard solution is titrated with a second standard solution.
    2.2 **Indirect titration in the narrower sense:** The substance to be determined is converted into a defined compound in a stoichiometric reaction and this is determined titrimetrically.
    2.3 **Substitution titration:** A substance is added to the sample solution with which the substance to be determined reacts stoichiometrically, releasing a component of the added substance. The released component is then determined by direct titration.

If the titration cannot be carried out directly, either because the reaction is too slow or side reactions occur or the end point cannot be well indicated, the target is often reached indirectly.

This book is divided into Chapters 3 and 4 based on the type of endpoint detection. Chapter 3 is subdivided based on the four reaction types. The type of titration is indicated in the analytical methods.

2 Practical basics of volumetric analysis

In titrimetry, standard solutions are usually prepared by weighing a certain mass of a substance, dissolving and filling to a defined volume. Dissolved samples are generally specified by volume. The standard solutions required to convert the substances to be determined are usually determined volumetrically. The **volume measurement** is therefore of particular importance.

In the **International System of Units** (Système International d'Unités, SI) [40], the **unit of volume** is cubic meter (m^3). Common fractions of this unit are cubic decimeters ($1 \text{ dm}^3 = 10^{-3} \text{ m}^3$), cubic centimeters ($1 \text{ cm}^3 = 10^{-6} \text{ m}^3$) and cubic millimeters ($1 \text{ mm}^3 = 10^{-9} \text{ m}^3$). A special name for cubic decimeters, **liter** (l or L), has been introduced [41–45]. In analytical chemistry, the volume is often specified in subunits of the liter, in milliliters ($1 \text{ ml} = 10^{-3} \text{ l}$) and microliters ($1 \text{ µl} = 10^{-6} \text{ l}$).

In the historical development of measurement, cubic decimeters and liters were not always of the same size. In 1790, when the metric system was to be introduced, a commission of scholars working on behalf of the French National Assembly proposed borrowing the unit of length from nature and linking the unit of mass to it. In 1795, the **basic unit of length** the **meter** was chosen as the basic unit of length and defined as the forty millionth part of the Earth's meridian passing through the Paris Observatory. The unit of **mass was derived** from the unit of length and called the **kilogram**. One kilogram was defined as the mass of 1 dm^3 of pure, air-free water at 4 °C, the temperature of its greatest density, under normal pressure ($1 \text{ atm} = 760 \text{ torr} = 1.01325 \cdot 10^5 \text{ Pa}$). The space occupied by 1 kg of water under these conditions was given the name **liter**.

After the length measurements on the meridian, **prototypes** were made from platinum sponge in the form of rods and cylinders to represent meters and kilograms and were confirmed as base units by the French government in 1799. These were known as the **original meter** (mètre étalon) and the original **kilogram** (kilogramme étalon). They were replaced in 1889 by more perfect prototypes made of a platinum-iridium alloy (90% Pt, 10% Ir). They are kept as international standards at the Bureau International des Poids et Mesures in the Pavillon de Breteuil in Sèvres near Paris. Further prototypes were issued in 1889 to those countries that had joined the **International Meter Convention**, which had been signed 14 years earlier, an intergovernmental treaty on metrology.

After the first platinum prototypes were made, measurements by Bessel 1840 shown that the Earth's meridian was slightly longer than the original definition of the meter had been based on, so the embodiment of the meter was too short by fractions of a millimeter. However, the definition of the meter according to the prototype was retained, but the idea that the unit of length should be a natural measure was abandoned. The situation was similar to the unit of mass. Later investigations proved that the kilogram prototype was 28 mg too heavy and that one kilogram of water did not exactly fill the space of one cubic decimeter, but a slightly larger volume. For many

https://doi.org/10.1515/9783111350127-002

years, the definition based on the unit of mass, the liter, was retained (1 kg of water takes up the volume of 1 l) and it was determined that $1\,l = 1.000028\ dm^3$.

Finally, in 1964, the 12th General Conference on Weights and Measures (Conférence Général des Poids et Mesures, CGPM) decided to discard the previous definition of the liter and to equate cubic decimeters and liters. The unit of volume, the liter, was thus based on length and no longer on mass. Since 1965, the following has therefore applied: $1\,l = 1\ dm^3$. This decision was accompanied by the recommendation not to use the unit liter for measurements of high precision (relative measurement uncertainty $< 5 \cdot 10^{-5}$ or $5 \cdot 10^{-3}$ %). This is intended to avoid possible ambiguities arising from the question of whether liters are meant according to the old or the new definition. However, this recommendation has no effect on the requirements of titrimetry.

2.1 Devices for volume measurement

2.1.1 Measuring devices

The measuring devices used for volumetric analysis are generally made of glass; plastic vessels are also available for special applications. The chemical and physical properties of the glass types used for production are adapted to the special requirements of the laboratory. A distinction is made between chemically resistant glassware that is alkali-earth alkali-silicate-based and shows relatively high expansion coefficients and borosilicate glassware with high chemical resistance and significantly lower expansion coefficients. Glasses in the first group are used for devices subject to less thermal stress. One example is the **AR glass®** (SCHOTT, Mainz) with a coefficient of linear expansion $\alpha = 9.5 \cdot 10^{-6}\ K^{-1}$. Of the borosilicate glasses, the glass **DURAN®** (SCHOTT, Mainz) is applied, with a coefficient of linear expansion $\alpha = 3.3 \cdot 10^{-6}\ K^{-1}$, corresponding to the international standard DIN ISO 3585 [46]. The minimal thermal expansion of DURAN also results in a high thermal shock resistance. The glass is highly resistant to water, neutral and acidic solutions, including concentrated acids and acid mixtures, as well as chlorine, bromine, iodine and organic substances. Water and acids only dissolve predominantly monovalent ions from the glass to a small extent. As a result, a very thin, low-porosity silica gel layer forms on the surface, which inhibits further attack. However, hydrofluoric acid, hot phosphoric acid and alkaline solutions attack the glass to an increasing extent with rising concentration and temperature.

As a wide variety of liquid volumes has to be measured during titrimetric procedures, measuring devices of different sizes and shapes are required. Depending on the intended use, four main types are needed: **volumetric flasks**, **graduated cylinders**, **pipettes** and **burettes**. The vessels bear markings or graduations made of enamel or diffusion colors, which are permanently bonded to the glass surface and are chemically resistant. After machine printing, the inks are baked on by controlled heating to a maximum of 400–550 °C (depending on the type of glass).

There are basically two types of measuring devices: those that have to be calibrated by **pouring in** and those that have to be calibrated by **pouring out**.

In a vessel adjusted for pouring in (adjustment "to contain", "In"), the mark precisely limits the volume to be measured, i.e. after filling up to the mark, a defined volume is contained in the vessel. If the liquid is poured out, a small residue always remains on the wall of the vessel due to wetting. It is therefore not possible to completely remove the measured liquid from the container.

A vessel adjusted for pouring out (adjustment "to deliver", "Ex") allows a defined volume of liquid to be removed. The space between two marks or between a mark and the drain tip of the measuring device exceeds the volume to be measured by just as much as remains on the glass wall and possibly in the drain tip after draining. Since the remaining liquid residue was taken into account when calibrating such a device, it contains a slightly larger volume than the specified nominal volume.

The type of adjustment is indicated on the container. The marking is **In (to contain)** or **Ex (to deliver)**. Measuring vessels made of **amber glass are available** for working with light-sensitive substances. This has a strong light absorption in the short-wave range; the absorption edge is approximately at a wavelength of 500 nm.

Certificate of conformity. In Germany, volumetric measuring devices that are available and used for measurements in commercial transactions and in the medical and pharmaceutical sector (production and testing of medicinal products) must be certified in accordance with the German Weights and Measures Ordinance (Eichordnung) of August 12, 1988. With the conformity mark

$\bowtie$ · = abbreviation of the manufacturer

the manufacturer or the verification authority, if requested, certifies that the device in question meets the requirements of the verification regulations and the relevant standards. Insofar as requirements for different accuracy classes are specified in the standards for volumetric instruments, the specifications of the conformity tests and certificates only apply to class A and AS instruments. The exact procedure for certifying conformity is described in standard DIN 12600 [47].

2.1.1.1 Volumetric flasks

Volumetric flasks are long-necked standing flasks (Fig. 2.1), which are manufactured with a flanged rim, usually with a ground joint and a fitting stopper. The volumetric flasks are available in a pear-shaped or conical design. In volumetric analysis, volumetric flasks are mainly used that are liquidtight sealable with a stopper made of less elastic plastic (e.g. polyethylene). Plastic stoppers are preferred to glass stoppers as they protect the flask from breakage if it falls over. The conical, trapezoidal cross-sectional design is chosen for small-volume volumetric flasks. Due to their larger standing surface, conical flasks stand more securely than pear-shaped flasks.

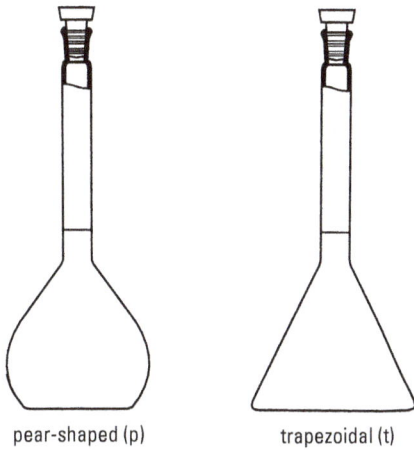

pear-shaped (p)   trapezoidal (t)   Fig. 2.1: Volumetric flask.

The volume is defined by a **ring mark** with a maximum line thickness of 0.4 mm drawn around the neck of the flask. Adjustment is made to the containing (In) for a reference temperature of 20 °C. The nominal volume is contained in the flask when the lowest point of the liquid meniscus and the upper edge of the front and rear parts of the ring mark are in the same plane. To be able to take such a parallax-free reading, the meniscus must be at eye level.

The **sizes of volumetric flasks** available on the market are shown in Tab. 2.1. As the uncertainty in the volume measurement depends on the neck width of the flask, the German Institute for Standardization (DIN) specifies a certain neck width for each flask size. There are also regulations for the minimum diameter of the flask and the base (standing surface). The distance between the ring mark and the neck base, the overall height and the wall thickness are also standardized [48]. The conical sleeve for the stopper is a standard ground joint [49], and the stoppers are manufactured in accordance with standards [50].

Tab. 2.1: Volumetric flasks and their error limits.

Nominal volume	Shape	Error limits			
		Class A		Class B	
		Narrow neck	Wide neck	Narrow neck	Wide neck
(ml)		(±ml)	(±ml)	(±ml)	(±ml)
1	t	0.025			0.05
2	t	0.025		0.05	
5	t, p	0.025	0.04	0.05	0.08
10	t, p	0.025	0.04	0.05	0.08
20	t, p	0.04	0.06	0.08	0.12

Tab. 2.1 (continued)

Nominal volume	Shape	Error limits			
		Class A		Class B	
		Narrow neck (±ml)	Wide neck (±ml)	Narrow neck (±ml)	Wide neck (±ml)
(ml)					
25	t, p	0.04	0.06	0.08	0.12
50	t, p	0.06	0.1	0.12	0.2
100	p	0.10		0.2	
200	p	0.15		0.3	
250	p	0.15		0.3	
500	p	0.25		0.5	
1,000	p	0.4	0.6	0.8	1.2
2,000	p	0.6		1.2	
5,000	p	1.2		2.4	

The error limits apply to water with a reference temperature of 20 °C.

With regard to the error limits, two **accuracy classes** are distinguished: **Class A** with narrow error limits corresponds to the German calibration regulations, for **class B** the limit deviations are twice as large. The error limits in Tab. 2.1 indicate the largest permissible deviations from the nominal volume. In addition to class A measuring vessels, most laboratory glassware manufacturers produce vessels that have smaller tolerances than those prescribed for class B. Their error limits are within 1.5 times according to the deviations of class A.

The **labeling** of the volumetric flask (Fig. 2.2) must be permanently affixed and must contain the following information:
- the numerical value for the nominal content with unit symbol (e.g. 250 ml),
- the code letter of the accuracy class (A or B, followed by the letter W for wide-neck flasks) and the error limit,
- the adjustment symbol In and the reference temperature (20 °C),
- the ground joint size for volumetric flasks with ground joint,
- the material (glass type),
- the name or trademark of the manufacturer and
- the conformity mark for conformity-certified class A volumetric flasks with identification number.

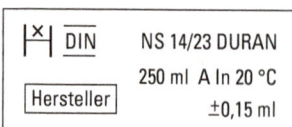

Fig. 2.2: Label of a volumetric flask (in Germany, "Hersteller" means manufacturer).

The previously required indication of the DIN association mark (see Fig. 2.2) is now often replaced by the inscription ISO 1042. It is also a manufacturer's declaration of conformity (requirements of the standard are met).

Volumetric flasks are mainly used to prepare standard solutions of a certain concentration and dilute solutions to a defined volume. The latter is necessary if not the entire sample solution is to be used for the analysis, but only an aliquot.

Plastic volumetric flasks for special applications are commercially available in selected sizes (25, 50, 100, 250, 500, 1,000 ml). The flasks made of polypropylene (PP) are sufficiently translucent for the position of the meniscus on the mark to be recognized. Flasks made of polymethylpentene (PMP) are transparent. Plastic flasks belong to accuracy class B.

2.1.1.2 Graduated cylinders

These measuring devices are upright cylinders with a base and pouring device, which are provided with a volume scale. They are manufactured in two versions: a high form and a low form. If the high form has a ground joint with a PE stopper instead of the pouring device, the measuring device is called a mixing cylinder (Fig. 2.3). The calibration of graduated cylinders is performed for containing (In).

Different volumes can be measured with such a device; however, the permissible error limit is greater. According to [51], two accuracy classes (A and B) are also distinguished for mixing cylinders and other graduated cylinders. Their error limits differ by a factor of 2. Graduated cylinders of the high form and mixing cylinders are available in both classes, whereas graduated cylinders of the low form are only in class B (Tab. 2.2). The error limits are the largest deviations at any point on the scale as well as the largest permissible differences between the deviations at any two points on the scale. Class A cylinders are supplied with a certificate of conformity. In addition to the overall height, the internal height from the base to the top graduation mark, the distance from the top of the scale to the top edge of the cylinder and the volume at the lowest graduation mark are standardized. For class A cylinders, the graduation must correspond to type II (main point ring graduation) and for class B cylinders to type III (line graduation) of standard ISO 384 [52]. The meniscus must be set so that its lowest point just touches the upper edge of the graduation line when observed without parallax.

The size of the graduated cylinder is selected according to the volume to be measured. Large graduated cylinders should not be used for small volumes because the reading uncertainty increases with the diameter of the cylinder. The graduated cylinder should therefore be filled to at least ⅕ of its nominal volume so that the relative uncertainty does not become too large.

Graduated cylinders made of plastic [53] can be used for filling and pouring out due to the lack of wetting. Graduated cylinders made of polymethylpentene (PMP) with main point ring graduation are available with conformity certification in accordance with the German calibration regulations.

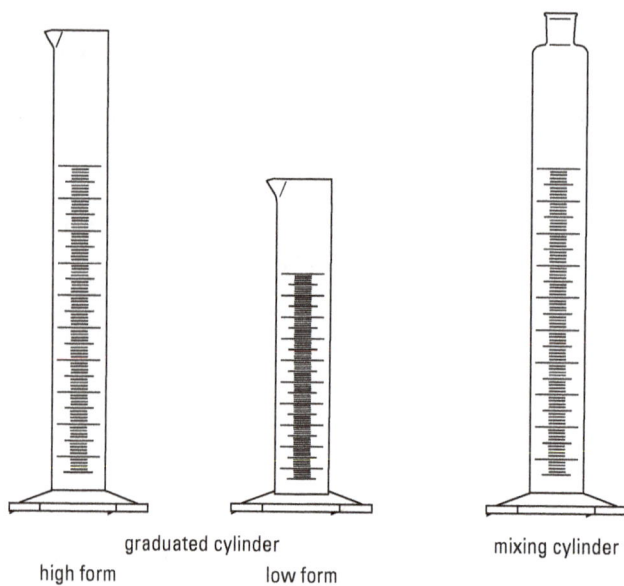

graduated cylinder   mixing cylinder

high form   low form

Fig. 2.3: Graduated cylinder.

Tab. 2.2a: Graduated cylinder: high form and mixing cylinder.

Nominal capacity ml	Subdivision ml	Error limits ± ml	
		Class A	Class B
5	0.1	0.05	0.1
10	0.2	0.1	0.2
25	0.5	0.25	0.5
50	1	0.5	1
100	1	0.5	1
250	2	1	2
500	5	2.5	5
1000	10	5	10
2000	20	10	20

The error limits apply to water with a reference temperature of 20 °C.

Tab. 2.2b: Graduated cylinder: low form.

Nominal capacity ml	Subdivision ml	Error limits ± ml
5	0.5	0.2
10	1	0.3
25	1	0.5
50	1 or 2	1
100	2	1
250	5	2
500	10	5
1000	20	10
2000	50	20

The error limits apply to water with a reference temperature of 20 °C.

2.1.1.3 Pipettes

Pipettes are glass suction tubes that end in a tip at the bottom. Their volume is marked with one or more marks so that a defined volume of liquid can be measured by drawing it up into the pipette and letting it flow out. The pipette is calibrated to deliver (Ex). A distinction is made between **volumetric pipettes** and **graduated pipettes** (Fig. 2.4).

Volumetric pipettes are glass tubes that are cylindrically widened in the middle and have a **ring mark** on the upper or both the upper and lower parts of the tube. They allow the measurement of a specific volume, which is indicated on the cylindrical extension.

Tab. 2.3a: Volumetric pipettes according to DIN, their error limits and delivery times.

Nominal volume (ml)	Error limits		Delivery time		
	Class A and AS (±ml)	Class B (±ml)	Class A without waiting time (s)	Class AS with waiting time (s)	Class B without waiting time (s)
0.5	0.005	0.010	10–20	6–10	4–20
1	0.008	0.015	10–20	7–11	5–20
2	0.010	0.02	10–25	7–11	5–25
5	0.015	0.03	15–30	9–13	7–30
10	0.02	0.04	15–40	11–15	8–40
20	0.03	0.06	25–50	12–16	9–50
25	0.03	0.06	25–50	15–20	10–50
50	0.05	0.10	30–60	20–25	13–60
100	0.08	0.15	40–60	25–30	25–60

Tab. 2.3b: Other class AS volumetric pipettes according to the manufacturer's specifications.

Nominal volume (ml)	Error limits (±ml)	Delivery times (s)
2.5	0.010	7–11
3	0.010	7–11
4	0.015	7–11
6	0.015	8–12
7	0.015	8–12
8	0.020	8–12
9	0.020	8–12
15	0.030	9–13
30	0.030	13–18
40	0.050	13–18

Delivery times and error limits apply to water with a reference temperature of 20 °C.

The commercially available **sizes of volumetric pipettes** are listed in Tab. 2.3a and b. To carry out the volume measurements as reproducibly as possible, the German and international standards specify guidelines on dimensions and handling [54]. These refer to the diameter of the tube and the cylindrical body, its length as well as the length of the upper suction and lower discharge tube, the total length of the pipette, the position of the ring mark as well as the discharge tip and delivery time. The classification was again divided into two **accuracy classes.** For **class A** pipettes, the drainage time for the liquid is extended by narrowing the tip opening so that the last amount of liquid already runs down the pipette wall during drainage. Their error limits comply with the German calibration regulations. **Class B** pipettes have considerably shorter draining times and error limits that are twice as large. **No waiting times** need to be observed for pipettes of either class after expiry. Class B pipettes are also available with error limits within 1.5 times the values for class A, i.e. with smaller tolerances than those required for pipettes in this class.

The use of automatic production machines and extensive automation in the calibration and graduation technology made it possible to manufacture pipettes that are within the error limits of the German calibration regulations (class A) and yet have short cycle times [55]. The designation **class AS** (accuracy class A, fast flow S) was introduced for them. This designation has been included in the German Weights and Measures Ordinance (Eichordnung) and in the regulations of the German Institute for Standardization (Deutsches Institut für Normung) [54]. As their error limits correspond to those of class A, these devices are conformity-certified. They require a **waiting time of 15 s** after expiry; for AS pipettes produced from 2007 onward, the waiting time is only 5 s. In practice, class AS pipettes have become established and have replaced class A pipettes with longer expiry times.

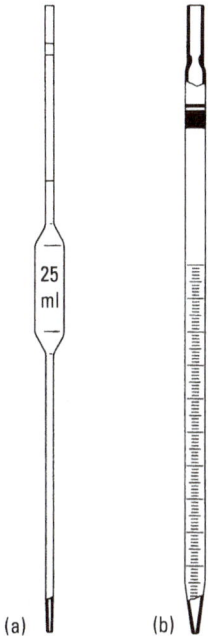

(a)   (b)   Fig. 2.4: Volumetric pipette (a) and graduated pipette (b).

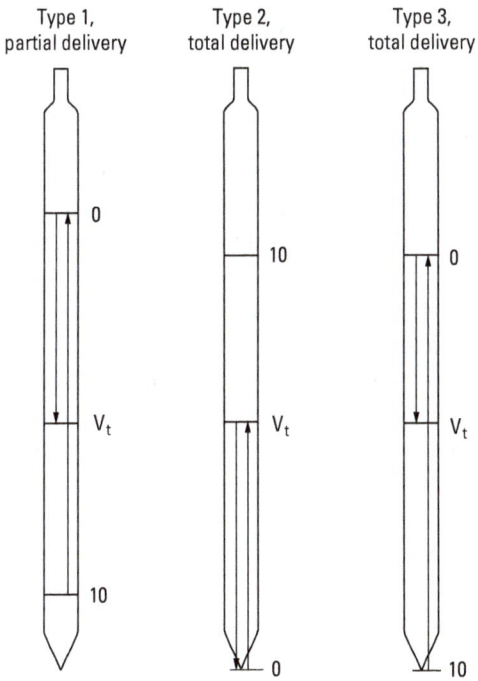

Fig. 2.5: Types of graduated pipettes.

Volumetric pipettes made of PP are translucent and unbreakable (nominal volumes 1/2/5/10/25/50 ml). Their error limits correspond to class B values.

In addition to volumetric pipettes with a ring mark in the suction tube, there are also pipettes available that have a **second ring mark** in the discharge tube [54]. The ring marks limit the volume indicated on the pipette. These devices are therefore not intended for complete drainage (nominal volumes 0.5/1/2/3/5/10/15/20/25/50 ml). The waiting time is 15 or 5 s, error limits and drain times are within the values specified for class AS pipettes. **Suction piston pipettes** have a cylinder with a glass piston firmly connected to the pipette at the aspiration end. They are available as volumetric pipettes with 1/2/5/10/20/25 ml capacity (for piston-operated pipettes, see p. 22 ff.).

Graduated pipettes are cylindrical, calibrated glass tubes that allow any volume of liquid to be measured within their nominal volume (Fig. 2.4). They are mainly used for measuring small volumes, even parts of whole milliliters.

They are also manufactured in the aforementioned accuracy classes **Class A** or **AS** and **Class B**. In classes A and AS, the error limits correspond to the regulations of the German Weights and Measures Ordinance, while those of class B are greater (see Tab. 2.4). Class A and B graduated pipettes are calibrated without waiting time, for class AS pipettes the waiting time is 15 or 5 s (depending on the time of manufacture, see label on the pipette).

With regard to the design, four different types of graduated pipettes have been defined [56], which differ in the scale arrangement (see Fig. 2.5): **Type 1** is the design for partial dispensing of the desired volume. The zero mark is located at the upper end of the pipette scale; the nominal volume is indicated by the lowest graduation mark, which is not identical to the pipette tip. The liquid is dispensed from the top to the desired graduation mark, the residual liquid remains in the pipette. **Type 2** is designed for complete liquid dispensing. The zero line is at the bottom, and the mark for the nominal volume is at the top. The liquid is aspirated up to the desired volume mark on the scale and allowed to drain completely. **Type 3** is also used to completely dispense the selected volume. Here, however, the zero line is at the top and the pipette tip represents the nominal volume. The pipette is adjusted so that you aspirate up to the top, set the meniscus to the zero mark, remove the liquid by draining to the desired mark and discard the rest. Finally, **type 4** is designed as a blow-out pipette for complete dispensing. The last drop of liquid in the tip is ejected by blowing it out. Blow-out pipettes only exist in accuracy class B, the other three types are manufactured in classes A, AS and B.

Table 2.4 provides an overview of the commercially available values, the scale divisions and the error limits. The latter apply to each point on the scale and also indicate the largest permissible difference between the deviations of any two points. The international standard DIN EN ISO 835 contains specifications on the diameter and total length of pipettes, on the length, arrangement and numbering of the scale and on the properties of the pipette tip. The expiry times for the four types are specified in an annex. The newly included type 2 has the advantage that the meniscus setting only needs to be made once during the pipetting process.

Tab. 2.4: Graduated pipettes and their error limits.

Nominal volume (ml)	Subdivision (ml)	Error limits	
		Class A and AS (±ml)	Class B (±ml)
0.1	0.01	0.006	0.01
0.2	0.01	0.006	0.01
0.5	0.01	0.006	0.01
1	0.01	0.007	0.01
1	0.10	0.007	0.01
2	0.02	0.010	0.02
2	0.10	0.010	0.02
5	0.05	0.030	0.05
5	0.10	0.030	
10	0.1	0.05	
20	0.1	0.1	
25	0.1	0.1	
25	0.2	0.1	

The error limits apply to water with a reference temperature of 20 °C.

The pipettes in the dimensions 0.1:0.001 (nominal volume in ml:subdivision in ml), 0.2:0.001 and 0.2:0.002 (rows 1–3 in Tab. 2.4) are available for complete draining and in adjusted in class AS (limit deviations ±1/±2/±2 µl) and in class B (±1.5/±3/±3 µl). They are emptied by rinsing.

They are also commercially available as graduated pipettes:

Blow-out pipettes in the dimensions 1:0.01/2:0.01/2:0.002/5:0.05/5:0.1/10:0.1/20:0.1/25:0.1 ml, which are divided up to the tip and correspond to class AS according to volume deviations and expiry times; the waiting time is 2 s; they are then blown out once briefly.

Suction piston pipettes for 1:0.01/2:0.02/5:0.05/10:0.1/20:0.1/25:0.1/50:0.2 ml capacity; **graduated pipettes made of plastic** (PP) without wetting for 1/2/5/10 ml nominal volume with 0.1 ml subdivision. The tolerances correspond to those of class B.

The **labels** on the pipettes contain information that has already been listed for the volumetric flasks (Fig. 2.6). The calibration mark on the pipettes is Ex. For graduated pipettes, the subdivision (scale value) should also be noted in addition to the nominal volume, e.g. 10:0.1 for a 10 ml pipette with 0.1 ml subdivision.

In addition, a **color coding** (color stripes) can also be applied. The color is internationally agreed upon for each pipette type (**color code**) and defined by a standard [57], but its application is not mandatory. Volumetric pipettes and graduated pipettes for complete drainage are marked with one or two stripes of the same width, while graduated pipettes for partial drainage often have an additional narrow stripe (see Tab. 2.5).

Measuring a volume of liquid with a pipette is carried out as follows: The pipette is immersed with its pointed end deep enough into the liquid to be measured and the liquid is drawn up to approximately 5 mm above the ring mark (zero mark). Then remove the pipette, wipe off the liquid adhering to the outside with a paper

1. Color-code
2. Manufacturers name or logo
3. Nominal volume (e.g. 25), for graduated pipettes incl. subdivision (e.g. 10:0.1)
4. Error limit (e.g. ±0.03)
5. Volume unit
6. Sign for certificate of conformity (⋈)
7. Designation of the standard
8. Reference temperature (20 °C), calibration (TD, Ex), waiting time (e.g. 15 sec.), class (e.g. AS)
9. Country of origin, further information (year of production, charge)

Fig. 2.6: Pipette label.

Tab. 2.5: Color code system (extract).

Volumetric pipettes		Graduated pipettes		
Nominal volume (ml)	Color code	Nominal volume (ml)	Subdivision (ml)	Color code
0.5	2× black	0.1	0.001	2× green
1	Blue	0.2	0.001	2× blue
2	Orange	0.2	0.002	2× white
3	Black	0.5	0.01	2× yellow
4	2× red	1	0.01	Yellow
5	White	1	0.1	Red
6	2× orange	2	0.01	2× white
7	2× green	2	0.02	Black
8	Blue	2	0.1	Green
9	Black	5	0.05	Red
10	Red	5	0.1	Blue
15	Green	10	0.1	Orange
20	Yellow	20	0.1	2× yellow
25	Blue	25	0.1	White
30	Black	25	0.2	Green
40	White			
50	Red			
100	Yellow			

towel, place the pipette vertically against the wall of a beaker held at an angle and lower the meniscus until its lowest point touches the upper edge of the ring mark and lies in the same plane as its front and rear parts (parallax-free reading). After the tip of the pipette has been wiped cautiously against the tube wall, the pipette has to be emptied. It is held vertically over the inclined receiving vessel again, with its tip rest-ing against the vessel wall (Fig. 2.7). The liquid is now allowed to drain completely or up to the desired mark and the pipette tip is wiped against the vessel wall. For pipettes of class AS, a waiting time of 5 or 15 s must be observed between draining and wiping. The pipette must not be blown out. The remaining liquid in the tip has been taken into account during adjustment.

A prerequisite for the correct measurement of the volume is that the measuring conditions specified during calibration are also adhered to during use. This means that the flow and waiting times must be the same, as must the temperature of the measuring liquid. The pipette must also be handled in the same way. The expiry time is determined by the inner diameter of the pipette tip, and it changes if the tip is damaged. To be able to check its integrity, a 1–2 mm wide ring mark in diffusion color can be attached to the tip. The waiting time depends on the pipette class. If the measuring temperature deviates from the reference temperature (20 °C), both the volume of the measuring device and that of the liquid will change. While the volume expansion of the glass is very small and generally does not need to be taken into account (see p. 8), the expansion of the liquid is considerably greater (see p. 50), so the measurement result can be distorted. When handling the pipette, it is important that it is held vertically and rests against the wall of the vessel. With normally adjusted graduated pipettes, several volumes must not be measured in succession from one filling. If a pipette is used for the first time to measure a specific liquid, it must be prerinsed twice with this liquid.

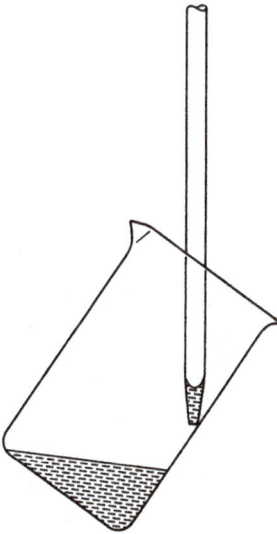

Fig. 2.7: Correct pipette position for delivery.

The previously common technique of drawing up the liquid into the pipette using the mouth, closing the aspiration opening with the moistened index finger and lowering the meniscus by carefully lifting the finger is now prohibited by occupational health and safety regulations [58]. This ban is intended to prevent vapors from aggressive liquids produced during the pipetting process from damaging the mucous membranes of the respiratory tract and from corrosive or toxic substances entering the mouth. This risk is particularly high if liquid residues are to be removed from containers so that a visual inspection of the pipette tip is not possible when aspirating with the mouth. If the liquid level drops below the pipette tip, air will flow into the pipette. The suction resistance suddenly decreases and the liquid shoots into the oral cavity. There is also a high risk of infection in clinical laboratories. For fundamental hygienic reasons, a general ban has been imposed.

Pipetting aids must therefore be used to draw up liquids. These devices are commercially available in a wide variety. They are attached to the suction tube of the pipette and can be operated with one hand. Of course, you have to get used to the devices. Some examples of pipetting aids are described below. The **pipette ball** made of rubber (e.g. **Peleus ball**, safety pipetting device according to Pels Leusden, D. B. P. No. 897930) is equipped with three valves (Fig. 2.8). After opening the air outlet valve A with thumb and forefinger, the rubber ball is compressed. The suction valve S (suction) is then opened until the desired liquid level is reached in the pipette. The pipette is then aerated and emptied using valve E (emptying). The **Pumpett** also works with a rubber bulb (B. Braun, Melsungen) (Fig. 2.8).

The **pi-pump**® (Glasfirn, Giessen; Roth, Karlsruhe) device is based on the piston pump principle. In a plastic housing, a piston is attached to a toothed rack which can be moved up or down with the thumb via a drive wheel. A stepped insertion cone made of soft plastic ensures a secure fit on the pipette, while a drain valve allows the liquid to drain freely. Three sizes are available (for pipettes up to 2 ml, up to 10 ml and up to 25 ml). Another model for pipettes up to 0.2 ml is manufactured without a drain valve (Fig. 2.9). The **macropipette controller** (BRAND, Wertheim) can be used for all volumetric and graduated pipettes from 0.1 to 100 ml (Fig. 2.10). The valve system allows easy compression of the suction bellow. The drawing up and dispensing of liquids can be sensitively controlled with a small lever. The pipetting lever is moved upward for drawing up. The further it is pushed upward, the stronger the suction force. Up to 50 ml can be drawn up in one go in 12 s. The meniscus is easily adjustable. If the pipetting lever is moved downward, the liquid drains off. When using blow-out pipettes, the rubber bladder has to be pressed additionally. A hydrophobic membrane filter (3 µm pore size) protects the system against penetrating liquids. As the liquid vapors only come into contact with silicone, PP and PTFE, very good chemical resistance is guaranteed. The entire device is resistant to steam sterilization (121 °C). For small pipettes (up to 1 ml), the **micropipetting aid** (BRAND, Wertheim) has been developed. An ejector is integrated into the device, with which disposable micropipettes can be ejected.

Electrically operated pipetting aids are suitable for pipetting larger series (battery or mains operation). The **accu-jet**, pipetting aid (BRAND, Wertheim) with a lithium battery, can be used for all volumetric and graduated pipettes from 0.1 to 100 ml (Fig. 2.11). All pipetting processes can be continuously controlled with two function buttons: slow

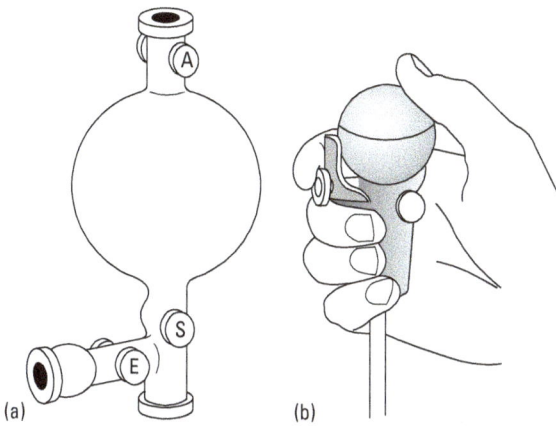

Fig. 2.8: Pipetting aids with rubber ball: (a) Peleus ball and (b) Pumpett.

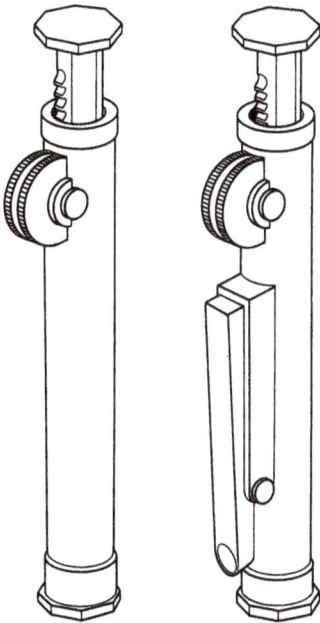

Fig. 2.9: Pipetting aids pi-pump with and without drain valve.

and fast drawing up or dispensing by running out or blowing out, the latter with motor support, as well as precise adjustment of the meniscus. The nonreturn valve integrated into the pipette adapter, together with a hydrophobic membrane filter (0.2 μm pore diameter), provides effective protection against the ingress of liquids. Liquid vapors are discharged to the outside by pressure equalization, which protects the inside of the device

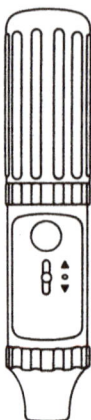

Fig. 2.10: **Macro**pipette controller .

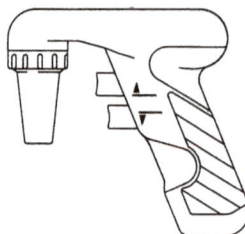

Fig. 2.11: Pipette controller **accu-jet**®.

from corrosion as far as possible. One battery charge is sufficient for 8 h of continuous pipetting; the charging socket is positioned so that work can continue while charging.

Several pipetting aids work in a similar way to the devices described: devices of the **pipetus series** (Hirschmann Laborgeräte, Eberstadt) **pipetus-junior** (manual), **pipetus-micro** (for pipettes up to 1 ml), **pipetus-akku** (with adjustable pipetting speed) and **pipetus-standard** (with power supply unit) as well as the PIPETBOY acu (INTEGRA Bioscience, Fernwald) and the manually operated **Easypet** (Eppendorf, Hamburg).

Solutions of highly volatile gases (NH_3 and SO_2) are not drawn up into the pipette, but are better pressed into the pipette using a rubber hand blower.

For storing pipettes, flat trays with notched edges are used for horizontal storage or racks in various designs for vertical insertion, as well as plastic baskets that can be used in conjunction with dishwashers.

Piston pipettes (microliter pipettes) are used to quickly pipette small volumes of liquid. A distinction is made between two types according to the functional principle: air cushion pipettes and positive displacement pipettes [59]. In the former, there is an air cushion between the piston of the pipette and the liquid to be pipetted. With positive displacement pipettes, the liquid is in direct contact with the pipette piston. A replaceable pipette tip made of PP or glass is attached to both types. A further distinguishing feature is the design as a piston-operated pipette with a fixed volume, which is used exclusively for dispensing its nominal volume, e.g. 100 µl, and as a variable volume pipette for dispensing a user-selectable, fixed usable volume, e.g. between 10 µl and 100 µl. With the adjustable microliter pipettes, the piston stroke volume is changed by turning a knob or wheel. The set volume can be read off digitally. Some designs are shown in Fig. 2.12.

The volume range usually covered by piston-operated pipettes is between 5 µl and 5 ml. For example, under the name Transferpette® (BRAND; Wertheim), there are 12

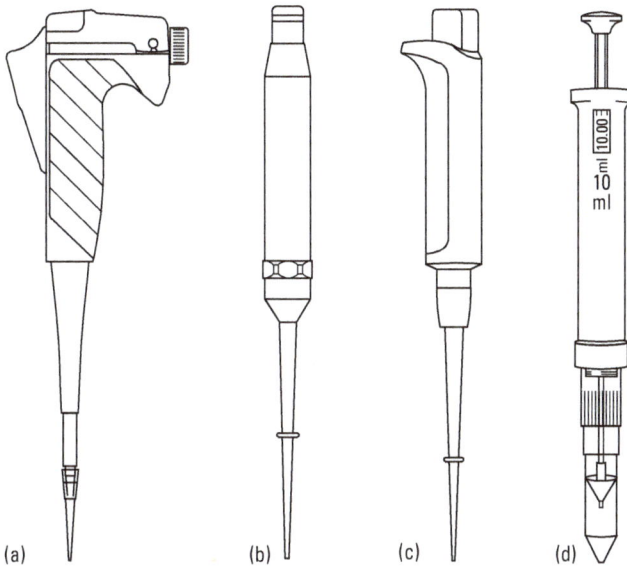

Fig. 2.12: Piston pipettes.
(a) Transferpette®, BRAND GmbH & Co, Wertheim; (b) Eppendorf®: Prototype of a piston-operated pipette; (c) Pipetman® P, Gilson International via ABIMED, Langenfeld; and (d) Transferpettor, BRAND GmbH & Co, Wertheim.

fixed models in the 5 μl to 2 ml range and 10 digital models, each of which covers a power of ten in volume, between 0.1 μl and 5 ml. As a rule, four variable devices are sufficient to cover the usual volume range. The relative deviation from the target value of the volume to be measured is specified as < 1%. This applies to water at 20 °C. An ISO standard specifies details on adjustment, labeling, limits for the measurement deviation and test methods [60].

The pipette shaft of the air cushion pipettes contains a cylinder in which an airtight piston is guided, as well as a system for limiting the stroke of the piston. The total stroke is divided by an intermediate stop. The pipetting volume is determined by the cylinder diameter and the stroke from the upper end stop to the intermediate stop. The further stroke to the lower end stop allows the complete dispensing of the liquid to be pipetted, including the adhering amount. A PP tip is placed on the lower end of the pipette, which alone absorbs the liquid so that it does not come into contact with the pipette itself. The tip is not wetted by aqueous solutions. However, to prevent carry-over errors, the tip is replaced with a new one when changing to a different pipetted liquid. After attaching the tip, press the pipetting button to the first stop before immersing the tip in the liquid. After immersion, the button is slowly allowed to slide back and the liquid is drawn up by the upward movement of the piston. The tip is wiped against the wall of the dispensing vessel. To empty, place it against the wall of the sample vessel held at an angle, press the pipette button slowly to the first stop and, after one to two seconds, continue to the end stop. In this position, the pipette has to be pulled out of the vessel by the wall. The button is then allowed to slide back slowly. The pipette must be held vertically during the pipetting process. No liquid should be aspirated without the tip attached, and the pipette should not be placed down with the tip filled. Air cushion pipettes are available as single-channel and multi-channel pipettes (with 8 or 12 channels). The latter are used for working in microtiter plates with 96 wells.

Maximum accuracy when pipetting with piston-operated pipettes is achieved by avoiding the influence of the air cushion. This is achieved by using the direct displacement principle. These devices can also be used to pipette viscous media, liquids with high vapor pressure and solutions that tend to foam. Figure 2.12d contains the example of the **Transferpettor** (BRAND) for 1 µl to 10 ml (fixed type) and from 100 µl as a variable type. The piston directly draws the liquid up and wipes it off completely when ejected. Glass capillaries are used for small volumes and plastic tips for volumes from 200 µl.

Further aspects in the differentiation of piston-operated pipettes are the presence of an integrated tip ejector, the ergonomic design of the pipette, the arrangement of the functional units (pipetting button, volume adjustment and tip ejector) and the possibility of adjusting the pipette to the respective working conditions with or without tools.

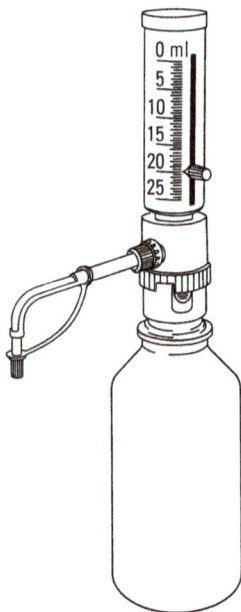

Fig. 2.13: Dispenser (Dispensette, BRAND, Wertheim).

Dispensers are piston-stroke devices that have been developed for the repeated, rapid dispensing of preselected and same-size liquid volumes from storage bottles – primarily for serial dispensing. Single-stroke dispensers enable a single dispensing, while multiple dispensers allow several dispensing per filling stroke. The latter save time as there is no need for repeated drawing up. The devices are attached or screwed directly onto a storage bottle. They are adjusted to drain (Ex) for water at 20 °C. Various designs are available: Manual dispensers with a fixed dispensing volume (e.g. 1/2/5/10 ml), those with adjustable volumes for different ranges between 0.05 µl and 100 ml, whereby an arrangement is used to sweep one power of 10 of the volume in each case. The volume set manually with a slide switch can be read off a scale on the piston shaft of the device or displayed digitally. Figure 2.13 shows an example. Battery-operated manual dispensers are also available. The manufacturer generally specifies accuracy as 0.5% (based on the final volume) and reproducibility as 0.1%. The ISO standard [60] specifies the limits for the systematic and random error of measurement for single-stroke dispensers from 10 µl to 200 ml and for multiple dispensers from 1 µl to 200 ml.

Dilutors are also piston-stroke devices that are used to prepare liquid mixtures with the aim of dilution. They are used to save time when analyzing large series. The dilution liquid is drawn into the cylinder from a storage vessel up to a defined volume, after which the preselected volume of sample liquid is taken up from a second vessel. This can be done by a second movement of the piston or by a second piston and cylinder system. During dispensing, the sample liquid is ejected first, followed by the dilution liquid. The two volumes are adjusted before the liquids are taken up so that the desired dilution ratio is achieved. The adjustment is to be carried out with water at a reference temperature of 20 °C. Regarding the sample liquid, it takes place for the drawing up (In), and for the dilution liquid to the expiry (Ex). The limits for the systematic and random error of measurement are specified in ISO standard 8655 [60] for nominal volumes of 5 µl to 1 ml for the sample liquid and 50 µl to 100 ml for the dilution liquid.

2.1.1.4 Burettes

Burettes are long cylindrical glass tubes with an adjustable outlet at the bottom. The cylindrical part of uniform diameter is graduated. Burettes are adjusted to drain (Ex).

A commonly used burette allows 50 ml to be drawn; it is divided into tenths of a milliliter. The first burettes were fitted with a short piece of rubber tubing at the lower end, into which a glass outlet tube was pushed, ending in a point. The closure was made with a **pinch clamp** (Mohr's burette). Such burettes are still on the market today. Bunsen used a small glass rod inserted into the rubber tubing instead of a stopcock. Later, **ground-glass stopcocks** were introduced, which are attached either vertically or laterally. The glass taps must be greased so that they are tight. Today, the glass taps have been replaced by **taps with polytetrafluoroethylene** (PTFE) **plugs. Fine-dosing valves** with a PTFE spindle are also on the market. When using plastic plugs, there is no risk of the measuring liquid and the glass wall being contaminated by tap grease. Figure 2.14 shows various designs.

For the collection of smaller volumes, **microburettes** with a capacity of 2, 5 or 10 ml are used (Fig. 2.15). They usually have graduations of 0.01 ml or 0.02 ml. The burette can be easily refilled from the storage vessel using a filling tube with a tap.

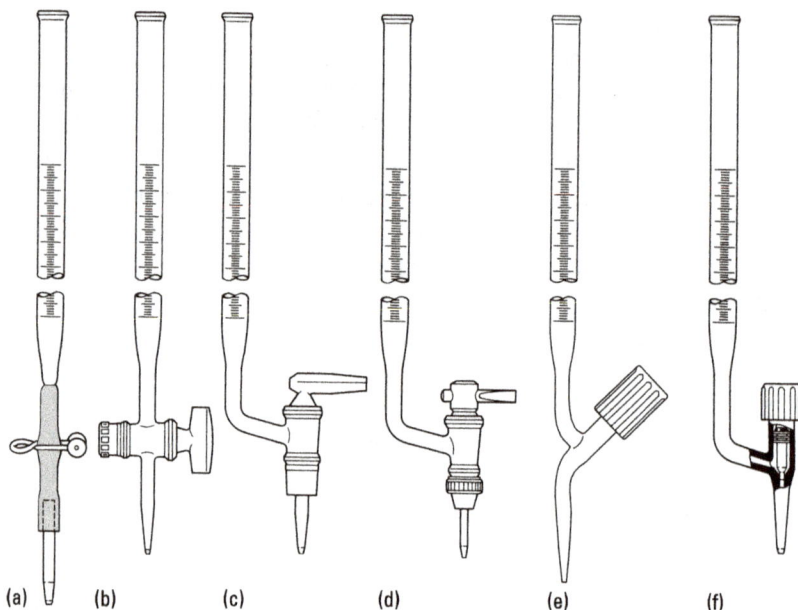

Fig. 2.14: Burette taps: (a) rubber tubing with pinch clamp, (b) straight stopcock, glass; (c) side tap, conventional shape, glass; (d) side tap, conventional shape, PTFE; (e) straight tap with PTFE valve spindle; and (f) lateral tap with PTFE valve spindle.

Burettes on storage bottles, known as **automatic burettes,** also allow convenient filling (Fig. 2.16). The storage bottle holds 1 or 2 l. The liquid can be pressed through the filling tube into the burette using a rubber hand blower, which is pushed onto the lateral nozzle. Adjustment to the zero mark is made automatically by independently lifting off the excess liquid (device according to Pellet). These burettes are manufactured with and without an intermediate stopcock.

Schilling titrators are particularly suitable for use in the plant and for field tests (Fig. 2.17). They are robust and can still be used to carry out precise titrations. They consist of a polyethylene storage bottle in a plastic base and a burette with a Pellet top, which are connected to each other by a holding clamp. The standard solution is fed to the burette from the storage bottle via a polyvinyl chloride riser tube. The burette is filled by applying light pressure to the bottle. The burette and outlet tube with tip are connected by plastic tubing which is closed in the burette holder by a spring. A push button allows fast titration (Fig. 2.18a), and a hinged microscrew allows the titration liquid to be added drop by drop (Fig. 2.18b). Versions are available with burettes of 10, 15, 25 and 50 ml capacity as well as storage bottles of 500 and 1,000 ml capacity.

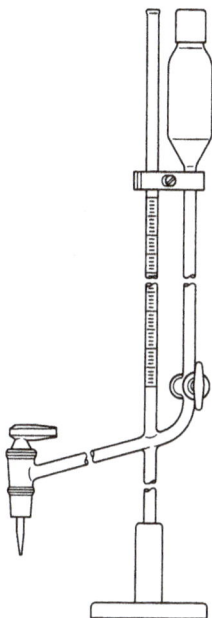

Fig. 2.15: Microburette.

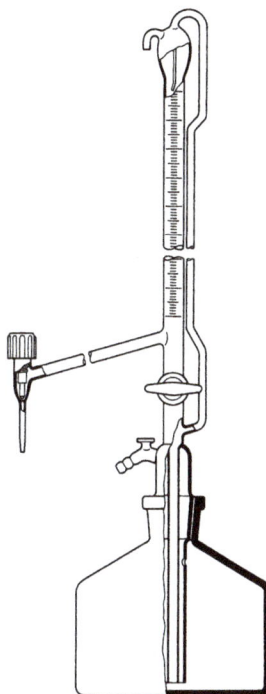

Fig. 2.16: Automatic burette.

Accuracy classes are also distinguished for burettes [61]: **Class A** (narrow error limits, ring graduation, no waiting time), **class AS** (narrow error limits, ring graduation at the main points, shortened elapsed time, 30 s waiting time) and **class B** (double error limits as for classes A and AS, line spacing, no waiting time). As already explained for pipettes (see p. 13), class AS burettes are now used instead of class A burettes. The maximum scale length, the minimum graduation distance as well as the length and width of the graduation marks are prescribed. Table 2.6 provides an overview of the standard sizes, error limits and expiry times.

As with the other measuring devices, device manufacturers offer burettes with tolerances smaller than those specified for class B; they are guaranteed to have 1.5 times the error limits of class A.

The error limit is the largest permissible deviation. It applies to both the total volume and each partial volume. However, to ensure that the relative deviation is not too large, the volume of liquid to be measured should be at least ⅕ of the nominal volume. The **flow time** is the time it takes for the burette to release continuously from the zero mark to the full-scale value when the burette tap is fully open.

In the case of volumetric flasks and pipettes, the **labels** have already been presented which must or may be affixed to volumetric glassware (see p. 10 and 17 f.). They apply analogously to burettes. The adjustment mark is Ex for class A and B

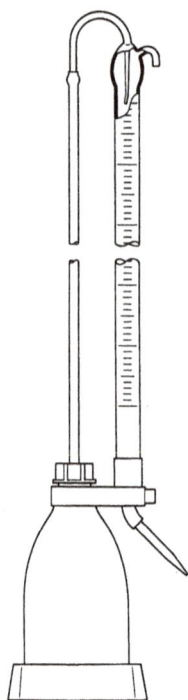

Fig. 2.17: Schilling titrator.

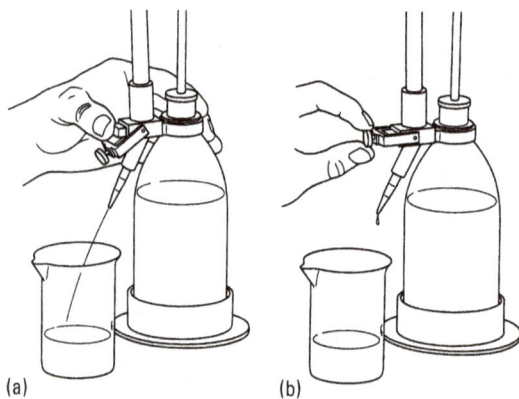

(a)   (b)

Fig. 2.18: Schilling titrator: (a) rapid and (b) dropwise titration.

burettes and Ex + 30 s for class AS burettes. The type of glass from which the burette is made is indicated by the glass type or by a color coding (borosilicate glass: green; soda-lime glass: black). The DIN association symbol has been replaced by the inscription ISO 385.

Tab. 2.6: Burettes and their error limits.

Nominal volume (ml)	Subdivision (ml)	Error limits	
		Class A and AS (ml)	Class B (ml)
1	0.01	±0.006	±0.01
2	0.01	±0.01	±0.02
5	0.01	±0.01	±0.02
5	0.02	±0.01	±0.02
10	0.02	±0.02	±0.05
10	0.05	±0.03	±0.05
25	0.05	±0.03	±0.05
25	0.10	±0.05	±0.10
50	0.10	±0.05	±0.10
100	0.20	±0.10	±0.20

Delivery times:

Nominal volume (ml)	Subdivision (ml)	Delivery times			
		Class A		Class B	
		(s max.)	(s min.)	(s min.)	(s max.)
1	0.01	20	50	20	50
2	0.01	15	45	10	45
5	0.01	20	75	20	65
5	0.02	20	75	20	65
10	0.02	75	95	40	95
10	0.05	75	95	45	75
25	0.05	70	100	30	70
25	0.10	35	75	30	70
50	0.10	50	100	40	100
100	0.20	60	100	30	100

Delivery times for class AS burettes:

Nominal volume (ml)	Subdivision (ml)	Delivery times	
		(s min.)	(s max.)
2	0.01	8	20
5	0.01	15	25
5	0.02	15	25
10	0.02	35	45
10	0.05	35	45

(continued)

Nominal volume (ml)	Subdivision (ml)	Delivery times	
		(s min.)	(s max.)
25	0.05	35	45
25	0.10	35	45
50	0.10	35	45

To use the burette, it is clamped vertically in a stand and filled using a small funnel. Briefly open the tap to expel the air from the lower constricted tube and the tap hole. The liquid should then be about 5 mm above the zero mark. Then carefully open the tap and drain the liquid down to the zero mark. A drop adhering to the drain tip is wiped off on the wall of a beaker. After this adjustment, the desired volume of liquid can be measured or the titration carried out. To do this, open the stopcock completely and allow the liquid to run down to about 5 mm above the selected graduation mark or into the solution to be titrated until shortly before the indicator changeover is reached. In contrast to pipettes, burettes are adjusted in such a way that they must drain freely and must not rest against the wall of a vessel held at an angle. After closing the stopcock, wait 30 s for class AS burettes and then drain the liquid to the desired mark or titrate drop by drop until the color changes. There is no waiting time for class A and B burettes. Then wipe the tip on the inside wall of the glass and rinse with a wash bottle. As with other volumetric instruments, the liquid level must be read at the point where a plane passing through the lowest point of the meniscus perpendicular to the axis of the burette intersects its walls. The eye must be at the same height as the meniscus to ensure freedom from parallax. This requirement can be easily checked if the burette has a ring graduation. In this case, the front and rear parts of the mark must coincide. Even if the graduation line is only halfway around the burette, the parallax error can be avoided. At the titration end point, the meniscus is usually not exactly aligned with a graduation mark. The graduation marks above and below the meniscus are then used as a guide and the volume between them is estimated. A 50 ml burette, for example, can be read to tenths of a milliliter (see Tab. 2.6); hundredths of a milliliter must be estimated.

Reflections in the liquid level often make it difficult to read the burette accurately. The **visor diaphragm** according to Göckel can be helpful. It consists of a piece of cardboard, the upper half of which is white and the lower half black. If the black and white dividing line is held slightly below the meniscus, the disturbing light reflections are reduced (Fig. 2.19a). It is also recommended to slit a corresponding piece of paper and pull it over the burette (Fig. 2.19b).

Reading burettes is easier if **Schellbach stripes** are used. This is a broad white stripe with a narrow blue stripe in the middle, which is printed on the back of the burette. The point of contact of the mirror images produced by the upper and lower surfaces of the meniscus appears as a constriction of the blue stripe in the form of two wedges of different widths (Fig. 2.20). The reading is taken in the plane in which the two wedge tips touch one another; however, it also requires the correct eye level.

The following **general rules** should be observed when **handling burettes:** Alkaline solutions should not be left in the burette as they attack the glass and a glass tap can seize up. Evaporation of the solutions and dusting of the burettes can be avoided by closing them with a suitable reagent glass or a commercially available PP burette cap. Glass taps on burettes must be greased. A tiny amount of tap grease (Vaseline applied above and below the hole) is sufficient for this purpose. It is advisable to apply a thin layer of grease to the plug and then wipe it off, ensuring that no grease gets into the tap hole. The grease remaining in the grinding bed is sufficient; it is wrong to over-grease the tap plug. The finish should look uniformly clear and the tap should turn easily. After emptying the burette, a seized tap plug is carefully loosened by immersing it in hot water. It is then carefully cleaned and greased again. The outlet tip must not be damaged during any of these operations, otherwise, the outlet speed and thus the measuring accuracy will change.

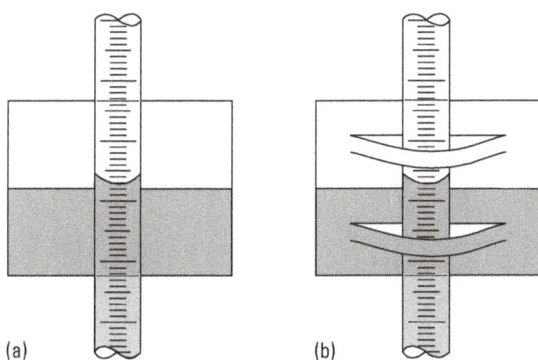

Fig. 2.19: Visor diaphragm.

Piston burettes contain a precision glass cylinder with a PTFE piston as the dosing unit. In the first **manual piston burettes,** of which Fig. 2.21 shows two examples, this component was mounted on a base frame with the reading scales and operating elements. A three-way stopcock allowed the standard solution to be removed from the storage vessel as the plunger slid back (filling) and, after turning the stopcock, dispensed into the solution provided (dosing, titration). By guiding the piston on a spindle, the volume could be measured as the spindle stroke. One turn of the spindle can be measured with a very high resolution according to the micrometer screw principle and therefore precisely. The reading was taken on a vertical coarse scale and a horizontal fine scale, the graduations of which corresponded to 0.1% of the total volume of 5, 10, 20 or 50 ml (Metrohm AG, Herisau, Switzerland, see Fig. 2.21a) or on a dial gauge with two

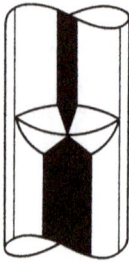

Fig. 2.20: Meniscus at the Schellbach stripe.

interlocking scales up to 5 ml with subdivision into 0.002 ml (formerly Ströhlein GmbH & Co. Düsseldorf; see Fig. 2.21b). Interchangeable units made it possible to exchange the standard solution and/or the cylinder volume. Parallax and tracking errors cannot occur with piston burettes. These devices are only of historical interest.

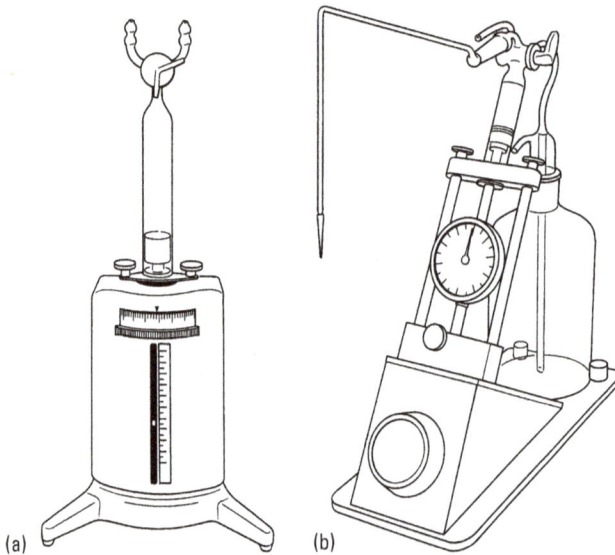

(a)   (b)

Fig. 2.21: Hand piston burettes: (a) Metrohm and (b) Ströhlein.

The development of equipment in the 1980s led from these simple piston burettes to manually operated **piston burettes with a digital display**. The precision of these burettes is still subject to the influence of individual working methods during piston guidance. Subsequently, battery-operated burettes were developed and are offered by various manufacturers today. They stand between the classic glass burettes and the expensive motorized burettes, which are used in **automatic titrators** where the addition of the standard solution is controlled by an electrical signal. The control signal results from the continuous measurement of an electrical or optical variable in the solution (see p. 322).

The Digital III burette (BRAND, Wertheim) should be mentioned as an example (Fig. 2.22). It is screwed onto a storage bottle containing the standard solution. A button is used to switch from "filling" to "titrating". Two handwheels allow the titration speed to be continuously adjusted from fast to drop by drop. An LCD display shows the volume of the standard solution consumed. The burette is equipped with fast calibration technology. According to the manufacturer, the lithium battery lasts for 60,000 titrations, without recharging. For the nominal volumes of 25 ml and 50 ml with a resolution of 0.01 ml, the accuracy is specified as 0.2% and the precision as a coefficient of variation of 0.1%.

The solar-powered digital burette Solarus® (Hirschmann Laborgeräte, Eberstadt) is powered by a built-in solar cell with 500 lux as a sufficient light intensity.

The metrological requirements for piston burettes are specified in ISO Standard 8655 [60].

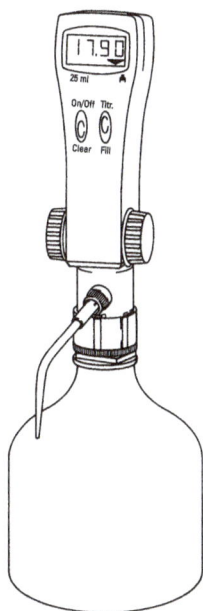

Fig. 2.22: Bottle-top burette Titrette® (BRAND, Wertheim).

Weight burettes offer the possibility of determining the mass of the standard solution required to reach the titration end point. For this purpose, the burette with the standard solution is weighed before the start and after the end of the titration. This methodological variant is called **weight titration**. The fact that it is not the volume of the standard solution but its mass that is determined seems to contradict the nature of volumetric analysis. However, its characteristic feature remains the same: The determination is made using the reagent, not the reaction product.

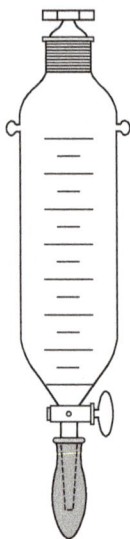

Fig. 2.23: Weight burette.

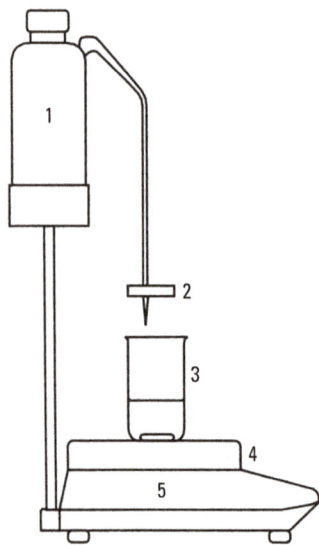

Fig. 2.24: Gravimetric titrator.
(1) storage container (standard solution); (2) dosing device; (3) beaker with stirring rod; (4) weighing pan with magnetic stirrer; and (5) precision balance with integrated computer.

Numerous devices were proposed as weight burettes in the last century. Their various design principles are illustrated by a few examples: Fig. 2.23 shows one designed by Friedman and LaMer [62a] with a filling opening and drain cock, which can be suspended in a pan balance. During titration, it is held by a stand clamp and used like a normal burette. The graduation is only for orientation, it is irrelevant to the actual measurement. The simplest and cheapest burette for weight titration is a polyethylene syringe bottle [62b], the outlet tip of which has been drawn out so that the drops are sufficiently small. Szebelledy and Clauder [62c] used an injection syringe with a capacity of 2 ml in a stand holder with a screw spindle as a weight burette for titrations on a microscale. An arrangement given by Rellstab [62d] allows the withdrawn liquid to be measured continuously during the titration using a top pan balance. The historical development of weight titration and the devices used for this purpose are described by Kratochvil and Maitra [62e].

The great advantage of weight titration is the improved precision of the results. While a reading uncertainty of 0.02 ml must be expected when reading a burette subdivided to 0.1 ml, the reading on a macroanalytical balance can be taken to 0.1 mg. Even if a weighing uncertainty of 1 mg is assumed, this only corresponds to a volume deviation of around 0.001 ml. This results in a precision for weighing titration that is at least an order of magnitude better. A further advantage is the elimination of influences listed below: volume changes due to temperature deviations; delayed flowing liquid when using viscous solutions; difficulties that can occur when reading the meniscus. One disadvantage is the more complicated handling when working discontinuously. The greater time required for weighing compared to volume measurement can be largely compensated or significantly reduced by using modern electronic weighing technology. Saur and Spahn [62 f] describe a measuring arrangement consisting of a microprocessor-controlled balance with an integrated magnetic stirrer in the weighing pan and a simple dosing device (Fig. 2.24). All work steps (tare weighing of the titration vessel with stirring rod, weighing of the sample, addition of reagents, dilutions, stirring during the titration, weighing out after addition of the standard solution) are carried out on the balance without changing location. The microprocessor system stores the masses of the sample provided and the standard solution consumed, calculates the values with the stored stoichiometric factor for the determination and displays the result of the titration. The titration values can be recorded using a measured value printer connected to the interface of the balance. A simple wash bottle or a storage vessel with hose and pinch clamp or solenoid valve can be used as a dosing unit.

About standard solutions for weight titrations, see p. 54.

2.1.2 Cleaning and drying

Measuring instruments, but also beakers, flasks, dishes, etc. used for quantitative analytical work must be completely clean. The measuring devices must be **free of grease.** This can be recognized by the fact that the glass wall is evenly wetted by the liquid. Uneven wetting is noticeable when the liquid film is torn, forming irregular streaks or droplets. The devices are most frequently contaminated by grease. A greasy film reduces the wetting of the walls and causes measuring errors in burettes and pipettes due to an uncertain increase in the volume dispensed. For this reason, a glass burette stopcock should only be greased very lightly; in particular, no grease should be applied to the bore of the stopcock plug (see p. 25). Taps with Teflon plugs, which do not need to be greased and are therefore widely used, are more favorable.

Laboratory equipment made of glass and plastic can be cleaned manually in an immersion bath or mechanically in a laboratory dishwasher. To ensure gentle cleaning, this should be carried out immediately after use at a low temperature (< 70 °C), low alkalinity (< pH 12) and a short contact time (approx. 2 h). Laboratory devices must never be treated with abrasive scouring agents or sponges, as this would damage the surface. Frequent cleaning of volumetric glassware with strongly alkaline media at high temperatures and long exposure times leads to volume changes due to glass erosion and destruction of the graduation. Water, diluted and concentrated acids and alkalis as well as detergents can be used for cleaning. The use of **detergents** has become increasingly popular because they can be handled safely, do not attack the equipment, are gentle on skin and clothing and do not contribute to the contamination of waste water with heavy metals, but also clean quickly and thoroughly as they dissolve many types of dirt and grease well. There are many detergent mixtures on the market that are offered specifically for cleaning laboratory equipment (e.g. Extran, Mucasol, RBS and Tween). By using these mixtures consisting of an optimum combination of organic alkali carriers, surfactants, complexing agents and corrosion inhibitors, it is even possible to dispense with cleaning using dichromate sulfuric acid with its high hazard potential for people and the environment. Dichromate sulfuric acid consists of sodium dichromate or potassium dichromate, in each case mixed with concentrated sulfuric acid.

According to the German accident prevention guidelines for laboratories [58], highly reactive cleaning agents (e.g. concentrated acids and dichromate sulfuric acid) may only be used if other agents have proven to be unsuitable. Before using them, check that the residues in the container do not react dangerously with the cleaning agent.

In addition to dichromate sulfuric acid, these previously frequently used highly reactive agents include chromium nitric acid (a solution of potassium dichromate in concentrated nitric acid), fuming nitric acid and a mixture of concentrated sulfuric acid and fuming nitric acid.

Alkaline solutions are also suitable as degreasing agents. Alcoholic and aqueous potassium hydroxide solution and sodium hydroxide solution were suggested, as well as alkaline permanganate solution, prepared by mixing saturated potassium permanganate solution with the same volume of sodium hydroxide solution (20% by mass). As alkaline solutions attack glass, they should not be used for cleaning measuring instruments.

However, treatment with a neutral saturated potassium permanganate solution (64 g/l) and subsequent posttreatment with concentrated hydrochloric acid has proved successful. The resulting chlorine causes particularly thorough oxidation in statu nascendi.

After treatment with cleaning agents, the vessels are carefully rinsed with tap water and then with deionized water. If alkaline solutions are used for cleaning, an intermediate rinse with acid is required after the first wash with water.

The measuring vessels can be dried by sucking a powerful stream of air through them using a vacuum pump. To protect them from dust, the air is filtered through a piece of filter paper that is sucked in at the open end. Rinsing the containers with ethanol or, better still, acetone beforehand speeds up the drying process. However, this only makes sense if these solvents do not contain any traces of dissolved fat. The time-consuming drying of pipettes and burettes can be avoided by rinsing the vessels several times with a few milliliters of the solution to be measured.

The question of whether glass measuring instruments can be heated without loss of measuring accuracy is answered in the negative in almost all practical books. Drying by heating, e.g. in a drying cabinet, is not recommended because this changes the vessel volume and the original volume is only restored very slowly or not at all. Glassware manufacturers, on the other hand, argue [55] that the thermal expansion of properly tempered glassware is completely reversible so that no change in volume can be detected after cooling. However, the glass must not be heated to such a high temperature that deformation occurs as a result of softening. Under this premise, no detrimental consequences are to be expected with dry glass. In the case of heating wet glass, the hydrolytic attack of the water on the glass surface at higher temperatures can lead to its roughening and thus to a change in the flow properties and impairment of the measuring accuracy. If a glass vessel is to be heated to higher temperatures, it should be pre-dried below 100 °C until the water has evaporated; only then should the temperature be increased. Some manufacturers state that their volumetric glassware can be heated up to 180 °C.

However, you should avoid thermal stress caused by sudden temperature changes. It is therefore advisable to place the glassware in a cold drying cabinet and heat it up slowly. After drying, the oven should be allowed to cool down slowly.

2.1.3 Testing of measuring devices

The most accurate method of measuring the volume of a liquid is weighing. This is why the calibration or adjustment of volumetric measuring devices is carried out by weighing water. **Calibration** is the determination of the deviation of the measured value from the specified target value (nominal volume) without interfering with the system. The adjustment of the measuring device takes place after calibration, using technical measures so that the required tolerance, i.e. the maximum deviation or limit deviation, is maintained or restored. The adjustment of glassware during production is fully automated using computer-controlled systems. In contrast, only the official institutes of the state are authorized to officially check the measuring device, an activity, in German called **Eichen.** The mass m corresponding to the volume V is calculated from the density ρ at the measuring temperature according to $m = \rho \cdot V$. For adjustment, the mass of the liquid corresponding to the nominal volume is filled into the measuring vessel and the ring mark is (re)set. You can also determine the mass of the liquid in the vessel when it is filled up to the mark by weighing and calculating the actual volume from this. The volume calculation takes into account the thermal expansion of the measuring vessel, and the mass determination takes into account the air buoyancy during weighing.

2.2 Solutions for volumetric analysis

As explained in the introduction (see p. 2), a standard solution used for titrimetric analyses contains a reagent in a known concentration as a titrator, which reacts quantitatively with the substance to be determined in the analysis solution during the titration. In principle, the concentration of the standard solution can be chosen at will, but it is advisable to adjust it to the expected amount of substance in the sample solution. If the concentration of the standard solution is too low, the volume required for the titration will be too large. If you use a solution that is too concentrated, only a small amount is used up to the equivalence point, so the relative reading error can assume large values.

2.2.1 Empirical solutions; standard solutions, according to an equivalent; usual standard solutions

When the development of volumetric analysis began in the first decades of the nineteenth century (see p. 331), the standard solutions were prepared in such a way that 1 l of the solution corresponded to one or a few grams of the substance to be determined.

Gay-Lussac in 1832 developed a method for titrating silver ions with a sodium chloride solution (see p. 139), of which 1 l displayed just 5 g of silver. Marguerite, who introduced potassium permanganate to volumetric analysis in 1846, described a "chamaeleon solution", of which 1 l corresponded to just 10 g of iron. The advantage of this solution was that when 1 g of an iron-containing substance was weighed out and a burette with a capacity of 100 ml was used, the percentage of iron by mass could be read directly from the burette after titration of the dissolved sample.

The inappropriateness of such **empirical solutions** became apparent when it was realized that not only iron, but also other substances such as hydrogen peroxide, nitrous acid, oxalic acid, etc. could be titrated with potassium permanganate solution. If one wanted to proceed as Marguerite did with iron, one would have to have a special standard solution ready for each of these substances, each corresponding to a certain mass of the substance to be determined. If only one solution had been used, complicated conversions would have had to be carried out each time. To make matters worse, different units of measurement were in use in different European countries at the time. The search was on for a generally applicable method for preparing the standard solutions that was independent of the respective measuring system, and this was found in the use of chemical units. It proved to be useful to measure solutions on **an equivalent basis.** Thus, **standard solutions, according to an equivalent, were introduced** at an early stage into analytical practice (from 1844). However, it was only after the publication of Mohr's *Lehrbuch der chemisch-analytischen Titrirmethode* (*sic*) **(1855) (i.e. textbook on the chemical-analytical titration method)** that their use became generally accepted and led to a generalization of volumetric analysis [167].

The concentration of a standard solution was based on the **equivalent weight.** A 1 normal (1 N) solution was defined as one that contained the equivalent weight (the equivalent mass) of the dissolved substance (in g) – also known as the **gram equivalent** contained in 1 l. In addition to **normality, molarity** was introduced as a further term for concentration. In a 1 molar (1 M) solution, as much of the substance was dissolved in 1 l as its **molecular weight** or the molar mass (in g) indicated. The term **gram molecule** was also commonly used for this term, which was abbreviated to **mol** for short. The corresponding abbreviation for gram equivalent was **Val.** Thus, 1 mol of $AgNO_3$ was understood to be 169.873 g of $AgNO_3$, 1 mol of H_2SO_4 was 98.07 g of H_2SO_4; in contrast, 1 Val of H_2SO_4 was 49.04 g of H_2SO_4. The molecular weight of polyatomic molecules is the **sum of the atomic weights** of the elements they contain, taking into account the stoichiometric composition. The equivalent weight of a substance is obtained by dividing its atomic or molecular weight by the **valence,** which it exerts in the reaction under consideration compared to the substance to be determined (stoichiometric valence) or, in the case of redox reactions, by the change in valence (change in oxidation number) that occurs during the reaction.

The term **valence** has undergone a differentiation in the course of the development of theoretical ideas about chemical bonding that does not allow a uniform application in all cases. While it was originally said that one atom of a monovalent substance can bind or replace one atom of hydrogen or half an atom of oxygen, today a number of valence-theoretical terms are distinguished, all of which can be summarized under the collective term "valence" [66]: The stoichiometric valence, the ionic valence (ionic charge), the oxidation number (oxidation state), the coordination number, the bonding strength (number of covalent bonds) and the formal charge. The ambiguous term "valence" is largely avoided in the following.

With the introduction of the International System of Units (SI) and its adoption into national law by the German Law on Units in Metrology [41], a changeover was made to a new conceptual basis, which did not lead to a change in the previously used numerical values, but to a new and unambiguous language. Since the concept of equivalence plays an important role in volumetric analysis, the elimination of the terms "molecular weight" and "equivalent weight", which do not fit into the SI system, with the individual mass specifications such as gram molecule and gram equivalent, has a particular impact in this area. The quantities and units now in use will therefore be explained in more detail below.

2.2.1.1 Amount of substance

In 1971, the 14th General Conference on Weights and Measures (CGPM) introduced the **base unit mol** was introduced into the International System of Units. The term **amount of substance** was defined for the associated physical quantity [41]. Although the name mole has thus been adopted from earlier times, it has been given a different definition. The term "amount of substance" has also acquired a very specific meaning. Amount of substance is no longer generally understood as a limited system of matter or a portion of substance [67, 68] (a sample of substance **is** an amount of substance), but as a measurable property of a portion of substance (a sample of substance **has** an amount of substance). The quantity, the "how much" of a substance portion, can be determined by specifying its properties of mass, volume or the number of particles of a certain type it contains. Of these possibilities, the last is of decisive importance for chemistry and has therefore led to the definition of the amount of substance [67, 69, 70].

As individual atoms or molecules cannot be counted directly, it is necessary in practice to select particle numbers large enough to allow measurement. These very large numbers of particles in macroscopic portions of substances are too inconvenient for practical calculations, so the introduction of a suitable **counting unit** for the number of particles seemed to make sense. This was done with the definition of the base unit mole. The mole is therefore no longer an individual unit of mass, but a counting unit and, by definition, the following applies:

– A mole is the amount of substance in a system that contains $6.02214076 \cdot 10^{23}$ of a specific individual particle [70a].

When using the base unit mole (unit symbol: mol), it must always be specified which individual particles are meant. They can be atoms, molecules, ions, electrons as well as other particles or groups of such particles with a precisely definable composition [42, 67]. What is to be regarded as a particle in a substance must be decided on a case-by-case basis and depends on the respective approach. For substances that are not made up of molecules but crystallize in ion lattices, for example, a group of particles with a defined composition is generally regarded as a single particle, e.g. $NaCl$, $BaSO_4$, $KMnO_4$, regardless of whether this combination of particles is present in the crystal as a building block or not.

In calculations, the amount of substance (quantity symbol: n) should be given by a **quantity equation** where the symbol of the underlying particle type is placed in brackets after the quantity symbol, e.g. $n(H_2SO_4) = 0.5$ mol. The amount of substance is a pure counting quantity; its unit mole is a universal counting unit that can be applied to all conceivable types of particles. It is based on the experimentally accessible number of particles per mole, which is known as the **Avogadro constant (molar number**

of particles) and has the value $N_A = 6.022 \cdot 10^{23}$ mol^{-1}. The amount of substance n (X) of any particle type X is related to the number of particles $N(X)$ by the relationship

$$n(X) = \frac{N(X)}{N_A}.$$

It records the number of particles contained in a portion of a substance in multiples of N_A. (The type of particle must also be specified for the number of particles N, as for the quantity of substance n.)

2.2.1.2 Equivalent particle

The concept of particles can also be applied to equivalents, which is particularly important for volumetric analysis. An **equivalent particle, or equivalent** for short is formally understood as the fraction $1/z^*$ of a particle X (atom, molecule, ion or atomic group) that is involved in the exchange of a positive or negative elementary charge in a particular chemical reaction. The equivalent particle has no real significance; the division is of pure abstract nature. Since a material decomposition is not meant by the division, the qualitative properties of the particle type are retained. The number of equivalents per particle X, the **equivalent number** z^*, is always an integer resulting from the reaction equation (equivalence relationship) or from the ionic charge. Its reciprocal value is placed in front of the particle symbol when specifying the quantity of equivalents and thus identifies the equivalent particle, e.g. $n(\frac{1}{2} Mg^{2+})$, $n(\frac{1}{2} H_2SO_4)$, $n(\frac{1}{5} KMnO_4)$, generally n $(^1/_z{}^* X)$.

The following types of equivalent particles can be distinguished:

- **Acid-base equivalent (neutralization equivalent)**: it characterizes an imaginary particle that can release or bind a proton in an acid-base reaction, e.g. HCl, $\frac{1}{2} H_2SO_4$, $\frac{1}{3} H_3PO_4$, NaOH, $\frac{1}{2} Ba(OH)_2$, $\frac{1}{3} Al(OH)_3$. The equivalent number z^* is equal to the number of H$^+$ or OH$^-$ ions released by the particle when the reaction is complete. If $z^* = 1$, the equivalent particle and particle X are identical ($\frac{1}{1}$ HCl = HCl, $\frac{1}{1}$ NaOH = NaOH).
- **Redox equivalent**: It identifies a particle that can accept or release an electron in a redox reaction, e.g. Fe^{2+}, $\frac{1}{5}$ KMnO$_4$, $\frac{1}{6}$ KBrO$_3$. The equivalent number z^* is equal to the amount of the difference between the oxidation numbers before and after the reaction of the atom that changes its oxidation number.
- **Ion equivalent** is to be considered as a part of an ion, carrying a positive or negative elemental charge, e.g. Na$^+$, $\frac{1}{2}$ Mg^{2+}, $\frac{1}{3}$ Al^{3+}, Cl$^-$, $\frac{1}{2}$ SO$_4^{2-}$. The equivalent number z^* *is* equal to the charge number of the ion involved in the reaction, e.g. during ion exchange or electrolytic deposition.

The amount of equivalents n $(^1/_z{}^* X)$, in short form also n (eq), is also given in the unit mole, which is used as a counting unit for all particles. This makes the unit Val superfluous and dispensable, as only one name should belong to each term. The principle of the law of units [41], according to which there must be a fixed conversion factor between two different units of a quantity, also speaks against its continued use. Val and mole are linked by variable conversion factors that depend on the associated reaction (e.g. $KMnO_4$ in acidic solution has the equivalent number $z^* = 5$, in neutral solution $z^* = 3$, see p. 157) and on the type of equivalent particle (for the oxidation of Fe^{2+} to Fe^{3+} $z^* = 1$, for the ionic equivalent of Fe^{2+} $z^* = 2$). If the previously common specification in Val is replaced by the substance quantity specification in moles on the basis of equivalent particles, the numerical value of the specification does not change (previously: 0.1 Val H_2SO_4, now: ($\frac{1}{2}$ H_2SO_4) = 0.1 mol).

For a given portion of sulfuric acid, the amount of substance in relation to H_2SO_4 molecules is

$$n(H_2SO_4) = 0.1\,\mathrm{mol}.$$

If they are related to equivalents, the amount of substance is

$$n(\tfrac{1}{2}\,H_2SO_4) = 2 \cdot 0.1\,\mathrm{mol}.$$

It follows from this:

$$2\,n(H_2SO_4) = n(\tfrac{1}{2}\,H_2SO_4).$$

The numerical value of the amount of substance n therefore doubles compared to the value related to H_2SO_4 molecules if equivalents are taken as a basis, because each H_2SO_4 molecule corresponds to 2 equivalents of $\frac{1}{2}$ H_2SO_4. In general, for a portion of substance, the relationship between the amount of substance of the particles X and the amount of substance of their equivalent particles is as follows

$$n\left(\frac{1}{z^*}X\right) = z^* \cdot n(X).$$

2.2.1.3 Molar mass

The reaction equations we use to describe chemical processes reflect the conversions on the basis of particles, e.g.

$$HCl + NaOH \longrightarrow NaCl + H_2O.$$

As the number of particles is proportional to the amount of substance related to the same type of particle, $n(X) = N(X)/N_A$, the equation states that 1 mol of HCl reacts with 1 mol of NaOH to form 1 mol of NaCl and 1 mol of H_2O. In practical work, however, the quantity of the reacted substances is generally not measured by the number of particles, but by their mass using the balance. Therefore, a quantity is required to convert the mass and the amount of substance of a portion of substance into each other. This quantity is the **molar mass M**, the quotient of the mass m and the amount of substance n:

$$M(X) = \frac{m}{n(X)}.$$

Its SI unit is kg/mol, the usual unit is g/mol. As it can be related to different types of particles, the symbol for the particle is placed in brackets after the size symbol, e.g.

$$M(S) \quad = \quad 32.06 \, g/mol \quad (atoms),$$

$$M(SO_4^{2-}) \quad = \quad 96.06 \, g/mol \quad (ions),$$

$$M(H_2SO_4) \quad = \quad 98.07 \, g/mol \quad (molecules),$$

$$M(½\, H_2SO_4) \quad = \quad 49.04 \, g/mol \quad (equivalents).$$

The relationship exists between the molar mass $M(X)$ of the particles X and the molar mass of their equivalents M (eq):

$$M\left(\frac{1}{z^*}X\right) = \frac{1}{z^*} \cdot M(X).$$

The molar mass is a constant for a given substance and a given type of particle; the values calculated for the individual substances can be found in tabular form, e.g. in [5]. The relationship between the molar mass M and the mass of a single particle m_T is described by the relationship

$$M = m_T \cdot N_A$$

(the mass of N_A molecules, atoms or equivalents is the molar mass).

To calculate the mass from the amount of substance and vice versa, the defining equation for the molar mass is rearranged according to the desired quantity:

$$m = M(X) \cdot n(X), \quad \text{respectively,} \quad n(X) = \frac{m}{M(X)}.$$

Example: What is the mass of a portion of potassium permanganate if the desired amount of substance is 0.1 mol equivalents ($z^* = 5$)?

$$n(⅕\, KMnO_4) \quad = \quad 0.1 \, mol,$$

$$M(⅕\, KMnO_4) \quad = \quad 31.607 \, g/mol,$$

$$m \quad\quad = M(⅕\, KMnO_4) \cdot n(⅕\, KMnO_4),$$

$$m \quad\quad = 31.607 \, g/mol \cdot 0.1 \, mol,$$

$$m \quad\quad = 3.1607 \, g.$$

Example: What is the amount of substance in a portion of oxalic acid with a mass of 4.5018 g?

$$m \quad = \quad 4.5018\,g,$$

$$M(C_2H_2O_4) \quad = \quad 90.035\,g/mol,$$

$$n(C_2H_2O_4) \quad = \quad \frac{m}{M(C_2H_2O_4)},$$

$$n(C_2H_2O_4) \quad = \quad \frac{4.5018\,g}{90.035\,g/mol},$$

$$n(C_2H_2O_4) \quad = \quad 0.05\,mol.$$

2.2.1.4 Content of solutions

The term **solution** is of central importance in volumetric analysis. A solution is a liquid, homogeneous mixture of substances (mixed phase) that contains one or more dissolved substances distributed in a solvent. If you want to make a statement about the quantitative composition of the solution, you specify its solute content. The term **content** should only be used if the statement is of a general (qualitative) nature (e.g. the lead content of an ore or the chloride content of water is to be determined). Content should be regarded as a generic term for the quantities **proportion** and **concentration**, which should be used when numerical values are given [71]. The quantitative composition of a solution can be described using the quantities mass m, amount of substance n or volume V. It should be noted that the amount of substance depends on the type of particle and the volume depends on the temperature.

Proportions are ratio quantities (dimension 1) in which one of the named quantities of a component i (m_i, n_i, V_i) is related to the same quantity of all components of a substance portion (m, n, V_0). The following dimensions have to be distinguished:

$$\text{Mass fraction} \qquad w_i = m_i/m,$$

$$\text{Mole fraction} \qquad x_i = n_i/n,$$

$$\text{Volume fraction} \qquad \varphi_i = V_i/V_0.$$

V_0 is the volume of all components before mixing; it is only identical to the total volume V of the mixture if no change in volume occurs during the mixing process (ideal mixture). The values of mass and amount of substance of the individual components, on the other hand, add up during mixing (dissolving).

The numerical values can be given in the form of decimal fractions, in percent or per thousand or with a unit in both the numerator and denominator; e.g. for 5% hydrochloric acid (conventional designation)

$$w(HCl) = 0.05 = 5\% = 50\%_0 = 0.05\,g/g = 50\,mg/g.$$

The symbols % and ‰ do not represent units, but stand for the powers of 10^{-2} and 10^{-3}. The previously common designations (percentages) should be replaced by the corresponding symbols w, x and φ because they pretend special units.

If a solution contains several dissolved substances, it is of course not possible to deduce the complete composition of the solution from the proportion of one component. However, the mass fraction w_i of a component can be determined independently of other solution partners because the total mass m of the solution portion can be determined directly. The same applies to the volume fraction φ_i for ideal mixtures. In contrast, the mole fraction x_i can only be specified if all components are known by type and quantity, because only then can the total mass n of the solution portion be calculated.

Concentrations are content quantities in which the quantity of component i is related to the volume V of the solution portion. A distinction is made between

$$\text{Mass concentration} \qquad \beta_i = \frac{m_i}{V},$$

$$\text{Molar concentration} \qquad c_i = \frac{n_i}{V},$$

$$\text{Volume concentration} \qquad \sigma_i = \frac{V_i}{V}.$$

Since the volume changes with the temperature, the three concentration variables are temperature-dependent; the pressure dependence of the volume of solutions, on the other hand, can be neglected.

Mass concentration β_i and mass fraction w_i of a component can be converted into each other via the density $\rho = \frac{m}{V}$ of the solution:

$$\beta_i = \frac{m_i}{V} = \frac{m_i \cdot \rho}{m} = w_i \cdot \rho.$$

Example: What is the mass concentration of "5% hydrochloric acid"?

$$w(HCl) \qquad = 0.05,$$
$$\rho(HCl\,solution) = 1.023\,g/ml\ (at\ 20\ °C),$$
$$\beta(HCl) \qquad = 0.05 \cdot 1.023\,g/ml = 0.0512\,g/ml,$$
$$\beta(HCl) \qquad = 51.2\,g/l.$$

Volume concentration σ_i and volume fraction φ_i are the same for ideal mixtures ($V_0 = V$). When specifying volume contents, the volume V of the solution is generally the reference value; e.g. the specification "40% alcohol" for an ethanol-water mixture refers to the volume concentration $\sigma(EtOH)$ and not the volume fraction (EtOH). Both quantities are not the same due to the nonideal behavior of alcohol-water mixtures.

The most important content quantity in volumetric analysis is the molar concentration c_i. It is usually meant when talking about the concentration. The use of the term "concentration" for molar concentration is not recommended [71]. To express the particle relation of the amount of substance and the quantities derived from it, we put the symbol for the particle type in brackets after the quantity symbol

$$c(X) = \frac{n(X)}{V}.$$

The SI unit of $c(X)$ is mol/m³; the usual unit is mol/l. The term **equivalent concentration** is also commonly used for the concentration of equivalents of substances [67]. There is a relationship between it and the concentration of substances in relation to whole particles

$$c\left(\frac{1}{z^*}X\right) = z^* \cdot c(X).$$

The use of the terms **molarity** and **normality** for molar concentration or equivalent concentration is no longer recommended [67]. Instead of the previous notation 0.05 M H_2SO_4 or 0.05 molar sulfuric acid is the specification

Sulfuric acid, $c(H_2SO_4) = 0.05$ mol/l   (size equation)

or

H_2SO_4, 0.05 mol/l   (short form).

Analogously, instead of 0.1 N H_2SO_4 or 0.1 normal sulfuric acid the indication
Sulfuric acid, $c(\frac{1}{2} H_2SO_4) = 0.1$ mol/l or
Sulfuric acid, $c(H_2SO_4) = 0.05$ mol/l or
H_2SO_4, 0.05 mol/l.

The conversion of mass concentration into molar concentration and vice versa is carried out using the molar mass of the solute according to

$$\beta_i = c(X) \cdot M(X) \qquad \text{respectively} \qquad c(X) = \frac{\beta_i}{M(X)}.$$

Example: What is the mass concentration of hydrochloric acid with a mass concentration of β (HCl) = 51.2 g/l?
With M (HCl) = 36.461 g/mol, the result is

$$c(HCl) = \frac{51.2\,g/l}{36.461\,g/mol} = 1.404\,mol/l.$$

The following examples are intended to explain the derivation of the equivalent number z^* from the reaction equation and to illustrate the relationship between the equivalent concentration and the mass concentration of standard solutions (former concentration designation in brackets).

1. The neutralization of potassium hydroxide solution by hydrochloric acid takes place according to the reaction scheme

$$KOH + HCl \rightarrow H_2O + KCl$$

or

$$OH^- + H^+ \rightarrow H_2O.$$

A KOH particle provides an OH^- ion that binds an H^+ ion; consequently, the equivalent number $z^* = 1$. A potassium hydroxide solution of the equivalent concentration $c(KOH) = 1$ mol/l (1 N) has the mass concentration

$$\beta(KOH) = c(KOH) \cdot M(KOH),$$

$$\beta(KOH) = 1 \, mol/l \cdot 56.105 \, g/mol = 56.105 \, g/l.$$

The same applies to a sodium hydroxide solution:

$$c(NaOH) = 1 \, mol/l \, (1 \, N) \text{ and } \beta(NaOH) = 39.997 \, g/l.$$

In contrast, the neutralization of barium hydroxide solution according to

$$Ba(OH)_2 + 2\,HCl \rightarrow 2 \, H_2O + BaCl_2$$

or

$$2\,OH^- + 2\,H^+ \rightarrow 2 \, H_2O$$

the equivalent number $z^* = 2$. For a $Ba(OH)_2$ standard solution of the equivalent concentration $c(\frac{1}{2} \, Ba(OH)_2) = 1$ mol/l (1 N) is obtained:

$$\beta(Ba(OH)_2) = c(\tfrac{1}{2} \, Ba(OH)_2) \cdot \tfrac{1}{2} M \, (Ba(OH)_2),$$

$$\beta(Ba(OH)_2) = 1 \, mol/l \cdot \frac{171.35}{2} g/mol = 85.67 \, g/l.$$

2. For the precipitation titration of chloride with silver ions according to

$$Ag^+ + Cl^- \rightarrow AgCl$$

is the equivalent number for both ion types $z^* = 1$. A silver nitrate standard solution of the equivalent concentration $c(AgNO_3) = 1$ mol/l (1 N) has the mass concentration

$$\beta(AgNO_3) = c(AgNO_3) \cdot M(AgNO_3),$$

$$\beta(AgNO_3) = 1 \, mol/l \cdot 169.873 \, g/mol = 169.873 \, g/l.$$

3. For a sodium thiosulfate standard solution used for titration of iodine according to

$$2\,S_2O_3^{2-} + I_2 \rightarrow S_4O_6^{2-} + 2\,I^-$$

is to be used, $z^* = 1$ must also be used as the basis for determining the equivalent particle, as one thiosulfate ion reduces one iodine atom (change in oxidation number by 1). If the equivalent concentration is to be $c(Na_2S_2O_3) = 1$ mol/l (1 N), then

$$\beta\,(Na_2S_2O_3) = c(Na_2S_2O_3) \cdot M(Na_2S_2O_3),$$

$$\beta(Na_2S_2O_3) = 1\,mol/l \cdot 158.11\,g/mol = 158.11\,g/l.$$

If the salt $Na_2S_2O_3 \cdot 5\,H_2O$ containing water of crystallization is used to prepare 1 l of standard solution, 248.19 g must be weighed in.

4. If you titrate with a potassium permanganate solution in a strongly acidic solution, a MnO_4^- ion takes up five electrons during the reduction, the oxidation number of the manganese changes from VII in the permanganate to II:

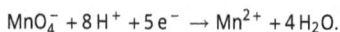

$$MnO_4^- + 8\,H^+ + 5\,e^- \rightarrow Mn^{2+} + 4\,H_2O.$$

The equivalent number of $KMnO_4$ is $z^* = 5$. However, if titrated in a weakly acidic or neutral solution, the oxidation number of the manganese only changes by 3 ($Mn^{VII} \rightarrow Mn^{IV}$):

$$MnO_4^- + 2\,H_2O + 3\,e^- \rightarrow MnO_2 + 4\,OH^-.$$

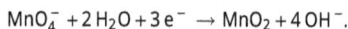

Thus, $z^* = 3$. The equivalent concentration of a potassium permanganate solution is of a potassium permanganate solution in the first case is $c(\frac{1}{5}\,KMnO_4) = 1$ mol/l (1 N), in the second case $c(\frac{1}{3}\,KMnO_4) = 1$ mol/l (also 1 N). The corresponding mass concentrations are

$$\beta(\tfrac{1}{5}\,KMnO_4) = c(\tfrac{1}{5}\,KMnO_4) \cdot \tfrac{1}{5}\,M(KMnO_4),$$

$$\beta(\tfrac{1}{5}\,KMnO_4) = 1mol/l \cdot \frac{158.034}{5}\,g/mol = 31.607\,g/l,$$

respectively

$$\beta(\tfrac{1}{3}\,KMnO_4) = c(\tfrac{1}{3}\,KMnO_4) \cdot \tfrac{1}{3}\,M(KMnO_4),$$

$$\beta(\tfrac{1}{3}\,KMnO_4) = 1\,mol/l \cdot \frac{158.034}{3}\,g/mol = 52.678\,g/l.$$

The former notation 1 N is not a clear indication of concentration because it does not express which equivalent is meant.

5. For a potassium dichromate standard solution of the equivalent concentration $c(eq) = 1$ mol/l (1 N), which serves as an oxidizing agent in acidic solution, $z^* = 6$ for the dichromate, since the oxidation number of both chromate atoms changes from VI to III, a dichromate ion consequently accepts six electrons during the reduction:

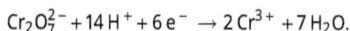

$$Cr_2O_7^{2-} + 14\,H^+ + 6\,e^- \rightarrow 2\,Cr^{3+} + 7\,H_2O.$$

The equivalent concentration is $c(\frac{1}{6}\,K_2Cr_2O_7) = 1$ mol/l (1 N), the mass concentration is

$$\beta(\tfrac{1}{6}\,K_2Cr_2O_7) = c(\tfrac{1}{6}\,K_2Cr_2O_7) \cdot \tfrac{1}{6}\,M(K_2Cr_2O_7),$$

$$\beta(\tfrac{1}{6}\,K_2Cr_2O_7) = 1\,mol/l \cdot \frac{294.185}{6}\,g/mol = 49.031\,g/l.$$

A potassium dichromate solution, $c(eq) = 1$ mol/l (1 N), which is not used for redox titrations, but for the precipitation of barium ions according to

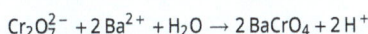

$$Cr_2O_7^{2-} + 2\,Ba^{2+} + H_2O \longrightarrow 2\,BaCrO_4 + 2\,H^+$$

is to be used, ¼ $K_2Cr_2O_7$ must be assigned as the ion equivalent ($z^* = 4$), since 1 dichromate ion is equivalent to 2 barium ions with $z^* = 2$. The equivalent concentration of a potassium dichromate solution for precipitation purposes is therefore c (¼ $K_2Cr_2O_7$) = 1 mol/l (again 1 N), its mass concentration

$$\beta(\tfrac{1}{4}\,K_2Cr_2O_7) = c(\tfrac{1}{4}\,K_2Cr_2O_7) \cdot \tfrac{1}{4}\,M(K_2Cr_2O_7),$$

$$\beta(\tfrac{1}{4}\,K_2Cr_2O_7) = 1\,mol/l \cdot \frac{294.185}{4}\,g/mol = 73.546\,g/l.$$

The last two examples clearly show that the equivalent number z^* of a substance and thus its equivalent concentration $c(eq)$ are not constant values, but can have different values for different reactions.

The advantages of standard solutions which are produced on the basis of equivalents:

– Equal volumes of solutions with the same equivalent concentration contain equivalent amounts of substance. A volume of 20 ml hydrochloric acid, $c(HCl) = 1$ mol/l, neutralizes exactly 20 ml potassium hydroxide solution, $c(KOH) = 1$ mol/l, or 20 ml barium hydroxide solution, $c(\tfrac{1}{2}\,Ba(OH)_2) = 1$ mol/l. This simplifies the calculations.

– One liter of such a standard solution shows the different substances with which it reacts under the same reaction conditions in the ratio of the molar masses of these substances related to equivalents. One liter of a potassium permanganate solution for titrations in a strongly acidic medium, $c(\tfrac{1}{5}\,KMnO_4) = 1$ mol/l, corresponds to ½ mol nitrous acid, 1 mol iron(II) or ½ mol oxalic acid and therefore also 23.507 g HNO_2, 55.845 g Fe and 45.018 g $H_2C_2O_4$.

The volume of the standard solution consumed in a titration on an equivalent basis in ml is equal to the numerical value of the mass fraction in percent of the substance to be determined in a substance sample, if the weight in g is selected so that it corresponds to one tenth of the mass of the substance sought in terms of equivalents. For example, if you want to determine the mass fraction $w(NaCl)$ in percent in a mixture of sodium chloride and sodium nitrate, weigh out 5.8443 g of sample ($\triangleq \tfrac{1}{10}\,M$ (NaCl)) and titrate with silver nitrate solution after dissolving, $c(AgNO_3) = 1$ mol/l. Since 1 l of the $AgNO_3$ solution corresponds to 58.442 g NaCl, 1 ml indicates 1% of the sample weight. However, it is better to use a solution with $(AgNO_3) = 0.1$ mol/l and 0.5844 g initial weight.

In practice, one generally does not work with standard solutions of the concentration $c(eq) = 1$ mol/l (1 N) at all, but with more dilute solutions, e.g. with $c(eq) = 0.1$ or 0.2 mol/l (0.1 N or 0.2 N), more rarely with $c(eq) = 0.5$ or 0.05 or 0.01 mol/l.

Table 2.7 summarizes some of the quantities frequently used in volumetric analysis in the currently valid or recommended designation (according to [67]).

Tab. 2.7: Usual designations and units in volumetric analysis.

Designations	Units
Mass m (X)	kg or g
Amount of substance n (X), n (eq)	mol
Molar mass M (X), M (eq)	g/mol
Mass concentration β (X)	g/l
Molar concentration c (X)	mol/l
Equivalent concentration c (eq)	mol/l
Mass fraction w (X)	g/g
Mole fraction x (X)	mol/mol

One mole according to the new definition denotes a portion of substance containing N_A particles as a unit of substance quantity. One mole according to the old definition (as the molecular weight of a substance in grams) is the mass of a portion of substance that also contains N_A particles. Despite the different definition of the term "mole", it refers to the same quantity and therefore to the same portion of substance. Therefore, the **numerical values** of the molar masses are equal to those of the former atomic, molecular and formula weights or equal to the relative atomic, molecular and formula masses.

A detailed treatment of the terms, relationships and calculations discussed here can be found in [72, 73].

2.2.2 Production of standard solutions

In principle, standard solutions are prepared as follows: Weigh the calculated portion of the desired substance, which results from the molar mass of the equivalents, on an analytical balance, transfer it quantitatively into a clean volumetric flask with a funnel attached, fill the flask to about three-quarters of its capacity with water at room temperature, bring the substance completely into solution while swirling vigorously, add water carefully – finally drop by drop – up to the ring mark and mix well after closing the flask to equalize the concentration.

As the flask is adjusted at 20 °C, an error is made if you fill up at a different temperature. The error can be avoided if the flask is tempered to (20 ± 0.2) °C using a circulation thermostat. The error can also be corrected or taken into account by calculation. If the temperature of the solution and flask is above 20 °C, the solution is too concentrated and would occupy a smaller volume at 20 °C. If you fill up at lower temperatures, the solution would be more diluted than desired.

When calculating the volume error it must be taken into account that the volume of the glass flask and that of the solution have different temperature dependencies; the volume of the solution changes much more with temperature. If γ_1 is the volume expansion coefficient of the glass ($\gamma_1 = 1 \cdot 10^{-5}$ K^{-1} for borosilicate glass; see p. 7 f.) and γ_2 that of the solution ($\gamma_2 = 20.6 \cdot 10^{-5}$ K^{-1} at 20 °C for water and aqueous solutions with $c \le 0.1$ mol/l, then the volume of the flask adjusted at 20 °C at the temperature t_M, at which the standard solution is prepared, is

$$V(K)_M = V(K)_{20}[1 + \gamma_1(t_M - t_{20})].$$

The solution has the same volume: $V(L)_M = V(K)_M$. At 20 °C, however, the solution in the flask would occupy the volume $V(L)_{20}$, which results from the relationship

$$V(L)_{20}[1 + \gamma_2(t_M - t_{20})] = V(K)_{20}[1 + \gamma_1(t_M - t_{20})]$$

to

$$V(L)_{20} = V(K)_{20} \frac{1 + \gamma_1(t_M - t_{20})}{1 + \gamma_2(t_M - t_{20})}$$

results. This expression allows the deviations ΔV of the solution volume from the adjustment volume of the piston to be calculated

$$\Delta V = V(K)_{20} - V(L)_{20},$$

if the flask is filled at a temperature t_M that deviates from the calibration temperature of 20 °C. Table 2.8 shows the deviations ΔV as a function of t_M for a 1 l borosilicate glass volumetric flask; they were calculated taking into account the temperature dependence of γ_2.

Example: The portion required to prepare 1 l of potassium bromate solution of equivalent concentration c(⅙ KBrO$_3$) with the mass m(KBrO$_3$) = 167.001/6 = 27.834 g is weighed and dissolved at 25 °C in a 1 l flask adjusted to 20 °C and filled up. At a temperature of 20 °C, the solution would have a volume of $V(L)_{20}$ = 1,000 − 1.11 ml = 998.89 ml. It is therefore too concentrated; it contains as much potassium bromate in 998.89 ml as it should contain in 1,000.0 ml. Its actual concentration is c(1/6 KBrO$_3$) = $\frac{1\,\text{mol}}{0.99889\,\text{l}}$ = 1.0011 mol/l.

The quotient that contains the actual concentration $c(X)_{\text{actual value}}$ in the numerator and the theoretical concentration $c(X)_{\text{target value}}$ in the denominator is the **titer** t of the solution (formerly **normal factor**):

$$t = \frac{c(X)_{\text{actual value}}}{c(X)_{\text{target value}}}.$$

The volume of a standard solution consumed during a titration must be multiplied by the titer *t in order to* obtain the consumption of a solution of the correct concentration.

If a standard solution with $t = 1.000$ is to be prepared, the volume to be taken from Tab. 2.8, 1.11 ml in the selected example, must be added to the solution.

The direct preparation of exact measured solutions by simple weighing is only possible if the substance to be weighed fulfills the following conditions:

- It must be analytically pure, i.e. have a composition exactly corresponding to its formula, or it must be possible to bring it easily and safely to the required high degree of purity by simple operations (recrystallization, drying).
- It must be easy to weigh accurately, that is, it must not be sensitive to oxygen or attract carbon dioxide and moisture from the air.
- The concentration of a freshly prepared standard solution must not change during prolonged storage.

From the following substances, for example, exact standard solutions can be produced by direct weighing: Sodium carbonate, sodium oxalate, sodium chloride, potassium bromate, potassium iodate, potassium dichromate, potassium hydrogen phthalate, calcium carbonate, etc. Therefore, they are suitable original titrimetric reagents (**primary standard**) that ca be applied for the production or the standardization of titrimetric standard solutions.

Tab. 2.8: Temperature correction for standard solutions: deviations ΔV of the solution volume V (L)$_{20}$ from the target volume (1 l).

t_M (°C)	ΔV (ml)	t_M (°C)	ΔV (ml)
10	+1.40	21	−0.20
11	+1.31	22	−0.41
12	+1.22	23	−0.64
13	+1.11	24	−0.87
14	+0.98	25	−1.11
15	+0.85	26	−1.36
16	+0.70	27	−1.60
17	+0.54	28	−1.89
18	+0.37	29	−2.17
19	+0.19	30	−2.46

For very high accuracy requirements, air buoyancy must be taken into account when weighing in (the molar masses are always calculated for weighing in a vacuum).

In all cases where the requirements to be met by original titrimetric reagents cannot be fulfilled, the standard solutions must be prepared indirectly. This applies to all acids (unknown water content of concentrated solutions) and alkalis (fluctuating carbonate content of alkali metal hydroxides) as well as to substances that can decompose in solutions (potassium permanganate, sodium thiosulfate). In the beginning, a solution whose concentration is slightly greater than the desired concentration is prepared by weighing out a coarse sample or, in the case of liquids, by measuring with a graduated cylinder, and the actual concentration is then determined by several titrations of weighed portions of a suitable original titrimetric reagent (adjustment of the standard solution). If the portions of substance to be weighed in for a consumption of 10 to 20 ml of the standard solution are too small, so that the weighing error is too large, the adjustment is carried out with aliquots by volume of a standard solution of the original titrimetric reagent, which has been prepared by weighing in a larger portion of the substance. Special care must be taken here, as the volume measurement is always subject to greater measurement uncertainty than weighing (see p. 35). After calculating the actual concentration, you can finally add enough water to the standard solution to obtain a solution of the desired concentration.

In volumetric analysis, however, it is not necessary to go to so much trouble and spend so much time to prepare a standard solution with a titer of $t = 1.000$. You can work just as well with solutions whose equivalent concentration is approximately $c(eq) = 1.0$ or 0.1 or 0.01 mol/l if you have determined the titer by careful titrations and always take it into account later.

If the titration is not carried out using an primary standard, but with another standard solution of known concentration (e.g. hydrochloric acid with sodium hydroxide solution or vice versa), its titer must be known very reliably, because the error is transferred to the titer of the solution to be adjusted (propagation of uncertainty).

It must be emphasized that the determination of the titer of a standard solution must be carried out with particular care. Every error made in the process will have the effect of a systematic deviation on all determinations that are carried out with the standard solution. Correct analysis results cannot be obtained with an incorrectly adjusted standard solution. A detailed treatment of further systematic errors that goes beyond the list given on p. 57 can be found, for example, in [15].

The random deviations that occur during titrations can be recognized by the scattering of the measured values when several identical samples are titrated one after the other with the same standard solution using the same burette according to the same working instructions. They are mainly caused by droplet errors, reading errors and procedural errors (fuzzy end point detection). They allow a statement to be made about the precision of the analysis results. To estimate them, titrate several samples depending on the precision requirements and calculate the standard deviation.

In order to save time during practical work, plastic ampoules are commercially available that contain **concentrates** for the preparation of all common standard solutions (trade names: Titrisol® and Fixanal®). The ampoule is placed on the volumetric flask, pierced with a glass rod and its contents are quantitatively rinsed into the flask, which then only needs to be filled up to the mark with water. The titer deviation of $t = 1.000$ is specified by the manufacturer as a maximum of $\pm 0.2\%$. In addition, ready-to-use **volumetric solutions** (maximum titer deviation $\pm 0.1\%$) are also available in plastic bottles.

Under the name Titripac®, ready-to-use volumetric solutions are available in 4 l and 10 l containers, which can be removed from the flexible inner bag of the packaging system via an integrated tap. This prevents contamination by air, CO_2 and microorganisms.

When used in the laboratory, the titer should be checked at certain intervals.

For **standard solutions used in weight titrations,** the amount of substance is not related to the volume but to the mass of the solution. The equivalent concentration $c(eq) = \frac{n(eq)}{V}$ is replaced by the amount of substance divided by the mass of the standard solution. This value is related to equivalents and defined as follows: $q(eq) = \frac{n(eq)}{m}$. The measurement of the standard solution is carried out using a balance and is independent of temperature. To prepare it, the dry substance or the contents of a concentrate ampoule is placed in a dry, balanced container that does not need to be graduated, e.g. a storage bottle, dissolved or diluted in water and mixed with water until a desired mass (e.g. 1 kg) is reached [62d].

2.3 Statement of the analysis result

2.3.1 Calculation of the analysis result

In titrimetry, the standard solutions are used according to the equivalent concentration. This requires knowledge of the equivalent number z^*, which results from the reaction equation (see examples on p. 41). The analysis result can be calculated from the concentration $c(eq) = \frac{n(eq)}{V}$ and the volume V_M of the standard solution in ml consumed during the titration as the measured value.

The amount of substance of the equivalents required for the titration is

$$n(eq)_M = c(eq)_M \cdot V_M.$$

(The index M stands for the standard solution, and the index S denotes the sample.) It is equal to the amount of substance of the equivalents, concerning the component to be determined in the sample:

$$n(eq)_M = n(eq)_P.$$

This means that $n(eq)_P = c(eq)_M \cdot V_M$.

If you want to convert the result into the mass m_P of the substance you are looking for, $n = \frac{m}{M}$, multiply by the molar mass of its equivalents:

$$m_P = M(eq)_P \cdot c(eq)_M \cdot V_M.$$

The product $M(eq)_P \cdot c(eq)_M$ in mg/ml is the factor F listed in tables (e.g. [5]) for volumetric analysis. It indicates the mass of the substance in mg, which corresponds to 1 ml of standard solution:

$$m_P = F \cdot V_M.$$

If the mass fraction of the substance being searched for in a solid sample, w_P is to be given as the result, it should be referred to the sample weight m_E

$$w_P = \frac{m_P}{m_E}.$$

$$w_P = \frac{F \cdot V_M}{m_E}.$$

If the mass concentration of the substance to be determined β_P in the sample solution provided (volume V_P) is required as the result, the following applies

$$\beta_P = \frac{m_P}{V_P}$$

$$\beta_P = \frac{F \cdot V_M}{V_P}.$$

Example: A volume of 20 ml sodium hydroxide solution of approximate concentration $c(NaOH) = 0.1$ mol/l is added, diluted with water to approximately 100 ml and titrated with sulfuric acid, $c(\frac{1}{2} H_2SO_4) = 0.1$ mol/l using methyl red as an indicator. The consumption was 20.84 ml. This results in the amount of substance presented as

$$n(NaOH) = 0.1 \, mmol/ml \cdot 20.84 \, ml = 2.084 \, mmol$$

and the concentration of sodium hydroxide to

$$c \, (NaOH) = \frac{2.084 \, mmol}{20 \, ml} = 0.1042 \, mmol/ml \, or \, mol/l.$$

If you want to use the titration to adjust the sodium hydroxide solution, this value indicates the concentration $c(NaOH)_{actual \, value}$. The desired concentration would be $c(NaOH)_{target \, value} = 0.1$ mol/l and the titer

$$t = \frac{c(NaOH)_{actual \, value}}{c(NaOH)_{target \, value}} = \frac{0.1042 \, mol/l}{0.1000 \, mol/l} = 1.042.$$

Example: The mass of sodium carbonate contained in a given soda solution, for the titration of which 42.6 ml of hydrochloric acid, $c(HCl) = 0.1$ mol/l, was used, is calculated as follows:

$$m_P(Na_2CO_3) = M(\frac{1}{2} \, Na_2CO_3) \cdot c(HCl)_M \cdot V(HCl)_M$$

$$m_P(Na_2CO_3) = 52.994 \text{ mg/mmol} \cdot 0.1 \text{ mmol/ml} \cdot 42.6 \text{ ml}$$

$$m_P(Na_2CO_3) = 225.75 \text{ mg}.$$

The factor $F = M(\frac{1}{2} Na_2CO_3) \cdot c(HCl)$ is 5.2994 mg/ml. If this soda was contained in a solid sample of 834.3 mg (initial weight), the soda content in the sample is

$$w(Na_2CO_3) = \frac{F \cdot V(HCl)_M}{m_E} = \frac{5.2994 \text{ mg/ml} \cdot 42.6 \text{ ml}}{834.3 \text{ mg}},$$

$$w(Na_2CO_3) = 0.2706 = 27.06\%.$$

Example: The mass concentration of a nitric acid solution is to be determined by titration with the sodium hydroxide solution set in the first example. A volume of 2 ml of the sample solution is added, diluted to 100 ml and titrated; sodium hydroxide consumption is 32.51 ml:

$$\beta(HNO_3) = \frac{M(HNO_3) \cdot c(NaOH)_M \cdot t \cdot V(NaOH)_M}{V(HNO_3)_P}$$

$$\beta(HNO_3) = \frac{63.013 \text{ mg/mmol} \cdot 0.1 \text{ mmol/ml} \cdot 1.042 \cdot 32.51 \text{ ml}}{2 \text{ ml}}$$

$$\beta(HNO_3) = 106.730 \text{ mg/ml or g/l}.$$

2.3.2 Uncertainty of a measured value

> Uncertainty is not a pleasant state, but certainty is an absurd one. (Voltaire, * November 21, 1694, †
> May 30, 1778, French philosopher and writer)

A measurement result can only be regarded as meaningful or scientific if, in addition to the measured value, it also contains the associated measurement uncertainty including the corresponding measurement unit:

$$c(NaOH) = (0.10 \pm 0.03) \text{ mol/l}.$$

Measurement uncertainty is defined as a nonnegative parameter that characterizes the dispersion of the individual values associated with a measurand based on the information used [73a].

In other words: The measurement uncertainty reflects a range in which the true value of the quantity sought is located with a certain probability. The uncertainty is therefore not a sign of deficiency, but describes the quality of the result.

According to the "Guide to the expression of uncertainty in measurement (GUM)", the measurement uncertainty can be determined using two methods [73b]:

Method A: Calculation using statistical methods from several independently measured values of a measurement repetition (see below),

Method B: Calculation based on information from other sources (see below).

The principles for calculating the measurement uncertainty, including instructions for practical application, were described in detail by M. Krystek [73c]. In addition, he compares the GUM methods with the traditional methods. The application of such calculations in analytical chemistry is described in detail in the EURACHEM/CITAC guide [73d].

2.3.2.1 Measurement deviations: accuracy and precision of a measurement

The **measured value** – here 0.10 mol/l – can be determined by a single measurement, but in analytical chemistry multiple determinations are common, i.e. the determination of individual measured values of a measurand under the same conditions.

The measured value is then the **arithmetic mean** $\bar{x}$ of the individual measured values x_i from n measurements

$$\bar{x} = \frac{x_1 + x_2 + x_3 + \cdots + x_n}{n} = \frac{1}{n}\sum x_i.$$

The measured value obtained in this way is an estimated value, which – like the individual values – is subject to an *error of measurement*, but is usually more reliable than an individual value.

The measurement deviation is calculated from the measured value and a reference value [3a].

A **reference value** can be a true value. The **true value** is an ideal value – a value that could only be determined under ideal conditions and is therefore unknown in principle. An agreed reference value, on the other hand, is known. For example, it can be a value that has been determined using a proven analytical method and may be part of the certificate of a certified reference material.

A basic distinction is made between two types of measurement error: **systematic measurement error** and **statistical measurement error**.

If the mean value of a measurement series does not correspond to the true value (x_w) or to the agreed reference value (x_{Ref}) and this is a unidirectional deviation, a systematic measurement **error (b)** is present:

$$\bar{x} \approx x_w + b \quad \text{or} \quad \bar{x} = x_{Ref} + b. \,[73c]$$

The approximately equal sign expresses the fact that the true value is not exactly known.

As a systematic error of measurement is unidirectional, it has both a magnitude and a sign. To calculate a measurement result, the measured value is corrected by the known systematic error.

$$\hat{x} = \bar{x} - b.$$

Systematic measurement deviations can be attributed to various causes. Without claim-
ing to be exhaustive, the following possibilities can be listed for the volumetric analysis:
-   Errors caused by equipment, e.g. by using incorrectly calibrated weighing and
    measuring equipment, test temperature deviating from the reference tempera-
    ture (usually 20 °C),
-   Methodological errors, e.g. due to incomplete reactions, occurrence of side reac-
    tions, co-detection of other components.

Systematic measurement deviations are decisive for the **trueness** of the result and
must therefore be recognized and corrected or eliminated.

The **statistical measurement deviation** results from the fact that the individual
values of a multiple measurement scatter to a greater or lesser extent around this
mean value. These deviations always occur, despite taking the greatest care and working under the most consistent conditions possible. They are caused by random influences that are independent of the will of the experimenter. This measurement deviation is therefore also referred to as random measurement error.

The extent of the scatter of the individual values can traditionally be determined
for normally distributed values using the least squares method proposed by C. F. Gauss.
One measure for this is the empirical standard deviation.

To calculate the **standard deviation** calculate the n deviations of the individual
values from the mean value $(x_i - \bar{x})$, square these differences, sum the squares $(x_i - \bar{x})^2$,
divide the sum by the number of measurements reduced by 1 and take the square root:

$$s = \sqrt{\frac{\sum (x_i - \bar{x})^2}{n-1}}.$$

The standard deviation indicates how far individual measured values differ from the
mean value and is therefore equal to the uncertainty of the individual values x_i:

$$u(x_i) = s.$$

The uncertainty of the measured value – i.e. the arithmetic mean of n individual
measurements – can then be calculated as follows for normally distributed values:

$$u(\bar{x}) = \frac{s}{\sqrt{n}}.$$

As this uncertainty, also known as **standard measurement uncertainty**, is estimated
from independent statistical single values of a multiple determination, this is **method
A** (see above) for determining the measurement uncertainty.

It is important to note that a small number of measured values do not have a normal distribution, but a *Student*'s *t*-distribution. In this case, a correction factor *t* must be taken into account:

$$u(\bar{x})_t = \frac{t \cdot s}{\sqrt{n}}.$$

The *t*-factor depends on the number of individual measurements and the **statistical certainty** (*P*) and can be taken from a table. More information on this, including the table, can be found in further literature [17, 18].

The measure of random deviations is often not the absolute value, but the **relative value** in relation to the mean value or its hundredfold value, the **coefficient of variation** v or **normalized coefficient of variation** v^* in percent:

$$v = \frac{s}{\bar{x}} \cdot 100\% = \frac{u(x_i)}{\bar{x}} \cdot 100\%,$$

$$v^* = \frac{s}{\sqrt{n} \cdot \bar{x}} \cdot 100\% = \frac{u(\bar{x})}{\bar{x}} \cdot 100\%.$$

In principle, random measurement deviations cannot be avoided. However, by increasing the number of individual measurements, these measurement deviations can be minimized by a factor of $\sqrt{n}$. The random error characterizes the **precision** of a measurement method.

The total measurement deviation of a measurement result is made up of the systematic and random measurement deviations and is reflected in the term **accuracy**. Accuracy is therefore a collective term for correctness and precision [73e].

Figure 2.25 shows mean values of four measurement series including standard measurement uncertainty. If the mean values of all measurement series are compared with the reference value x_{Ref}, it can be seen that the results $\bar{X}_1$ and $\bar{X}_2$ (first and second cases) are correct – the reference value x_{Ref} lies within the range of random deviations. In cases 3 and 4, however, there is such a large systematic difference that x_{Ref} lies outside the range of the standard measurement uncertainty – $\bar{X}_3$ and $\bar{X}_4$ are to be judged as incorrect and therefore unusable. A comparison of the standard measurement uncertainties shows that the measured values in cases 1 and 3 are more precise than those in cases 2 and 4. The first measurement result can be considered accurate due to its high trueness and precision. More detailed considerations on measurement uncertainty and the evaluation of measurement results can be found in the following sources [65, 73d].

Errors caused by a lack of care when measuring or abnormal device behavior, such as incorrect calculations, incorrect purity of the reagents, incorrect reading of the measured value, contamination of the balance, etc., are regarded as **blunder error**. These errors are generally avoidable when working carefully. When working routinely in the laboratory, such errors are usually easy to recognize and must be ex-

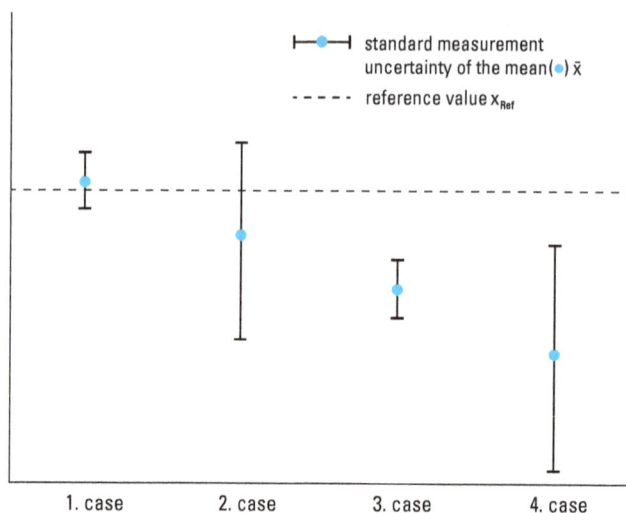

Fig. 2.25: Precision and trueness of measurement results.

cluded. When gaining initial experience in the laboratory, a comparison between the determined value and the reference value helps to recognize errors. Results with gross errors must not be included in the evaluation.

2.3.2.2 Determination of uncertainties from other sources

In some cases, it is not possible or is very time-consuming to repeat a measurement. In this case, the uncertainties cannot be determined statistically (**method B, see above**). For this purpose, the uncertainties of the output variables are first estimated from other sources.

Suitable sources are, for example, the limit deviations of the measuring devices used, the manufacturer's specifications on the purity of the substance used and so on.

The **limit deviations** (Δx), also known as the error limit, specify the guaranteed or agreed maximum possible deviation of the measured value. The read or measured value is then almost 100% certain to be within the specified limit range.

If it is expected that the individual values within this range will have larger and smaller deviations of the displayed value with equal probability, then a so-called rectangular distribution exists. If it can be assumed that smaller deviations occur more frequently than the larger ones, then a so-called triangular distribution exists. The associated standard uncertainties can be estimated from this:

Rectangular distribution: $u(x) = \dfrac{\Delta x}{\sqrt{3}}$,

Triangular distribution: $u(x) = \dfrac{\Delta x}{\sqrt{6}}$.

The values of the volumes measured using volumetric devices are usually triangularly distributed. When calculating the standard uncertainties of a sample weight, the degree of purity and the molar mass, it is recommended to use the equation of the rectangular distribution [18, 73d].

2.3.2.3 Propagation of standard uncertainties

If a final result is formed by mathematical operations and the standard uncertainties of the incoming variables are known, the standard uncertainty of the final result can be estimated mathematically.

The interaction of all these individual uncertainties is referred to as the propagation of uncertainty (formerly propagation of error). Standard uncertainties calculated in this way are called **combined standard uncertainties**.

The propagation of standard uncertainties is usually calculated using the general Gaussian law concerning the propagation of error. If the individual variables are independent of each other – not correlated – the uncertainty can be estimated using the following mathematical operations.

When *adding* or *subtracting*, the squares of the respective individual errors are added together:

$$\text{for} \quad y = x_1 + x_2 \quad \text{or} \quad \text{for} \quad y = x_1 - x_2 \quad \text{apply}$$

$$u^2(y) = u^2(x_1) + u^2(x_2).$$

This equation shows that **the standard uncertainty for burettes is always doubled** to estimate the combined standard measurement uncertainty, as this occurs both when zeroing and when reading the value:

$$u^2(y) = 2u^2(x).$$

Example: During a titration, the burette is first set to zero. After the titration, the volume used is read from the burette, which is 15.10 ml.

According to the manufacturer, the limiting deviation of the burette is ±0.05 ml. As no further information is given, a triangular distribution is assumed. The standard uncertainty can then be calculated as follows:

$$u(V) = \frac{0.05 \text{ ml}}{\sqrt{6}}.$$

The total uncertainty is then calculated as follows:

$$u(V)_{\text{consumption}} = \sqrt{u^2\left(V_{\text{zero-point position}}\right) + u^2\left(V_{\text{after titration}}\right)} = \sqrt{\left(\frac{0.05}{\sqrt{6}}\right)^2 + \left(\frac{0.05}{\sqrt{6}}\right)^2} = 0.03\,\text{ml},$$

$$V_{\text{consumption}} = (15.10 \pm 0.03)\,\text{ml}.$$

When *multiplying* or *dividing*, the squares of the relative uncertainty are added up. This applies to

$$y = x_1 \cdot x_2 \quad \text{or} \quad y = x_1 \div x_2$$

$$\left(\frac{u(y)}{y}\right)^2 = \left(\frac{u(x_1)}{x_1}\right)^2 + \left(\frac{u(x_2)}{x_2}\right)^2.$$

This equation can be used, for example, to calculate the concentration uncertainty of a solution:

Example: An EDTA solution is to be prepared. For this purpose, 37.2239 g ($\Delta m_{\text{scales}} = \pm 0.0001$ g) of dried EDTA dihydrate disodium salt with a purity of 99.9 ± 0.1% is weighed and dissolved in demineralized water in a 1 l volumetric flask (narrow neck, class A, $\Delta V = \pm 0.4$ ml) (see p. 238).
The substance concentration can be calculated using the following formula:

$$c(\text{EDTA}) = \frac{m(\text{Na}_2\text{C}_{10}\text{H}_{14}\text{N}_2\text{O}_8 \cdot 2\,\text{H}_2\text{O}) \cdot \text{purity level}}{M(\text{Na}_2\text{C}_{10}\text{H}_{14}\text{N}_2\text{O}_8 \cdot 2\,\text{H}_2\text{O}) \cdot V(\text{solution})} = \frac{37.2239\,\text{g} \cdot 0.999}{372.2392\,\frac{\text{g}}{\text{mol}} \cdot 1,000.0\,\text{ml}} = 0.09989995\,\text{mol/l}.$$

The associated uncertainty is then made up of the uncertainty of the balance, the uncertainty of the volumetric flask and the uncertainty of the EDTA purity.
The molar mass also contains an uncertainty (372.2392 ± 0.0009) g/mol. However, its share is small compared to the other uncertainties and can be neglected in the case of large molar masses, which is the case with the primary standards:

$$\frac{u(c(\text{EDTA}))}{c(\text{EDTA})} = \sqrt{2\left(\frac{u(m)}{m(\text{EDTA})}\right)^2 + \left(\frac{u(V)}{V(\text{solution})}\right)^2 + \left(\frac{u(\text{purity level})}{\text{purity level}}\right)^2} =$$

$$\sqrt{2\left(\frac{\Delta m_{\text{scales}}}{\sqrt{3} \cdot m(\text{EDTA})}\right)^2 + \left(\frac{\Delta V_{\text{volumetric flask}}}{\sqrt{6} \cdot V(\text{solution})}\right)^2 + \left(\frac{\Delta\,\text{purity level}}{\sqrt{3} \cdot \text{purity level}}\right)^2} =$$

$$\sqrt{2\left(\frac{0.0001\,\text{g}}{\sqrt{3} \cdot 37.2239\,\text{g}}\right)^2 + \left(\frac{0.4\,\text{ml}}{\sqrt{6} \cdot 1,000.0\,\text{ml}}\right)^2 + \left(\frac{0.1\%}{\sqrt{3} \cdot 99.9\%}\right)^2} = 0.00060057,$$

$$u(c(\text{EDTA})) = 0.09989995\,\text{mol/l} \cdot 0.00060057 = 0.00006\,\text{mol/l}.$$

2.3.2.4 Extended uncertainty
At this point, it should be emphasized once again that the uncertainties used to calculate the combined standard uncertainty are standard uncertainties. Therefore, no statis-

tical certainty (P) is associated with the combined standard uncertainty. In analytical chemistry, a multiple of the standard uncertainty, the so-called expanded uncertainty, is specified for this reason:

$$U(\bar{x}) = k_P \cdot u(\bar{x}).$$

In most cases, it is advisable to set the coverage factor k_P equal to 2, which corresponds to a statistical confidence (P) of approximately 95% for normally distributed values [73h]. In the case of *t-distributed* data, k_P should be replaced by the corresponding *t-factor*.

In the example above, the expanded safety $U(c(EDTA)) = 0.00012$ mol/l. The EDTA concentration of the EDTA solution can then be specified as follows:

$$c(EDTA) = (0.09990 \pm 0.00012) \, mol/l.$$

The EDTA concentration lies in the range from 0.09978 mol/l to 0.10002 mol/l ([c − U, c + U]) with a 95% probability. This range is called the **confidence interval**. This topic is explained in detail in [17, 18], for example.

2.3.3 Outlier test as a statistical test method

If individual values of a multiple determination are available, these are usually checked for possible outliers – especially if an extremely high or low value is present. Various tests have been developed for this purpose. Most of these tests (Grubbs, Nalimov, Dean-Dixon, also known as the Q-test, etc.) are applied to normally distributed values. Many analytical results are normally distributed values; whether this is really the case can be checked statistically.

For normally distributed individual values, the International Organization for Standardization [73f] recommends the **outlier test** according to Grubbs-Beck [73g]:

$$g_{calculated} = \frac{|x_i - \bar{x}|}{s},$$

where x_i is the measured value to be tested, $\bar{x}$ is the mean value and s is the standard deviation of the sample.

If $g_{calculated}$ is greater than $g_{crit.}$ (table value, see Tab. 2.9), then the tested measured value is an outlier with the previously defined statistical certainty (P). If the tested value is recognized as an outlier, it is clearly marked as such in the measurement series and removed from further calculations. The mean value and the corresponding standard deviation are calculated again from the remaining values.

Example: Analysis of an aqueous sample containing aluminum resulted in the following measured values: 10.5 mg/l; 13.5 mg/l; 11.5 mg/l; 10.7 mg/l; 9.6 mg/l and 9.9 mg/l. The second measured value (see Fig. 2.26) appears to be an outlier.

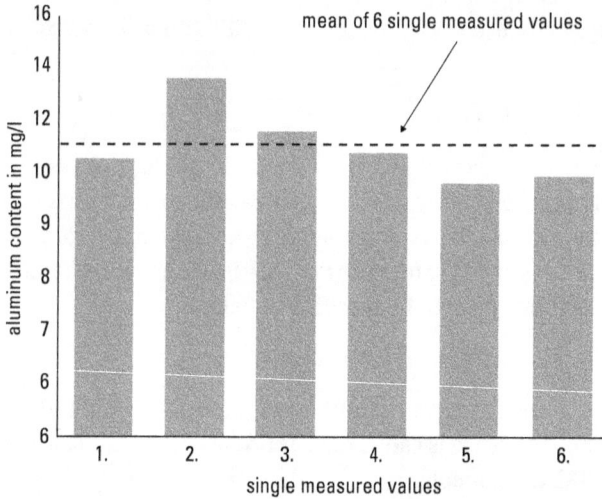

Fig. 2.26: Individual measured values and arithmetic mean of the analysis of an aqueous sample containing aluminum.

It is now to be investigated whether this is actually an outlier with a 95% statistical certainty (P).

As already described, the mean value and the standard deviation are calculated for this purpose:

$$\bar{x} = 10.9500 \text{ mg/l and } s = 1.4139 \text{ mg/l (unrounded values).}$$

This results in $g_{calculated} = 1.804$. The table value $g_{crit.}$ for $P = 95\%$ and 6 individual measurements ($n = 6$) then corresponds to 1.8871. Since the calculated value is smaller than the tabular value, the second measured value is not an outlier with a 95% statistical certainty and must therefore not be removed from the measurement series.

Tab. 2.9: Critical values ($g_{crit.}$) of the two-sided Grubbs-Beck outlier test, with n being the number of individual measurements and P the statistical reliability.

n	$P = 90\%$	$P = 95\%$	$P = 99\%$	n	$P = 90\%$	$P = 95\%$	$P = 99\%$
3	1.1531	1.1543	1.1547	17	2.4748	2.6200	2.8940
4	1.4625	1.4813	1.4963	18	2.5040	2.6516	2.9325
5	1.6714	1.7150	1.7637	19	2.5312	2.6809	2.9680
6	1.8221	1.8871	1.9728	20	2.5566	2.7082	3.0008
7	1.9381	2.0200	2.1391	21	2.5804	2.7338	3.0314
8	2.0317	2.1266	2.2744	22	2.6028	2.7577	3.0599
9	2.1096	2.2150	2.3868	23	2.6239	2.7803	3.0866
10	2.1761	2.2900	2.4821	24	2.6439	2.8016	3.1117
11	2.2339	2.3547	2.5641	25	2.6629	2.8217	3.1353
12	2.2850	2.4116	2.6357	26	2.6809	2.8408	3.1577
13	2.3305	2.4620	2.6990	27	2.6981	2.8589	3.1788
14	2.3717	2.5073	2.7554	28	2.7145	2.8762	3.1989
15	2.4090	2.5483	2.8061	29	2.7301	2.8927	3.2179
16	2.4433	2.5857	2.8521	30	2.7451	2.9085	3.2361

2.3.4 Rounding up the final result

If a final result is calculated from the measured values, the question often arises as to how exactly this must be specified or – in other words – how many valid digits should be rounded [73h].

All valid digits are referred to as **significant digits**. All digits not equal to zero and the zeros within a numerical value or at the right end of a number are significant. For example, the result of measurement method A has four significant digits and that of measurement method B has three.

Leading zeros, on the other hand, are never significant, as these zeros can disappear if the decimal places are shifted.

Further examples on the subject of significant positions are:

267.8 g   has four   significant digits; the value can also be written as $2.678 \cdot 10^2$ g, for example.

100 ml   has three significant digits; it can also be expressed as 0.100 l.

100.00 ml has five   significant digits, much more precise volume specification than in the second example.

0.001 l   has one   significant digit; it can also be specified as 1 ml.

If the numerical value of the measurement uncertainty after the leading zeros begins with a digit from 3 to 9, the measurement uncertainty must be stated with one significant digit. If the numerical value after the leading zeros begins with the number 1 or 2, the measurement uncertainty must be stated with two significant digits.

Furthermore, the numerical value of the measurement uncertainty must always be rounded up and the numerical value of the measured value is rounded up or down according to the arithmetic rules.

Example: Analysis of an aqueous sample for total iron content using two independent measurement methods yielded the following results:

	β(Fe) in g/l, unrounded result	β(Fe) in g/l, rounded result
Measuring method A:	0.20496231 ± 0.0007611	0.2050 ± 0.0008
Measuring method B:	0.21134 ± 0.01632011	0.211 ± 0.017

The number of significant digits reflects the accuracy of the experiment. If you compare the relative measurement uncertainty of the two measurement methods from the example above, you can see that measurement method A is significantly more accurate at around 0.4% than measurement method B at around 8%.

In most cases, higher accuracy is accompanied by increased measurement effort and the associated higher costs. If the result is displayed with too few significant digits, fewer than were achieved with the measurement method – e.g. 205 mg/l instead of 205.0 mg/l – then the accuracy is lower than it actually is and the measurement effort was therefore not worthwhile.

The indication of too many significant digits, on the other hand, pretends an accuracy that does not correspond to the facts – e.g. 211.34 mg/l instead of the correct 211 mg/l – and is therefore inadmissible and to be regarded as irresponsible:

> The lack of mathematical education is revealed by nothing so conspicuously as by excessive sharpness in arithmetic. (C. F. Gauss, * April 30, 1777, † February 23, 1855, German mathematician)

If the **measurement uncertainty is not known,** which usually occurs during training when solving exercises, the final result is rounded based on the significant digits of the values specified in the exercise.

In the case of *multiplication* or *division,* the number of significant digits in the result corresponds to the lowest number of significant digits in the specified values. For example,

$$c(\text{Pb}) = \frac{n(\text{Pb})}{V(\text{Solution})} = \frac{1.3469 \text{ mol}(5 \text{ sign. digits})}{250 \text{ ml}(3 \text{ sign. digits})} = 5.3876\frac{\text{mol}}{1} = 5.39\frac{\text{mol}}{1}(3 \text{ sign. digits}).$$

For *addition* and *subtraction,* however, the number with the fewest decimal places is decisive. Care must be taken to ensure that all summands have the same unit and – in the case of powers – are represented with the same exponent. When converting values beforehand, no additional significant digits may be created. The result is rounded so that it has the same number of decimal places as the number with the fewest decimal places (in this case 23.4 g):

$$588.1 \text{ mg} + 23.4 \text{ g} + 30.65 \; 10^{-2}\text{g} = 0.5881 \text{ g} + 23.4 \text{ g} + 0.3065 \text{ g} = 24.2946 \text{ g} = 24.3 \text{ g}.$$

Additions and *subtractions* are the only arithmetic operations in which the number of significant digits can change:

Example 1: Increasing the number of significant digits from 2 to 3:

$$10.0 \text{ ml} + 5.0 \text{ ml} = 15.0 \text{ ml (and not 15 ml)}.$$

Example 2: Reduction of the significant digits from 3 to 2:

$$18.9 \text{ mmol} - 10.5 \text{ mmol} = 8.4 \text{ mmol (and not 8.40 mmol)}.$$

It is extremely important to note that rounding the intermediate results often leads to a considerable error. To avoid this, the final result and its measurement uncertainty should be calculated in one calculation step or, if this is very laborious, with the help of a calculation program such as Excel.

3 Volumetric analysis with chemical end point determination

3.1 Acid-base titrations

In acid-base titrations, protons are transferred between the reaction partners. They are used to determine acids or bases. Either an acidic sample solution is converted with an equivalent amount of base by titrating with a standard solution containing the base (**alkalimetry**) or, conversely, the unknown amount of a base is determined by titration with a standard acid solution (**acidimetry**).

3.1.1 Theoretical foundations

3.1.1.1 Acids and bases

The ideas about the nature of acids and bases have undergone many changes in the history of the development of chemistry. The first steps toward clarifying chemical concepts were taken in the age of chemiatria the period that replaced medieval alchemy. Even then, the classification of substances into acids, bases (alkalis) and salts was introduced. In the second half of the seventeenth century, it was recognized that **salts** were made up of two components, one alkaline and one acidic [167]. Boyle (1627–1691) characterized the **acids** because of their properties: they have a dissolving power for many substances, they turn the blue color of plant juices (violet, cornflower, litmus juice) into red and they lose their properties when combined with **alkalis**. The latter were not yet clearly defined; a distincion was made between alkalis as oxides or hydroxides on the one hand and carbonates on the other in 1755 by Black. Even then, these groups of substances were referred to as **bases**. An exact definition of bases was not made until the nineteenth century. Because various nonmetal oxides form acidic solutions with water, Lavoisier (1743–1794) regarded oxygen as the carrier of the acidic properties of a substance and gave it the corresponding name (oxygenium = acid former). Davy then recognized in 1816 that hydrogen determines the acidic character of a substance, as there are also acids that do not contain oxygen (e.g. hydrohalic acids). Liebig finally discovered in 1838 that only hydrogen compounds in which the hydrogen can be replaced by metals can be considered acids. The reaction of an acid with a base was called **neutralization** by Liebig. He referred to compounds containing only one metal and one acid residue as **neutral salts** (e.g. NaCl and Na_2SO_4), those that also contain other hydrogen that can be replaced by metal are called **acid salts** (e.g. $NaHSO_4$ and NaH_2PO_4).

In the years 1884–1887, Arrhenius suggested the **theory of electrolytic dissociation**. According to this theory, freely moving ions are present in electrolyte solutions (solutions of acids, bases and salts that conduct an electric current). The ions are not

https://doi.org/10.1515/9783111350127-003

created by splitting the dissolved molecules under the effect of an electric field – as was previously assumed – but when the compounds dissolve in water. The decomposition into individual ions caused by the solvent water is called **electrolytic dissociation**. Ostwald's work on ionic equilibria in aqueous solutions further expanded the theory.

According to the definitions of the classical theory of Arrhenius and Ostwald acids are compounds that split off **hydrogen ions** (H^+) in aqueous solution (e.g. $HCl \rightarrow H^+ + Cl^-$) and bases are substances that dissociate with the release of **hydroxide ions** (OH^-) in aqueous solution (e.g. $NaOH \rightarrow Na^+ + OH^-$). The stoichiometric reaction of an acid and a base produces salt and water, e.g.

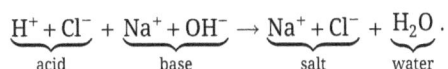

$$\underset{\text{acid}}{\underline{H^+ + Cl^-}} + \underset{\text{base}}{\underline{Na^+ + OH^-}} \rightarrow \underset{\text{salt}}{\underline{Na^+ + Cl^-}} + \underset{\text{water}}{\underline{H_2O}} \,.$$

The cations of the base and the anions of the acid are not involved in this process, **neutralization**, so salt formation is of secondary importance. The base cations and acid anions remain dissociated in the solution and form the crystalline salt when the solvent evaporates (easily soluble salts) or the crystalline salt is already formed during neutralization (poorly soluble salts). The essence of neutralization as defined by Arrhenius is the combination of the hydrogen ions of the acid with the hydroxide ions of the base to form water:

$$H^+ + OH^- \rightleftharpoons H_2O.$$

The reversal of neutralization, the reaction of a salt with water to form a base and an acid, was called **hydrolysis** of the salt. The classical theory thus explained the acidic or basic reaction of solutions of salts containing either a weak base or a weak acid (see p. 79). Here are two examples:

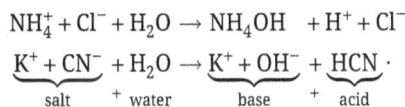

$$NH_4^+ + Cl^- + H_2O \rightarrow NH_4OH + H^+ + Cl^-$$
$$\underset{\text{salt}}{\underline{K^+ + CN^-}} + \underset{\text{water}}{\underline{H_2O}} \rightarrow \underset{\text{base}}{\underline{K^+ + OH^-}} + \underset{\text{acid}}{\underline{HCN}} \,\cdot$$

The Arrhenius-Ostwald theory can be used to describe many reactions in aqueous solutions. Using the dissociation constants of acids and bases made it possible to quantify their strength for the first time. The electrical conductivity of electrolyte solutions and their anomalies of osmotic pressure and freezing point depression could be explained. However, the theory has various weaknesses. It is limited to aqueous solutions, the acid-base reaction is restricted to electrically neutral molecular acids and bases such as HCl, H_2SO_4, KOH and $Ca(OH)_2$. It does not take into account that free protons (H^+) cannot occur in aqueous solutions. To interpret the basic behavior of compounds that do not contain OH^- groups (NH_3, organic bases), compounds had to be assumed (e.g. NH_4OH) that do not exist.

In 1923, Brønsted and Lowry independently proposed a more comprehensive definition for acids and bases which focuses on function rather than constitution. According to this definition, an acid is a compound that can release protons (**proton donor**), whereas a base is able to absorb protons (**proton acceptor**). Acids and bases are collectively referred to as **protolytes**. The definition is illustrated by the schematic reaction equation:

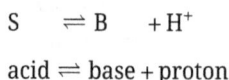

$$S \rightleftharpoons B + H^+$$

$$\text{acid} \rightleftharpoons \text{base} + \text{proton}$$

When a proton is released, the acid changes into a base (**corresponding** or **conjugate base**) and vice versa; when a proton is absorbed, the base changes into an acid (**corresponding** or **conjugate acid**). Acid and conjugate base are assigned to each other in a protolyte system, and they form a **corresponding acid-base pair**. Examples of such acid-base pairs are

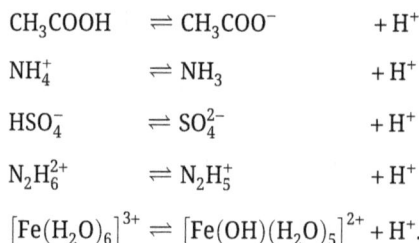

$$CH_3COOH \rightleftharpoons CH_3COO^- \quad\quad + H^+$$

$$NH_4^+ \rightleftharpoons NH_3 \quad\quad + H^+$$

$$HSO_4^- \rightleftharpoons SO_4^{2-} \quad\quad + H^+$$

$$N_2H_6^{2+} \rightleftharpoons N_2H_5^+ \quad\quad + H^+$$

$$\left[Fe(H_2O)_6\right]^{3+} \rightleftharpoons \left[Fe(OH)(H_2O)_5\right]^{2+} + H^+.$$

As you can see, acids can be neutral molecules, positively charged cations or negatively charged anions; the same applies to bases. Accordingly, a distinction is made between **neutral acids** (e.g. HCl, H_2SO_4 and H_3CCOOH) and **neutral bases** (e.g. NH_3, NH_2OH and PH_3), **cationic acids** (e.g. NH_4^+, $N_2H_6^{2+}$ and $[Fe(H_2O)_6]^{3+}$) and cationic **bases** (e.g. $N_2H_5^+$ and $[Fe(OH)(H_2O)_5]^{2+}$) and **anionic acids** (e.g. HSO_4^-, $H_2PO_4^-$ and HS^-) and **anionic bases** (e.g. H_3CCOO^-, SO_4^{2-} and OH^-).

In aqueous solution, the alkali metal hydroxides form the base OH^-, one special anion base among many.

A Brønsted acid always has one more positive charge unit than its corresponding base; however, the charge does not play a role in the acid or base effect.

The reaction equations listed are only of a formal nature, as free protons cannot occur in solutions due to their high reactivity, which is due to the small radius of the particle (approx. 10^{-3} pm) and the resulting extraordinarily high positive charge density. It follows that an acid can only release its proton if a base is present to absorb it. An **acid-base reaction** therefore requires two corresponding acid-base pairs:

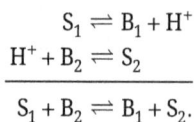

$$S_1 \rightleftharpoons B_1 + H^+$$
$$\underline{H^+ + B_2 \rightleftharpoons S_2}$$
$$S_1 + B_2 \rightleftharpoons B_1 + S_2.$$

According to Brønsted, the acid-base reaction is a **proton transfer (protolysis)**. The term "dissociation" of acids and bases is replaced by that of protolysis.

An acid can only protolyze in a solvent that can absorb protons, and a base can only protolyze in a solvent that can release protons. If, for example, ammonia is dissolved in water, the water reacts as an acid and the protolysis can be described by the reaction:

$$NH_3 + H_2O \rightleftharpoons NH_4^+ + OH^-$$

The H_2O molecule releases a proton and changes into the corresponding base OH^-. When gaseous HCl dissolves in water, however, water behaves as a base and absorbs a proton from the acid HCl:

$$HCl + H_2O \rightleftharpoons H_3O^+ + Cl^-.$$

The resulting acid H_3O^+ is the **oxonium ion**. The individual H_3O^+ ion has an extremely short lifetime in aqueous solution. After only around 10^{-13} s, it transfers one of its three protons to a neighboring H_2O molecule. The positive charge is not fixed, and it is distributed symmetrically among the three protons, which can therefore form three stable hydrogen bonds to neighboring H_2O molecules. The hydrated oxonium ion is also known as the **hydronium ion**, $H_9O_4^+$ ($H_3O^+ \cdot 3\ H_2O$); it is particularly stable compared to other possible hydrates such as $H_5O_2^+$ or $H_7O_3^+$. The hydroxide ion (OH^-) is also hydrated, preferably as $H_7O_4^-$ ($OH^- \cdot 3\ H_2O$). The secondary hydration is not important for the following considerations. The simplified notation H_3O^+ is therefore used.

As the reactions of water with NH_3 and HCl show, the H_2O molecule behaves either as an acid or as a base, depending on the reactant. Protolytes that can both absorb and release protons are called **ampholytes**. Other examples of ampholytes are the ions HSO_4^-, $H_2PO_4^-$ and HPO_4^{2-}:

$$HSO_4^- + H^+ \rightleftharpoons H_2SO_4$$
$$HSO_4^- \rightleftharpoons SO_4^{2-} + H^+$$
$$H_2PO_4^- + H^+ \rightleftharpoons H_3PO_4$$
$$H_2PO_4^- \rightleftharpoons HPO_4^{2-} + H^+$$
$$HPO_4^{2-} \rightleftharpoons PO_4^{3-} + H^+.$$

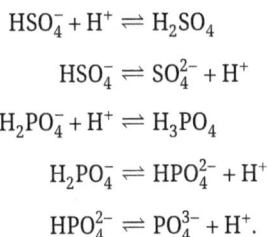

According to Brønsted electrolytes are divided into protolytes (acids and bases) and salts. Salts are all compounds that are made up of ions, e.g. NaCl, but also metal hydroxides such as NaOH and metal oxides such as Na_2O. The ions are formed when the salts dissolve in water can behave as protolytes (such as OH^- in NaOH solutions), but do not have to (such as Na^+ and Cl^- in NaCl solutions). Salts are not protolytes, but they can provide acids and bases.

If a solution of a practically completely protolyzed acid (e.g. HCl) is added to the solution of an equally protolyzed base (e.g. NaOH), the proton transfer from the H_3O^+ ions to the OH^- ions takes place; water is formed according to the following reaction:

$$H_3O^+ + Cl^- + Na^+ + OH^- \rightleftharpoons 2\,H_2O + Na^+ + Cl^-.$$

The acid-base reaction

$$H_3O^+ + OH^- \rightleftharpoons 2\,H_2O$$

corresponds to neutralization in the way Arrhenius described it. It proceeds at high speed and with a positive heat tint (exothermic reaction). If acid and base are used in equivalent quantities, neither H_3O^+ nor OH^- ions are present in excess after the reaction; complete neutralization occurs. From the rate constant of the above reaction ($k = 1.3 \cdot 10^{11}$ l/mol s at 25 °C), it can be calculated that in solutions with a concentration of 0.1 mol/l, 99.9% of the reactants have already reacted after $7.7 \cdot 10^{-8}$ s. Regardless of the type of acid and base used – provided they are subject to almost complete protolysis – the same amount of heat $\Delta H° = -57$ kJ/mol (heat of neutralization) is always released, proving that all neutralizations are based on the same reaction.

The term "hydrolysis" is no longer required in Brønsted theory. Since ions can also be acids and bases by definition, hydrolysis is nothing other than the acid or base effect of ionic acids or bases. Thus, the acidic reaction of an NH_4Cl solution is explained by the protolysis equilibrium

$$NH_4^+ + H_2O \rightleftharpoons NH_3 + H_3O^+$$

and the basic reactions of sodium acetate-, Na_2CO_3- and KCN-solution through the protolysis reactions:

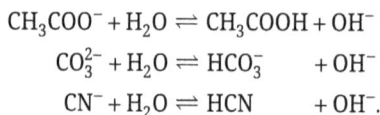

$$\begin{aligned}
CH_3COO^- + H_2O &\rightleftharpoons CH_3COOH + OH^- \\
CO_3^{2-} + H_2O &\rightleftharpoons HCO_3^- \quad + OH^- \\
CN^- + H_2O &\rightleftharpoons HCN \quad + OH^-.
\end{aligned}$$

Metal cations, such as Fe^{3+}, are hydrated in aqueous solution. The protolysis reaction

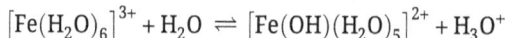

$$\left[Fe(H_2O)_6\right]^{3+} + H_2O \rightleftharpoons \left[Fe(OH)(H_2O)_5\right]^{2+} + H_3O^+$$

explains why the solution of a metal salt reacts acidically. The coordination of the H_2O molecule to the Fe^{3+} ion reduces the electron density at the oxygen and thus facilitates the detachment of a proton. How strongly acidic the metal salt solution reacts depends on the metal ion. For example, the hydrated Fe^{3+} ion is more acidic than the hydrated Al^{3+} ion; with the hydrated Na^+ ion, the acidic character is practically not noticeable at all.

From the explanations of the Brønsted theory, it is clear that the processes of dissociation of acids and bases, neutralization and hydrolysis, which have different names according to the classical theory of Arrhenius and Ostwald of acids and bases, are to be regarded as similar from the point of view of proton transfer. For Brønsted, the proton occupies the central position; the theory therefore applies not only to water but also to all solvents in which protons can be transferred (**prototropic solvents**), e.g. for liquid ammonia $2\,NH_3 \rightleftharpoons NH_4^+ + NH_2^-$, anhydrous acetic acid $(2\,CH_3COOH \rightleftharpoons CH_3COOH_2^+ + CH_3COO^-)$, anhydrous sulfuric acid $(2\,H_2SO_4 \rightleftharpoons H_3SO_4^+ + HSO_4^-)$, etc. It is also not limited to the liquid state. An acid-base reaction in the gas phase is, for example, the conversion of gaseous HCl and gaseous NH_3 to NH_4Cl.

An extension to other ionizing (ionotropic) solvents that do not contain protons is provided by the **solvent theory** given by Cady (1928). This and other acid-base theories of Lewis (1923), Bjerrum (1951), Ussanović (1939) and Pearson (1963), which each describe certain types of reaction from a different perspective and create organizing principles, will not be discussed in detail here. (For more details and further literature referencessee [74]).

3.1.1.2 Autoprotolysis of water

Even the purest water contains ions, as the – albeit very low – electrical conductivity of water proves (specific electrical conductivity $\kappa = 4.3 \cdot 10^{-6}$ S/m at 18 °C). The ions are formed by autoprotolysis of the water (**self-dissociation of the water**) according to

$$2\,H_2O \rightleftharpoons H_3O^+ + OH^-.$$

Autoprotolysis can be regarded as the transfer of protons from one water molecule to another. The equilibrium of the reaction is almost entirely on the left-hand side, i.e. there are very few ions present in pure water (10^7 l of water at 24 °C contains 1 mol each of H_3O^+ and OH^- ions).

If the **law of mass action** is applied to the autoprotolysis of water it results in the following relationship for the equilibrium state

$$\frac{c(H_3O^+) \cdot c(OH^-)}{c^2(H_2O)} = K_c$$

respectively

$$c(H_3O^+) \cdot c(OH^-) = K_c \cdot c^2(H_2O).$$

As the ion concentrations $c(H_3O^+)$ and $c(OH^-)$ in pure water are very small, they can be neglected compared to the equilibrium concentration of the water $c(H_2O)$:

$$c(H_2O)_{total} = c(H_2O) + c(H_3O^+) + c(OH^-).$$

This makes $c(H_2O)$ practically equal to the total concentration of water $c(H_2O)_{total}$:

$$c(H_2O)_{total} \approx c(H_2O).$$

In diluted solutions, the total concentration of water differs only slightly from that in pure water. Therefore, $c(H_2O)$ can be regarded as constant with a good approximation and included in the constant K_c.

At 25 °C

$$c(H_2O) = \frac{\rho_{25}(H_2O) \cdot 1000}{M(H_2O)},$$

with $\rho_{25}(H_2O)$ = 0.997043 g/ml, $M(H_2O)$ = 18.0152 g/mol and the conversion factor 1000 ml/l results in

$$c(H_2O) = 55.34 \, mol/l.$$

You get

$$K_c \cdot c^2(H_2O) = c(H_3O^+) \cdot c(OH^-) = K_w,$$

where K_w is the **ion product of water**.

Its experimentally determined value is K_w = 10^{-14} mol²/l² at 24 °C. The product of the mass concentrations of hydrogen and hydroxide ions is not only constant in pure water, the equation also applies to dilute aqueous solutions. In water and in solutions in which the ions formed from the dissolved substance are not subject to protolysis, the following applies

$$c(H_3O^+) = c(OH^-) = \sqrt{K_w} = 10^{-7} mol/l.$$

The autoprotolysis of water is an endothermic process and therefore increases with increasing temperature. The ion product is therefore temperature-dependent. This dependence is shown for temperatures between 0 and 60 °C in Tab. 3.1. For practical reasons, the negative decadic logarithm of the numerical value is often used instead of K_w, which is referred to as the pK_w value. The pK_w values are also contained in Tab. 3.1.

When applying the law of mass action to a chemical equilibrium reaction in the way described by Guldberg and Waage using concentrations, one obtains the **stoichiometric equilibrium constant** K_c. This quantity should only depend on the temperature, but not on the concentrations of the reactants or other ions present in the solution. However, as a result of electrostatic interactions between anions and cations, which are greater than the ions that are in the solution, a concentration dependence of K_c is observed. As a result of the attractive forces between the ions, their concentrations appear to be lower than they actually are. A concentration-independent quantity that only depends on the temperature for a given chemical equilibrium is the **thermodynamic equilibrium constant** K^0. It is obtained by replacing the ion concentrations c with their **activities** a in the equilibrium state in the expression for the law of mass action; for the autoprotolysis of water, for example, the result is

$$\frac{a(H_3O^+) \cdot a(OH^-)}{a^2(H_2O)} = K^0$$

The activity of a substance (ion) is a thermodynamic quantity that formally has the meaning of a corrected concentration. It is defined in such a way that the law of mass action is always strictly valid when it is used [36–39]. Concentration c and activity a are linked to each other by a correction factor, the **activity coefficient** f:

$$a = f \cdot c.$$

It records the deviations of the real solutions from the ideal behavior in which there are no interactions between the ions. The numerical value of f depends on the ionic charge number and the concentration of the ion in question as well as the ionic charges and the concentrations of the other ions present in the solution.

Tab. 3.1: Temperature dependence of the ion product K_w, the pK_w and pH values [6].

t (°C)	$K_w \cdot 10^{-14}$	pK_w	pH
0	0.1139	14.9435	7.4718
10	0.2920	14.5346	7.2673
15	0.4505	14.3463	7.1732
20	0.6809	14.1669	7.0835
25	1.008	13.9965	6.9983
30	1.469	13.8330	6.9165
40	2.919	13.5348	6.7674
50	5.474	13.2617	6.6309
60	9.610	13.0171	6.5086

With increasing dilution, the value of f increases and reaches the value 1 when the sum of the concentrations of all ions present in the solution becomes zero, expressed mathematically:

$$\lim_{\sum c \to 0} f = 1.$$

This hypothetical limiting case is achieved in the pure solvent. In real solutions, $f < 1$; but even in sufficiently dilute solutions ($c < 10^{-3}$ mol/l), $f \approx 1$ can be set, so that ion concentrations can be used as a good approximation instead of activities. If the deviations of the f-values from 1 are neglected in more concentrated solutions, only approximate calculations are possible. For more precise calculations, the exact form of the law of mass action with the activities must be used. In such cases, the value of the activity coefficient can be calculated using the method developed by Debye and Hückel (1923). According to this theory, there is a relationship between f and the **ionic strength** I in dilute electrolyte solutions. The ionic strength I is understood to be the quantity

$$I = \frac{1}{2} \sum_i c_i \cdot z_i^2$$

with the concentration c_i and the ionic charge number z_i of the ionic species i. For example, in a solution containing 0.01 mol/l Na_2SO_4 and 0.05 mol/l KCl, the ionic strength $I = \frac{1}{2}\,(2 \cdot 0.01 \cdot 1^2 + 0.01 \cdot 2^2 + 0.05 \cdot 1^2 + 0.05 \cdot 1^2) = 0.08$.

From the ionic strength, approximate formulas derived from the complete Debye-Hückel relationship between f and I, the activity coefficient f_i of an ionic species can be estimated, namely in aqueous solutions at 25 °C for $I \leq 0.001$ according to

$$\lg f_i = -0.509 \cdot z_i^2 \cdot \sqrt{I}$$

and for $I < 0.1$ according to

$$\lg f_i = -0.509 \cdot \frac{z_i^2 \cdot \sqrt{I}}{1 + \sqrt{I}}.$$

Table 3.2 shows some approximate values calculated according to these formulas for three different ionic charge numbers as a function of the ionic strength I. More precise f-values for numerous ions were calculated by Kielland [75] by including the effective radii for the hydrated ions in the calculation; numerical values in [11].

Tab. 3.2: Calculated activity coefficients f for different ionic strengths I.

I	f		
	$z = 1$	$z = 2$	$z = 3$
0	1.00	1.00	1.00
0.001	0.96	0.86	0.72
0.002	0.95	0.82	0.64
0.005	0.93	0.73	0.50
0.01	0.90	0.65	0.38
0.02	0.86	0.56	0.27
0.05	0.81	0.42	0.15
0.1	0.75	0.32	0.08

Although the formulas make it possible to calculate the individual activity coefficient of a single ion type, only mean activities $a_\pm$ and mean activity coefficients $f_\pm$ of ion pairs of cation and anion can be determined experimentally, as only solutions containing both cations and anions can be produced. As a first approximation

$$\lg f_\pm = -0.509 \cdot z_+ \cdot z_- \cdot \sqrt{I}.$$

For ionic strengths $I > 0.1$, the f-values are individually dependent on the ions that contribute to the ionic strength. They can no longer be estimated, but must be determined experimentally.

For volumetric analysis practice, it is usually sufficient to use concentrations instead of activities. For this reason, concentration and activity are generally equated in the following considerations. Of course, this only applies with sufficient approximation to appropriately diluted aqueous solutions.

3.1.1.3 Hydrogen ion concentration and pH value

The term "hydrogen ion concentration" goes back to the way Arrhenius described it. However, it is still used today instead of the term "oxonium ion concentration" for the quantity $c(H_3O^+)$; in the same way, the term "hydrogen ion" is used when referring to the H_3O^+ ion.

We had seen that in pure water and in aqueous solutions in which no protolysis reaction with water occurs, the concentrations of H_3O^+ and OH^- ions are the same. Their numerical value is determined by the ion product of the water and is 10^{-7} mol/l at 24 °C. In acidic solutions the concentration of hydrogen ions is greater than that of hydroxide ions,

$$c(H_3O^+) > 10^{-7} > c(OH^-),$$

in basic (alkaline) solutions there are more OH^- ions than H_3O^+ ions,

$$c(OH^-) > 10^{-7} > c(H_3O^+).$$

The ion product K_w is constant at a given temperature, but the two concentrations $c(H_3O^+)$ and $c(OH^-)$ in the equation

$$c(H_3O^+) \cdot c(OH^-) = K_w$$

can be varied within very wide limits by adding acids or bases to the water. Using the definition equation for the ionic product, the OH^- concentration present in the solution can be calculated for any H_3O^+ concentration and the corresponding H_3O^+ concentration can be calculated for any OH^- concentration. Thus, in a sodium hydroxide solution of concentration $c(NaOH) = 0.01$ mol/l, $c(OH^-) = 10^{-2}$ mol/l and $c(H_3O^+) = 10^{-14}/10^{-2} = 10^{-12}$ mol/l.

To characterize the acidic nature (acidity) of a solution, the negative decadic logarithm of its numerical value is usually used instead of the H_3O^+ concentration, whose numerical value can extend over many powers of 10, for ease of use. This simplified notation was introduced in 1909 by Sörensen, S. P. L. [76], who called the value **Wasserstoffexponent (hydrogen exponent)** and used the symbol **pH** for it. He defined:

$$pH = -\lg c(H^+).$$

Following the introduction of the concept of activity and taking into account the definition of acids established by Brønsted the current definition is as follows: the pH value is the logarithm of the numerical value of the H_3O^+ activity multiplied by (−1):

$$pH = -\lg a(H_3O^+) \approx -\lg c(H_3O^+).$$

As already mentioned, the activities are usually not used in volumetric analysis. In strongly acidic and strongly alkaline solutions, measured and calculated pH values therefore differ.

The pH value is dimensionless because it is not the concentration as a physical quantity, the product of numerical value and unit, that can be logarithmized, but only its numerical value. If $c(H_3O^+) = 10^{-2}$ mol/l in a solution, its pH value is 2. If the H_3O^+ concentration is $3.7 \cdot 10^{-8}$ mol/l, the following applies: $pH = -lg\,(3.7 \cdot 10^{-8}) = -(lg\,3.7 + lg\,10^{-8}) = -(0.57 - 8) = 7.43$. On the other hand, $pH = 5.8$ means that $c(H_3O^+) = 10^{-5.8}$ mol/l $= 10^{0.2} \cdot 10^{-6}$ mol/l $= 1.58 \cdot 10^{-6}$ mol/l.

In neutral solutions is $pH = 7$, in acidic solutions $pH < 7$ applies, in alkaline solutions $pH > 7$. The higher the pH value is, the more alkaline the solution is, the lower the pH is, the more acidic it is.

In a similar way, the alkaline character of a solution (the basicity or alkalinity) the negative decadic logarithm of the numerical value of the hydroxide ion concentration is defined as the **pOH value**:

$$pOH = -lg\,a(OH^-) \approx -lg\,c(OH^-).$$

From the equation for the ionic product of water, logarithmizing and multiplying by (−1) gives the relationship

$$pH + pOH = pK_w = 14.$$

In aqueous solutions, the normal pH scale ranges ranges from 0 to 14, which means that the H_3O^+ concentration in all solutions containing between a maximum of 1 mol/l H_3O^+ ions and a maximum of 1 mol/l OH^- ions can be expressed by positive numbers between 0 and 14. For solutions with pH values < 0 (overacidic solutions) and those with pH values > 14 (overalkaline solutions), the pH value can no longer be used to infer the H_3O^+ concentration.

The following diagram illustrates the relationships between the ion concentrations $c(H_3O^+)$ and $c(OH^-)$ as well as the pH and pOH values:

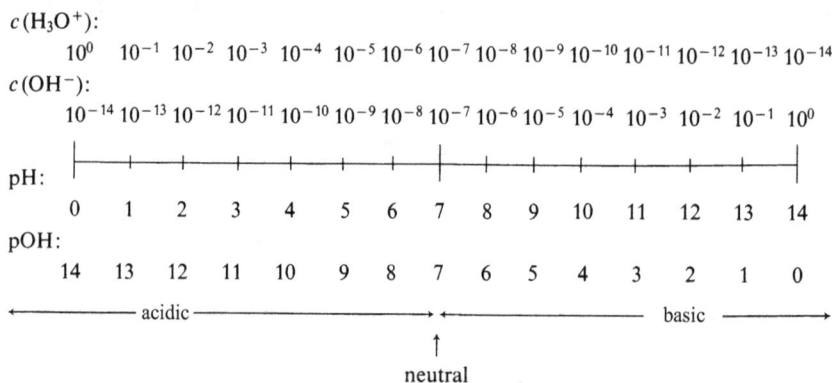

$c(H_3O^+)$:

10^0 10^{-1} 10^{-2} 10^{-3} 10^{-4} 10^{-5} 10^{-6} 10^{-7} 10^{-8} 10^{-9} 10^{-10} 10^{-11} 10^{-12} 10^{-13} 10^{-14}

$c(OH^-)$:

10^{-14} 10^{-13} 10^{-12} 10^{-11} 10^{-10} 10^{-9} 10^{-8} 10^{-7} 10^{-6} 10^{-5} 10^{-4} 10^{-3} 10^{-2} 10^{-1} 10^0

pH:

0 1 2 3 4 5 6 7 8 9 10 11 12 13 14

pOH:

14 13 12 11 10 9 8 7 6 5 4 3 2 1 0

←————— acidic —————→   ←————— basic —————→

↑
neutral

3.1.1.4 Strength of acids and bases

The strength of an acid can be characterized by its tendency to release protons, that of a base by its tendency to absorb protons. An absolute measure of the strength would be the equilibrium constant of the reaction

$$HA \rightleftharpoons A^- + H^+,$$

which, however, cannot take place unless a reaction partner absorbs or supplies the protons (see p. 69). The strength of acids and bases can therefore only be determined relatively, in relation to another base or acid. The usual reference substance chosen is water, which is particularly suitable as an ampholyte.

Due to the protolysis equilibrium of an acid HA or a base B in water,

$$HA + H_2O \rightleftharpoons H_3O^+ + A^- \qquad B + H_2O \rightleftharpoons BH^+ + OH^-$$

a distinction is made between strong acids or bases, in which the equilibrium is largely on the side of the protolysis products (right), and weak acids or bases, in which it is more or less on the side of the starting materials (left). A quantitative measure of the strength of the protolytes is obtained by applying the law of mass action to the equilibria:

$$\frac{c(H_3O^+) \cdot c(A^-)}{c(HA) \cdot c(H_2O)} = K_c, \qquad \frac{c(BH^+) \cdot c(OH^-)}{c(B) \cdot c(H_2O)} = K'_c.$$

If the concentration of water $c(H_2O) = 55.34$ mol/l, which is constant in dilute solutions, is included in the equilibrium constants, the following expressions are obtained:

$$\frac{c(H_3O^+) \cdot c(A^-)}{c(HA)} = K_a, \qquad \frac{c(BH^+) \cdot c(OH^-)}{c(B)} = K_b,$$

where K_a is the **acid constant** and K_b is the **base constant**; according to the Arrhenius theory, the quantities were referred to as **dissociation constants of the acid** or **base**, respectively. The decadic logarithms of their numerical values multiplied by (−1) are called the **acid exponent** or **base exponent**:

$$pK_a = -\lg K_a, \quad pK_b = -\lg K_b.$$

An acid (base) is stronger the greater its acid constant K_a (base constant K_b) or the smaller its acid exponent pK_a (base exponent pK_b).

If we consider a corresponding acid-base pair in water:

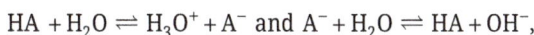

$$HA + H_2O \rightleftharpoons H_3O^+ + A^- \text{ and } A^- + H_2O \rightleftharpoons HA + OH^-,$$

so is

$$K_a = \frac{c(H_3O^+) \cdot c(A^-)}{c(HA)} \qquad \text{and} \qquad K_b = \frac{c(HA) \cdot c(OH^-)}{c(A^-)}.$$

The product of the protolysis constants

$$K_a \cdot K_b = \frac{c(H_3O^+) \cdot c(A^-) \cdot c(HA) \cdot c(OH^-)}{c(HA) \cdot c(A^-)},$$

$$K_a \cdot K_b = c(H_3O^+) \cdot c(OH^-),$$

$$K_a \cdot K_b = K_w.$$

is the ion product of water. With the corresponding exponents, the relationship is

$$pK_a + pK_b = pK_w.$$

You can see that the stronger an acid is, the weaker its corresponding base is, and the weaker an acid is, the stronger this base is. If you know the acid constant, you can easily calculate the corresponding base constant (and vice versa).

Table 3.3 lists some important acid-base pairs in order of decreasing acidity. The values apply to diluted aqueous solutions (approx. 0.1–0.01 mol/l) at 25 °C.

Tab. 3.3: Acid and base exponents of corresponding acid-base pairs [6, 20].

pK_a	Acid	Base	pK_b
~-9	$HClO_4$	ClO_4^-	~23
~-8	HI	I^-	~22
~-6	HBr	Br^-	~20
~-3	HCl	Cl^-	~17
~-3	H_2SO_4	HSO_4^-	~17
-1.74	H_3O^+	H_2O	15.74
-1.32	HNO_3	NO_3^-	15.32
1.25	$H_2C_2O_4$	$C_2O_4H^-$	12.75
1.76	$H_2SO_3 (SO_2 + H_2O)$	HSO_3^-	12.24
1.99	HSO_4^-	SO_4^{2-}	12.01
2.23	H_3PO_4	$H_2PO_4^-$	11.77
2.22	$[Fe(H_2O)]_6^{3+}$	$[Fe(H_2O)_5(OH)]^{2+}$	11.78
3.17	HF	F^-	10.83
3.25	HNO_2	NO_2^-	10.75
3.75	$HCOOH$	$HCOO^-$	10.25
4.28	$C_2O_4H^-$	$C_2O_4^{2-}$	9.72
4.75	CH_3COOH	CH_3COO^-	9.25
4.85	$[Al (H_2O)]_6^{3+}$	$[Al (H_2O)_5(OH)]^{2+}$	9.15
6.35	$H_2CO_3 (CO_2 + H_2O)$	HCO_3^-	7.65
7.00	H_2S	HS^-	7.00
7.20	HSO_3^-	SO_3^{2-}	6.80
7.21	$H_2PO_4^-$	HPO_4^{2-}	6.79
9.28	H_3BO_3	$H_2BO_3^-$	4.72
9.25	NH_4^+	NH_3	4.75
9.32	HCN	CN^-	4.68

Tab. 3.3 (continued)

pK_a	Acid	Base	pK_b
9.61	$[Zn(H_2O)]_6^{2+}$	$[Zn(H_2O)_5(OH)]^+$	4.39
10.25	HCO_3^-	CO_3^{2-}	3.75
11.65	H_2O_2	HO_2^-	2.35
12.32	HPO_4^{2-}	PO_4^{3-}	1.68
12.75	$H_2BO_3^-$	HBO_3^{2-}	1.25
12.92	HS^-	S^{2-}	1.08
13.80	HBO_3^{2-}	BO_3^{3-}	0.20
15.74	H_2O	OH^-	-1.74

Acids and bases have been categorized according to their strength as follows:

very strong acid (base)	pK_a (pK_b) $\leq$ -1.74
strong acid (base)	pK_a (pK_b) = -1.74 to 4.5
weak acid (base)	pK_a (pK_b) = 4.5 to 9.5
very weak acid (base)	pK_a (pK_b) = 9.5 to 15.74
extremely weak acid (base)	pK_a (pK_b) $\geq$ 15.74.

The acid constant of the oxonium ion is given by

$$H_3O^+ + H_2O \rightleftharpoons H_2O + H_3O^+,$$

where the H_2O to the left of the equilibrium sign is a molecule of the solvent water to which the proton is transferred, and the H_2O to the right represents the base corresponding to H_3O^+ to form

$$K_a = \frac{c(H_2O) \cdot c(H_3O^+)}{c(H_3O^+)} = c(H_2O),$$

$$K_a = 55.34 \, mol/l,$$

$$pK_a = -1.74.$$

The acid constant of water is obtained in the same way according to

$$H_2O + H_2O \rightleftharpoons OH^- + H_3O^+$$

to

$$K_a = \frac{c(OH^-) \cdot c(H_3O^+)}{c(H_2O)} = \frac{K_w}{c(H_2O)},$$

$$K_a = \frac{10^{-14} \, \text{mol}^2/\text{l}^2}{55.34 \, \text{mol}/\text{l}} = 1.80 \cdot 10^{-16} \text{mol}/\text{l},$$

$$pK_a = 15.74.$$

These two pK_a values limit the range in which acids and bases can occur in aqueous solutions. Stronger acids protolyze completely in water. Their solutions only contain the H_3O^+ ion as the acid and the anion as the corresponding base. Therefore, H_3O^+ is the strongest acid in diluted aqueous solutions. Similarly, OH^- is the strongest base that can occur in aqueous solutions. Stronger bases are protonated, and they contain the corresponding extremely weak acid in addition to the base OH^-. This fact means that all acids and bases with pK_a or pK_b values ≤ -1.74 are equally strong in diluted aqueous solutions, i.e. their solutions of the same concentration always have the same pH value (**leveling effect** of water).

If you want to determine the relative strength of such acids or bases, you must carry out the measurements in a different solvent that is more acidic (e.g. anhydrous acetic acid) or more basic (e.g. anhydrous ammonia) than water. The pK_a values listed in Tab. 3.3 for the very strong acids and bases were determined by measuring the solvent-dependent equilibrium constants and then related to the water system. They are to be regarded as approximate values and are for guidance only.

A distinction is made between **monovalent (monobasic)** and **polyvalent (polybasic) acids** according to the number of protons that an acid can release. Similarly, we speak of a **polyvalent (polyacid) base** if it can absorb several protons or supply OH^- ions. The protolysis of polyvalent acids and bases takes place in stages. The law of mass action can be applied to each **protolysis stage** (Arrhenius term: dissociation stage), whereby several acid or base constants are obtained. They become smaller from stage to stage (their pK values become larger), as the example of trivalent phosphoric acid shows (see Tab. 3.3):

$$H_3PO_4 + H_2O \rightleftharpoons H_3O^+ + H_2PO_4^-; \quad K_{a1} = 6.92 \cdot 10^{-3} \text{ mol}/\text{l}, \; pK_{a1} = 2.16,$$

$$H_2PO_4^- + H_2O \rightleftharpoons H_3O^+ + HPO_4^{2-}; \quad K_{a2} = 6.17 \cdot 10^{-8} \text{ mol}/\text{l}, \; pK_{a2} = 7.21,$$

$$HPO_4^{2-} + H_2O \rightleftharpoons H_3O^+ + PO_4^{3-}; \quad K_{a3} = 1.97 \cdot 10^{-13} \text{ mol}/\text{l}, \; pK_{a3} = 12.32.$$

The reason for the decrease in the K_a or K_b values is the increase in the negative charge after each proton release or the increase in the positive charge after each proton absorption (for bases). The electrostatic attraction or repulsion makes the release or absorption of the next proton more difficult. As the reaction equations show, every ion that occurs as an intermediate product in the protolysis equilibria can react as an acid and as a base, i.e. it is an ampholyte.

3.1.1.5 Calculation of pH values

For the following calculations, formulas are used that result from generally permissible simplifications. With their help, the calculations can be carried out quickly: Although the results must be regarded as approximate values, in most cases they fully satisfy the requirements of quantitative analysis. Deviations of the solutions from the ideal behavior are not taken into account.

Very strong acids and bases: As water has a leveling effect, solutions of very strong acids of the same concentration also have the same pH values. The same applies to very strong bases. Protolytes with K_a or K_b values > 55.34 mol/l (pK < –1.74) are to be regarded as completely protolyzed for concentrations < 1 mol/l. The hydrogen ion concentration is equal to the total concentration c_0 of the acid (S) or base (B):

$$c(H_3O^+) = c_0(S), \qquad \text{respectively,} \qquad c(OH^-) = c_0(B),$$
$$pH \quad = -\lg c_0(S) \qquad\qquad\qquad pH \quad = pK_W + \lg c_0(B).$$

Example: A hydrochloric acid with a concentration of 0.01 mol/l has a pH value of 2, a caustic soda solution with a concentration of 0.02 mol/l has a pH value of 12.30.

If the solution of a very strong acid or base is diluted to such an extent that $c(H_3O^+)$ or $c(OH^-) \leq 10^{-6}$ mol/l, the autoprotolysis of the water must be taken into account when calculating the pH (see p. 73).

If both protolysis constants of a divalent acid or base are very large, the approximation is valid:

$$c(H_3O^+) = 2 \cdot c_0(S) \qquad \text{or} \quad c(OH^-) = 2 \cdot c_0(B),$$
$$pH \quad = -\lg (2 \cdot c_0(S)) \quad \text{or} \quad pH \quad = pK_W + \lg (2 \cdot c_0(B)),$$

Example: The pH value of a Ba(OH)$_2$ solution with a concentration of 0.04 mol/l is

$$c(OH)^- = 2 \cdot c(Ba(OH)_2),$$
$$c(OH)^- = 0.08 \, mol/l,$$
$$pH \quad = 14 - 1.10,$$
$$pH \quad = 12.90.$$

In solutions containing several very strong acids or bases, protolysis takes place independently of each other so that the hydrogen or hydroxide ion concentration is the sum of the total concentrations of the individual protolytes:

$$c(H_3O^+) = \sum_i c_{0i}(S_i) \qquad\qquad c(OH^-) = \sum_i c_{0i}(B_i)$$

respectively

$$\text{pH} \quad = -\lg \sum_i c_{0i}(S_i) \qquad\qquad \text{pH} \quad = pK_w + \lg \sum_i c_{0i}(B_i).$$

Example: The pH value of an acid mixture with 0.01 mol/l HBr and 0.02 mol/l HCl is

$$c(H_3O^+) = c_0(HBr) + c_0(HCl),$$

$$c(H_3O^+) = 0.01 \, \text{mol}/l + 0.02 \, \text{mol}/l = 0.03 \, \text{mol}/l,$$

$$\text{pH} \quad = 1.52.$$

Weak acids and bases: With weak protolytes ($pK \geq 4.5$), the protolysis equilibrium is far to the left so that the equilibrium concentration of the protolysis products of an acid $c(H_3O^+) = c(A^-)$ is very small compared to $c(HA)$. In solutions that are not too dilute, the equilibrium concentration $c(HA)$ can therefore be set equal to the total concentration $c_0(S)$ for the sake of simplicity.

From

$$\frac{c(H_3O^+) \cdot c(A^-)}{c(HA)} = K_a,$$

$c^2(H_3O^+) = K_a \cdot c_0(S)$ and

$$(H_3O^+) = \sqrt{K_a \cdot c_0(S)}$$

$$\text{pH} \quad = \frac{1}{2}pK_a - \frac{1}{2}\lg c_0(S).$$

The same applies to bases:

$$c(OH^-) = \sqrt{K_b \cdot c_0(B)},$$

$$c(H_3O^+) = \frac{K_w}{\sqrt{K_b \cdot c_0(B)}},$$

$$\text{pH} \quad = 14 - 0.5 \, pK_b + 0.5 \lg c_0(B).$$

Whether the simplification made is permissible depends not only on the constant K_a or K_b but also on c_0. The calculation shows that for $pK = 4$ and $c_0 = 0.001$ mol/l the pH value only deviates by -0.07 U from that calculated using the more complicated formula. For larger pK values and larger concentrations, the error becomes smaller.

Example: What is the pH value of acetic acid with a concentration of 0.1 mol/l? With $K_a = 1.78 \cdot 10^{-5}$ mol/l ($pK_a = 4.75$):

$$pH = 0.5 \cdot 4.75 - 0.5 \cdot \lg 0.1,$$

$$pH = 2.88.$$

Example: What is the pH value of a sodium acetate solution with a concentration of 0.01 mol/l? The Na^+ ion is not a protolyte, and the acetate ion is a weak base with $pK_b = 9.25$. The result is

$$pH = 14 - 0.5 \cdot 9.25 + 0.5 \cdot \lg 0.01,$$

$$pH = 8.38.$$

Example: What is the pH value of an NH_4Cl solution with a concentration of 0.1 mol/l? NH_4^+ is a weak acid with $pK_a = 9.25$, Cl^- does not protolyze. Consequently

$$pH = 0.5 \cdot 9.25 - 0.5 \cdot \lg 0.1,$$

$$pH = 5.13.$$

Example: What is the pH value of a KCN solution, $c(KCN) = 0.02$ mol/l? Only the cyanide ion is subject to protolysis, it is a weak base with $pK_b = 4.60$. The following applies

$$pH = 14 - 0.5 \cdot 4.60 + 0.5 \cdot \lg 0.02,$$

$$pH = 10.85.$$

For divalent protolytes, the more the K_{a1} and K_{a2} values differ, the less noticeable the second protolysis stage becomes.

Example: What is the pH value of a saturated hydrogen sulfide solution if 2.46 l of H_2S dissolve in 1 l of water at 22 °C? The molar volume $V_m = 22.4$ l/mol gives the amount of substance $n(H_2S) = 2.46$ l: 22.4 l/mol = 0.11 mol; the concentration of the saturated solution is therefore $c(H_2S) = 0.11$ mol/l. The two acid constants of H_2S differ by five orders of magnitude; therefore, only the first one needs to be used:

$$pH = 0.5 \cdot 6.92 - 0.5 \cdot \lg 0.11,$$

$$pH = 4.42.$$

3.1.1.6 Buffer solutions

An aqueous solution whose pH value changes only slightly when certain quantities of very strong acids or very strong bases are added or when it is diluted is called a buffer solution. The pH-stabilizing effect of such buffer solutions is based on the fact that they are able to intercept added H_3O^+ or OH^- ions. For this purpose, a buffer solution must contain a base and an acid. It is therefore prepared by mixing a weak acid with its corresponding base or from a weak base and its corresponding acid. Depending on the acid or base selected, the solution is able to buffer in a certain pH range.

The pH value of common buffer solutions is between 2 and 12. An overview of buffer mixtures and their preparation can be found in [5, 77, 79].

If a weak acid HA and its salt, which contains the corresponding base A^-, are mixed in a stoichiometric ratio of 1:1, the concentrations of the HA molecules and the A^- ions become practically equal, as the addition of the salt results in equilibrium:

$$HA + H_2O \rightleftharpoons H_3O^+ + A^-$$

is shifted to the left and hardly any A^- ions are formed from the acid HA. The pH value of such a solution becomes equal to the pK_a value of the weak acid, as in the relationship (**buffer formula**, **Henderson-Hasselbalch equation**)

$$pH = pK_a + \lg \frac{c(A^-)}{c(HA)}$$

$c(A^-) = c(HA)$.

Although the addition of an acid or base changes the concentrations of HA and A^-, their ratio does not change significantly.

Example: A volume of 100 ml of hydrochloric acid, $c(HCl)$ = 1 mol/l, is added to 500 ml of a buffer solution containing 1 mol of acetic acid and 1 mol of sodium acetate. The H_3O^+ ions in the hydrochloric acid react almost completely with the acetate ions to form acetic acid molecules. The quantity of acetate ions thus becomes (1 – 0.1) mol, that of acetic acid (1 + 0.1) mol. The pH value is now

$$pH = pK_a + \lg \frac{0.9}{1.1},$$

$$pH = 4.66.$$

The addition of hydrochloric acid has reduced it from 4.75 to 4.66, i.e. only by ΔpH = 0.09. If the same amount of HCl solution was added to an unbuffered solution of pH = 4.75, the pH value would be around 1, as the added H_3O^+ ions are not bound by any base.

When sodium hydroxide solution, $n(NaOH)$ = 0.1 mol, is added to the buffer solution, the OH^- ions react with the acetic acid to form water and acetate ions. The pH value of the solution is only slightly increased to pH = 4.84, as can be shown in a similar way.

The measured pH values in a buffer system usually deviate slightly from the values calculated according to the buffer formula, as concentrations are used in the calculations instead of activities.

Buffer solutions can be prepared from commercially available buffer substances according to the literature (see [5, 77]). For buffer solutions with integer pH gradations between pH = 1.00 and pH = 13.00 as well as for standard buffer solutions [79], ampoules with concentrates as well as ready-to-use solutions are commercially available, e.g. under the name CertiPUR® (Merck, Darmstadt) as a Titripac® (Merck, Darmstadt) container (see p. 53).

Buffers are often required in analytical practice, e.g. if ions are to be separated by precipitation at certain pH values or if reactions in the course of which H_3O^+ ions are

released or consumed are to take place at a constant pH value, as in complex formation titrations with aminopolycarboxylic acids (see Section 3.4).

3.1.2 Titration curves

The importance of the ion product for acid-base titrations results from the following considerations: the H_3O^+ concentration of a hydrochloric acid is $c(H_3O^+) = 0.01$ mol/l. If a drop
of sodium hydroxide solution is added to a measured volume of this solution, the OH^-
concentration increases and with it the value of the ionic product $K_W = c(H_3O^+) \cdot c(OH^-)$.
However, this larger ion product no longer corresponds to the equilibrium state in the
solution. Therefore, H_3O^+ ions and OH^- ions combine to form nondissociated water until
the original equilibrium value of the ion product of approximately 10^{-14} mol^2/l^2 is reached
again. The H_3O^+ ion concentration decreases in the process. If more alkali is added, the
number of H_3O^+ ions constantly decreases and the number of OH^- ions increases. The
value of K_W, which is initially exceeded with each addition, returns to the equilibrium value of 10^{-14} mol^2/l^2 each time. The procedure described corresponds to the
process in a titration. In the course of the titration, the point is finally reached at
which the H_3O^+ and OH^- concentrations have become equal. This is the **neutral point**:
$c(H_3O^+) = c(OH^-) = 10^{-7}$ mol/l. If more alkali is added, the OH^- ions increasingly predominate and the solution becomes more and more alkaline.

3.1.2.1 Titration of strong acids and bases

On example of the titration of hydrochloric acid of concentration $c(HCl) = 0.01$ mol/l
with sodium hydroxide solution, we will calculate how the H_3O^+ concentration and
thus the pH value of the solution changes after each addition of reagent. The simplify-
ing assumption should be made that the volume of the starting solution does not
change during the titration. This requirement is met very closely if sodium hydroxide
solution with a concentration of $c(NaOH) = 1$ mol/l is used for the titration, which is
taken from a microburette. Table 3.4 shows how many milliliters of sodium hydroxide
solution were added to 100 ml of hydrochloric acid, what proportion of the hydrochloric acid was converted, the H_3O^+ concentration, the pH value and the degree of
titration τ after each reagent addition. The **degree of titration** is the ratio of the
equivalent amount $n_R(eq)$ of reagent added to the sample solution to the initial equivalent amount $n_0(eq)$ of the substance to be determined in the sample solution:

$$\tau = \frac{n_R(eq)}{n_0(eq)}.$$

At the start of a titration, $\tau = 0$, as $n_R(eq) = 0$. If the pH value in the template is plotted
against the degree of titration, the **titration curve** is obtained (Fig. 3.1). Its characteristic
curve shows that the pH value increases slowly at first, then faster and finally abruptly as

the degree of titration increases, i.e. with increasing addition of OH⁻ ions. Thereafter, it only increases slowly and ever more slowly. The curve has an inflection point. This point, at which the pH value increases the most, i.e. a certain small addition of OH⁻ ions causes the greatest relative change in the H_3O^+ ion concentration, is the **equivalence point** of the system. At this point, just as much alkali has been added as is required for the complete conversion of the hydrochloric acid provided, which corresponds to the degree of titration $\tau = 1$. Figure 3.1 shows the titration curves for the titration of hydrochloric acid with sodium hydroxide solution at three different HCl concentrations. The shape of the curve is the same in all three cases, but the pH jump at the equivalence point is greater and the more concentrated the used acid is. The aim of every titration is to determine the equivalence point of the respective titration system as accurately as possible.

Tab. 3.4: Titration of 100 ml hydrochloric acid (0.01 mol/l) with sodium hydroxide solution (1 mol/l).

Sodium hydroxide added (ml)	Hydrochloric acid converted (%)	Degree of titration (τ)	$c\,(H_3O^+)$ (mol/l)	pH
0.000	0	0.000	10^{-2}	2
0.900	90	0.900	10^{-3}	3
0.990	99	0.990	10^{-4}	4
0.999	99.9	0.999	10^{-5}	5
1.000	100	1.000	10^{-7}	7
1.001		1.001	10^{-9}	9
1.010		1.010	10^{-10}	10
1.100		1.100	10^{-11}	11
2.000		2.000	10^{-12}	12

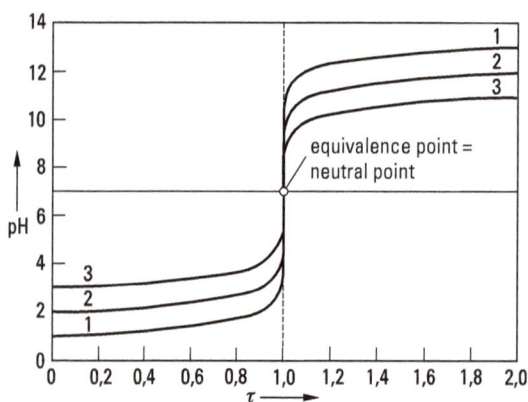

Fig. 3.1: Titration of hydrochloric acid with sodium hydroxide solution (1 mol/l): (1) c(HCl) = 0.1 mol/l, (2) c(HCl) = 0.01 mol/l and (3) c(HCl) = 0.001 mol/l (τ degree of titration).

If a strong base is used and titration is carried out with a strong acid, the titration curve (Fig. 3.1) is reversed. Starting from a high pH value, a low pH value in the acidic region is finally reached after a sudden decrease. The considerations for the titration of an acid can be transferred analogously to that of a base.

3.1.2.2 Titration of weak acids and bases

If a strong or very strong acid is titrated with an equally strong base (or vice versa), the equivalence point and the **neutral point** (pH = 7) coincide. However, when titrating a weak acid with a strong base, the equivalence point is above the neutral point in the alkaline range and when titrating a weak base with a strong acid, the equivalence point is below the neutral point in the acidic range. To explain this, the titration of 100 ml of an acetic acid of concentration $c(CH_3COOH)$ = 0.1 mol/l with caustic soda of concentration $c(NaOH)$ = 10 mol/l will be discussed as an example.

Table 3.5 provides information on the course of the titration; the titration curve is shown in Fig. 3.2.

In this case, the equivalence point is not at the neutral point (pH = 7), but at pH = 8.88, in the alkaline region. Although acetic acid and caustic soda are present in an equivalent ratio at the equivalence point, the number of OH^- ions in the solution is greater than that of H_3O^+ ions. The reason for this lies in the fact that the acetate ion is a weak base and protolyzes.

Tab. 3.5: Titration of 100 ml acetic acid (0.1 mol/l) with sodium hydroxide solution (10 mol/l).

Sodium hydroxide added (ml)	Acetic acid converted (%)	Degree of titration (τ)	$c\,(H_3O^+)$ (mol/l)	pH
0.000	0	0.000	$1.32 \cdot 10^{-3}$	2.88
0.100	10	0.100	$1.60 \cdot 10^{-4}$	3.80
0.500	50	0.500	$1.78 \cdot 10^{-5}$	4.75
0.900	90	0.900	$1.98 \cdot 10^{-6}$	5.70
0.990	99	0.990	$1.80 \cdot 10^{-7}$	6.75
0.998	99.8	0.998	$3.56 \cdot 10^{-8}$	7.45
0.999	99.9	0.999	$1.78 \cdot 10^{-8}$	7.75
1.000	100	1.000	$1.35 \cdot 10^{-9}$	8.88
1.001		1.001	$1.01 \cdot 10^{-10}$	10.0
1.002		1.002	$5.01 \cdot 10^{-11}$	10.3
1.010		1.010	$1.01 \cdot 10^{-11}$	11.0

Figure 3.3 shows the influence of the acid constants of different strong acids on the course of the titration curves when the titration is carried out with a strong base. These are titrations of acid solutions of concentration 0.1 mol/l with sodium hydroxide solution of concentration 10 mol/l. You can see:

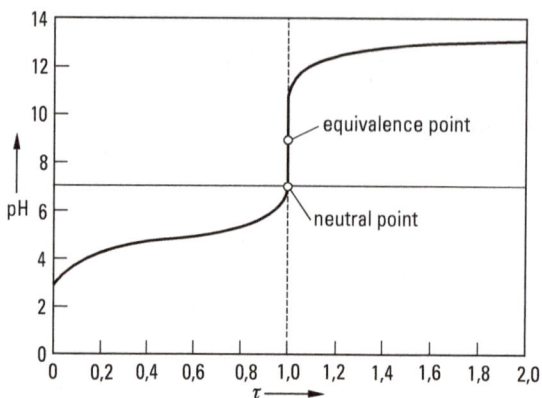

Fig. 3.2: Titration of acetic acid (0.1 mol/l) with sodium hydroxide solution (10 mol/l) (τ degree of titration).

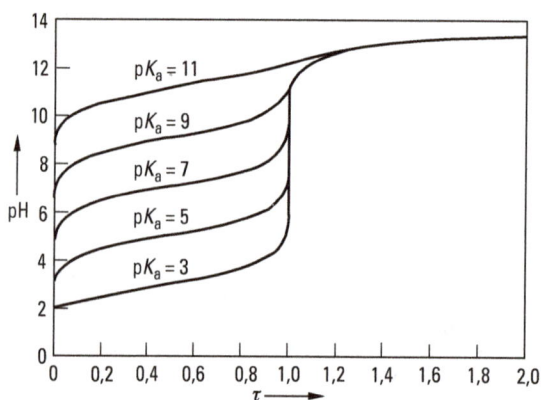

Fig. 3.3: Titration of acids of different strengths, $c(eq) = 0.1$ mol/l, with sodium hydroxide solution (τ degree of titration).

- The stronger the titrated acid, the greater the sudden decrease in the hydrogen ion concentration (increase in pH value) near the equivalence point.
- The position of the equivalence point ($\tau = 1$) deviates more from the neutral point (pH = 7) and shifts further into the alkaline range the weaker the titrated acid is; the transition area between the clearly acidic and the clearly alkaline reaction of the solution becomes wider the smaller the acid constant is, i.e. the greater the pK_a value of the titrated acid is.

The transition region becomes even more blurred when titrating a weak acid with a weak base, i.e. the change in the hydrogen ion concentration near the equivalence point is even smaller. Figure 3.4 shows the titration of an acetic acid of concentration 0.1 mol/l

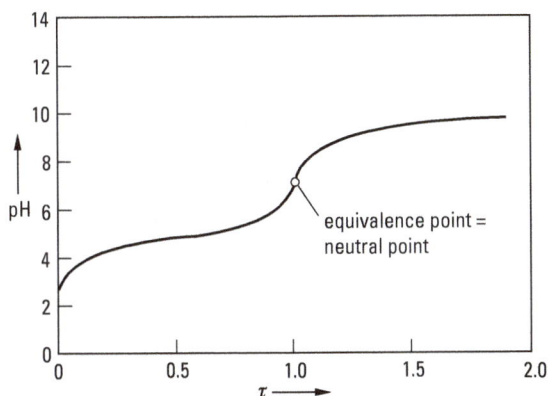

Fig. 3.4: Titration of acetic acid with ammonia solution (τ degree of titration).

with ammonia solution of the same concentration. Here the equivalence point is at the neutral point which is due to the fact that the numerical values of the acid constant of acetic acid and the base constant of ammonia are the same ($K_a(CH_3COOH) = K_b(NH_3) = 1.78 \cdot 10^{-5}$ mol/l).

3.1.3 Acid-base indicators

All acid-base titrations result in a more or less large pH jump at the equivalence point, but this cannot be easily observed visually, as the standard solutions used, the substances to be determined and the salts formed during the titration are usually colorless. Auxiliary substances, so-called **acid-base indicators** or **neutralization indicators**, are therefore added to the sample solution. Acid-base indicators are organic dyes (azo dyes, nitrophenols, phthaleins, sulfophthaleins, triphenylmethane dyes, etc.), which are themselves weak acids or bases and show a color change when they are protonated or deprotonated. If the indicator is selected correctly, this color change occurs in the area of the pH jump at the equivalence point indicating the end of the titration.

3.1.3.1 Indicator transition

As mentioned at the beginning, acid-base indicators are weak acids or bases; the relationships derived in Section 3.1.1 can therefore be applied to them. If the indicator is a weak acid, there is always an equilibrium in solution between the indicator acid HInd and the corresponding base Ind⁻, which have different colors, e.g. HInd could be yellow and Ind⁻ blue:

$$H_2O + HInd \rightleftharpoons H_3O^+ + Ind^-.$$
$$\text{yellow} \qquad\qquad \text{blue}$$

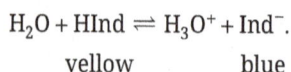

If the solution of a strong acid is mixed with a few drops of this indicator solution, the above equilibrium is shifted very far to the left, almost only HInd is present, the yellow color of which is the **borderline color of the indicator in the acidic range**. During the titration with a base, the H_3O^+ concentration decreases and the equilibrium shifts to the right until finally almost only the base Ind^- is present, whose blue color is the borderline color of the indicator in the basic range.

At a very specific pH value, the **transition point**, $pH_{1/2}$, of the indicator, $c(HInd) = c(Ind^-)$, the solution has a mixed color of the two limiting forms, i.e. green in the case under consideration. The transition point results from the pK_a value of the indicator acid when the law of mass action is applied to the above equilibrium:

$$pH = pK_a(HInd) - \lg\frac{c(HInd)}{c(Ind^-)}.$$

Since, by definition, $c(HInd) = c(Ind^-)$ at $pH_{1/2}$, it follows that $\lg 1 = 0$:

$$pH_{1/2} = pK_a(HInd).$$

The **transition range** is the pH range in which the color change of an indicator between the two borderline forms is visually recognizable. It extends approximately over an interval of $\Delta pH = 2$: at $pH = pH_{1/2} - 1$, approximately 90% of the indicator is present as HInd and 10% as Ind^-, at $pH_{1/2} + 1$, 90% as Ind^- and 10% as HInd. If the percentage of a form is plotted against the pH value, the S-shaped **transition curve of the indicator** is obtained.

The size of the transition range differs for the individual indicators. This is due to the different sensitivity of the human eye to the boundary colors and their different color intensity.

The turnover ranges of the indicators known today cover the entire pH range from strongly acidic to strongly alkaline so that a suitable indicator is available for almost every alkali or acidimetric determination.

Table 3.6 summarizes some important indicators.

A large number of other indicators are described in the literature; summaries can be found in relevant tables [5, 80–84].

A distinction is made between **single-color indicators** where only one boundary form is colored, and **two-color indicators** where both border forms are colored.

Using the two indicators methyl orange and phenolphthalein as examples, the reaction scheme underlying the color change will be explained formally:

Tab. 3.6: Acid-base indicators.

Indicator	Border colors: acidic–alkaline	Handling range (pH)	Working solution
Thymol blue	Red–yellow	1.2–2.8	0.04% in 20% ethanol
Methyl orange	Red–yellow-orange	3.1–4.4	0.04% in water
Bromocresol green	Yellow–blue	3.8–5.4	0.1% in 20% ethanol
Methyl red	Red–yellow	4.4–6.2	0.1% in ethanol
Litmus	Red–blue	5.0–8.0	0.2% in ethanol
Bromocresol purple	Yellow–purple	5.2–6.8	0.1% in 20% ethanol
Bromothymol blue	Yellow–blue	6.0–7.6	0.1% in 20% ethanol
Neutral red	Red–yellow	6.8–8.0	0.3% in 70% ethanol
Phenolphthalein	Colorless–red	8.2–9.8	0.1% in ethanol
Thymolphthalein	Colorless–blue	9.3–10.5	0.04–0.1% in 50% ethanol
Epsilon blue	Orange–violet	11.6–13.0	0.1% in water

Methyl orange is the sodium salt of 4-dimethylaminoazobenzene-4'-sulfonic acid, which is readily soluble in water:

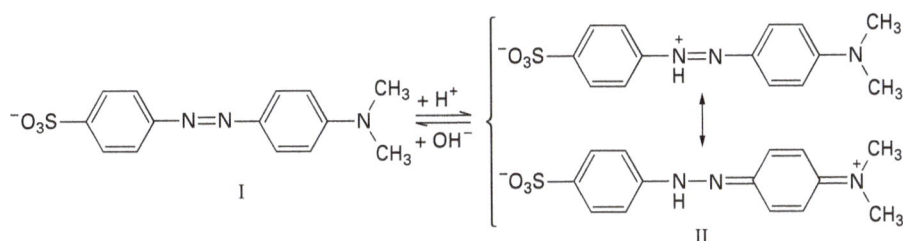

In neutral or alkaline solution, the anion I is present, which is yellow-orange in color. If the H_3O^+ concentration in the solution increases during a titration, I changes to the red form II as the amino group takes on the structure of an ammonium derivative (NR_4^+), the benzene ring changes to a quinone ring and a proton is bound to the azo group (lower formula). However, the formula only describes one of the possible limiting distributions of electrons in the molecule; it is one of the **limiting structures**. The π electrons occurring in conjugated double bonds (several double bonds separated by single bonds) cannot be localized. The second representable limiting distribution of electrons is given by the upper formula II. These bond ratios in conjugated π-electron systems are referred to as **mesomerism** (polymethin dye, cyanine). The actual structure of the dye cannot be specified; it lies between the boundary structures of the mesomeric compound (intermediate state). Mesomerism results in an energetic stabilization of the molecule.

Methyl red is the 4-dimethylaminoazobenzene-2'-carboxylic acid:

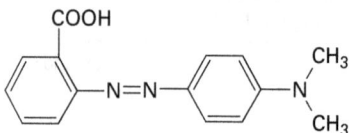

The color change from the alkaline border color yellow to the acidic border color red can be explained in a similar way for this azo dye.

Phenolphthalein, 4',4"-dihydroxydiphenylphthalide, is a colorless lactone (I) that turns deep red in weak alkaline solution. The color change can be formulated as follows:

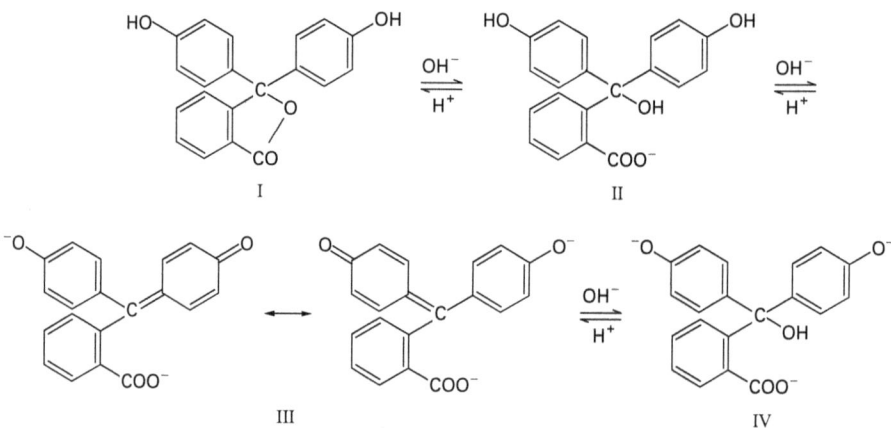

First, the addition of alkaline solutions opens the lactone ring (II), allowing one molecule of water to be split off. A quinoid ring is formed and a system of conjugated double bonds is formed over the entire molecule, for which only the boundary structures of a mesomeric compound can be indicated (III), which has a red color (polymethin dye). With further addition of OH⁻ ions, the molecule is converted into a colorless triple negatively charged anion (IV) in very strongly alkaline solutions.

The color change of an indicator is due to the different light absorption properties of the indicator forms involved in the chemical equilibrium. When light quanta are absorbed, an electron excitation occurs, namely a transition of an electron from the highest occupied to the lowest unoccupied molecular orbital, whereby the $\pi \rightarrow \pi^*$ transitions are of essential importance. The π electron system of the indicator molecule is changed by protolysis or deprotolysis. The difference in the excitation energy E ($E = h \cdot \nu$ with the Planck constant h and the frequency ν of the absorbed light) between the two molecular orbitals is greater in a benzoid system stabilized by mesomerism than in a quinoid system stabilized in the same way:

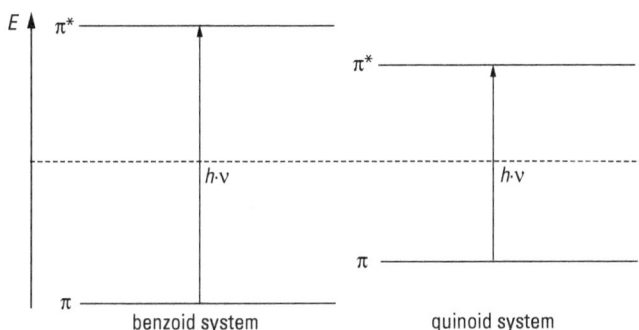

As a result, a benzoid system absorbs at higher frequencies v or at shorter wavelengths λ ($v = c/\lambda$, c = speed of light) than a quinoid system (**bathochromic effect of the latter**). Further details can be found in textbooks on spectroscopy [85–87] or organic chemistry [88].

The indicators methyl orange and phenolphthalein date back to the early days of volumetric analysis; they are still frequently used today. It should be noted that methyl orange cannot be used in hot solutions. This restriction does not apply to phenolphthalein. In strongly alkaline solutions, phenolphthalein, like all phthaleins, is rearranged and thus decolored.

With some indicators, including methyl orange, the color change is not easy to detect. In such cases, it is advisable to work with reference solutions.

Indicator mixtures have also proven their worth which have a stronger contrasting effect of the boundary colors and also make it easier to identify the transition color. A distinction is made between **mixed indicators** and **contrast indicators**. The former contain two indicator dyes with approximately the same transition range but different colors, while contrast indicators consist of an indicator dye and a pH-indifferent dye.

The following percentages refer to "mass fractions in %":

Methyl orange-indigocarmine contrast indicator: 25 ml of 0.1% methyl orange solution is mixed with 25 ml of a 0.25% indigocarmine solution and stored in a dark bottle. Acidic solutions are colored violet, neutral gray and slightly alkaline green. Titrate to a gray shade (pH = 4.1).

Neutral red-methylene blue contrast indicator: Mix 25 ml of a 0.1% neutral red solution in ethanol with 25 ml of a 0.1% ethanolic methylene blue solution and store in a dark bottle. Only a few drops are required for titration. Acidic solutions are colored violet-blue, and alkaline ones green. The color change occurs at pH = 7.0. According to Kolthoff [89], this indicator is suitable for the direct titration of acetic acid with ammonia solution without a reference solution.

Phenolphthalein-α-naphtholphthalein mixed indicator: Mix 50 ml of a 0.1% phenol-phthalein solution in 50% ethanol with 25 ml of a 0.1% solution of α-naphtholphthalein in 50% ethanol. A few drops of this mixture color acidic solutions slightly pink, alkaline solutions violet. At pH = 9.6, the color is green, making the Kolthoff indicator suitable for titration of phosphoric acid up to the second stage.

Bromocresol green-methyl red mixed indicator: Mix 75 ml of 0.1% bromocresol green solution and 25 ml of 0.2% methyl red solution, both in 96% ethanol. In the pH range 4–6, the mixed indicator changes from orange-red (mixed color of yellow and red, the two boundary colors of the indicators in the acidic range) to gray (mixed color of green and orange-red) to green (mixed color of blue and yellow, the two boundary colors of the indicators in the alkaline range). The color change is very sharp and easily visible to the eye.

In any case, make sure that only a few drops of the indicator solution are added to the sample to be titrated. As the indicator itself is a corresponding acid-base pair, it consumes standard solution.

For further indicator mixtures see [5, 80–83].

The turnover interval of an indicator is influenced by the ionic strength in the solution, namely when the activity coefficients of the two limiting forms of the indicator are influenced to different degrees. This is the case, for example, if one form is a neutral molecule and the other is an ion. This phenomenon is known as the **salt effect**. It is based on the difference between the thermodynamic and stoichiometric acid constants of the indicator system (see Section 3.1.1).

3.1.3.2 Indicator selection

A large number of indicators are available for the practical implementation of alkalimetric and acidimetric determinations, whereby the most varied turnover ranges occur. The following considerations apply to the question of which indicator is most suitable for a given problem:

The task of every analytical determination is to find the equivalence point. As has been shown, this does not always coincide with the neutral point, but can lie more in the acidic or more in the alkaline region as a result of protolysis reactions. The indicator is now selected so that its transition point is as close as possible to the equivalence point of the system to be titrated.

The titration curve of a strong acid with a strong base is characterized by a steep jump and thus by a relatively narrow transition region near the equivalence point. In addition, the equivalence point coincides with the neutral point. In this case, all indicators that transition close to pH = 7 are useful. However, since the error of a determination does not normally need to be less than 0.1%, all those indicators can also be used whose transition points lie in the range that is covered by 0.1% more or less base addition. These limits, which are just about permissible, are pH = 4 and pH = 10 when titrations in the concentration range of 0.1 mol/l are involved. This means that all indicators from methyl orange to phenolphthalein can be used. When titrating in the concentration range of 0.01 mol/l, the transition points must not fall below or exceed the pH limits of 5 and 9. Methyl orange may then no longer be used, and methyl red is just about permissible (see Fig. 3.5).

If a weak acid is titrated with a strong base, the equivalence point is in the alkaline range – in the example of acetic acid with a concentration of 0.1 mol/l with sodium hydroxide solution at pH = 8.88. The indicator transition points that are still permissible are pH = 7.75 and pH = 10 (see Tab. 3.5). An indicator that changes in the acidic range can no longer be used. Phenolphthalein can be selected as a suitable indicator (see Fig. 3.5).

Conversely, when titrating a weak base with a strong acid, the equivalence point is in the acidic range. For example, the permissible indicator transition points for the titration of an ammonia solution with a concentration of 0.1 mol/l with hydrochloric acid are at pH values of 4.0 and 6.25, while the equivalence point is at pH = 5.13. All indicators whose transition points lie between those of methyl orange and bromocresol purple can therefore be used. Under no circumstances should phenolphthalein be used.

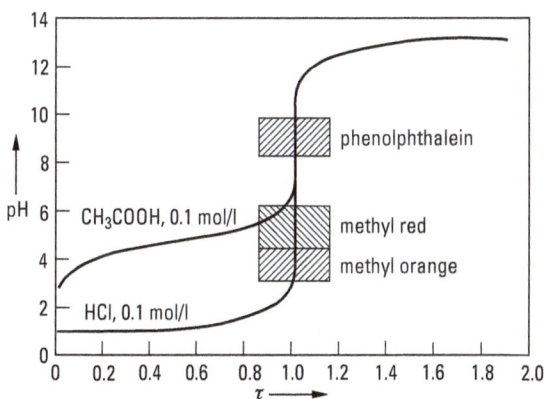

Fig. 3.5: Titration curves of acetic acid and hydrochloric acid and turnover ranges of the three indicators phenolphthalein, methyl red and methyl orange.

Finally, the titration curves of weak acids with weak bases only show a slight change in the H_3O^+ ion concentration at the equivalence point. If one wanted to titrate with an error of 0.1%, the indicator to be used should only have its transition point between pH = 6.96 and pH = 7.04. Almost only neutral red would be an option. But even with this indicator, no sharp color change can be achieved due to the slow change in the H_3O^+ ion concentration at the equivalence point. It would be necessary to work with the aid of neutral reference solutions containing the indicator in the same concentration as the solutions to be titrated. Nevertheless, only an approximate result could be expected. A mixed indicator can also be used.

The importance of selecting the right indicator can be seen by observing the course of the titration curve of a polybasic acid. If, for example, phosphoric acid is titrated with sodium hydroxide solution, the pH value in the solution at the first equivalence point, at which sodium dihydrogen phosphate is formed, is 4.7. If, however, the solution is titrated to disodium hydrogen phosphate, the pH value is 9.9. If one of these neutralization stages is to be correctly recorded titrimetrically, the indicator must be selected appropriately.

This results in the following four **basic rules** for the practice of acid-base titrations:

– Strong acids and strong bases can be titrated with each other using all indicators that convert between methyl orange and phenolphthalein.
– Weak acids can only be titrated with strong alkalis using indicators that convert in the weak alkaline range (e.g. phenolphthalein).
– Weak bases can only be titrated with strong acids using indicators that convert in the weakly acidic range (e.g. methyl orange or, better, methyl red).
    – Titrations of weak bases with weak acids and vice versa only give inaccurate results. However, if they cannot be avoided, only very few indicators can be used, which must be determined specifically for each case. The indicator can also only be titrated with the aid of a reference solution that has previously been brought to the pH value of the desired equivalence point and that has the same indicator concentration as the solution to be titrated.

A so-called **titration error**, which refers to the fact that the transition point of the indicator and the equivalence point of the titration do not correspond exactly, can be caused by the choice of an unsuitable indicator, but also by neglecting the dependence of the indicator transition on a number of factors, namely the temperature, the dilution of the solutions, the neutral salt content, the indicator concentration and others. Particular attention should be drawn to the increase in error with increasing dilution. Further details and the methods for calculating the titration error can be found in more detailed descriptions [9, 15, 80, 89].

3.1.4 Practical applications

3.1.4.1 Adjustment of acids

The most important acids used as standard solutions in acidimetric titrations are hydrochloric acid and sulfuric acid and secondarily nitric acid and perchloric acid. For some purposes, oxalic acid is suitable. These acids are usually used as solutions with a concentration of 0.5, 0.2 or 0.1 mol/l, based on equivalents.

To prepare the standard solution of a strong acid, you can start with the pure concentrated solutions available commercially, measure the density with a hydrometer and take the concentration of the substance corresponding to the density or the mass fraction in % from a table [5]. If you know the content of the concentrated acid, you can calculate the volume in milliliters that needs to be placed in a volumetric flask and diluted to the mark by adding water to obtain a solution of the desired concentration.

Example: Concentrated hydrochloric acid has the density $\rho_1 = 1.190$ g/ml (20 °C), and its concentration is $c_1 = 12.50$ mol/l. To prepare $V_2 = 1$ l of hydrochloric acid with approximately $c_2 = 0.1$ mol/l from this, let flow

$$V_1 = \frac{V_2 \cdot c_2}{c_1} = \frac{1,000 \text{ ml} \cdot 0.1 \text{ mol}/\text{l}}{12.50 \text{ mol}/\text{l}} = 8.00 \text{ ml}$$

from a burette into a 1-l volumetric flask and fill up to the mark (shake well!). As errors of up to 1% can occur with this method, an exact **titration** is still required.

The result of the adjustment, which should be carried out using one of the methods discussed below, is used to calculate the **titer t** of the solution.

Oxalic acid solutions of a precisely defined concentration can be prepared directly by dissolving a calculated quantity of the purest, air-dry oxalic acid dihydrate and making up the solution. Such a preparation is obtained by passing a stream of air over the finely powdered oxalic acid dihydrate, which has previously passed through a mixture of anhydrous (dried at 100 °C) and hydrated oxalic acid. The water content of the air stream then corresponds exactly to the water vapor partial pressure of the dihydrate.

Various primary standards are available for the precise adjustment of the acid solutions. These are sodium carbonate, sodium oxalate, potassium hydrogen carbonate and mercury oxide (Incze [92]).

Adjustment with sodium carbonate: The anhydrous salt must correspond exactly to the formula Na_2CO_3, and it must not contain any sodium hydroxide or sodium hydrogen carbonate. It must also be free of chloride, sulfate and anhydrous. Such a preparation can be produced as follows: Prepare a saturated solution of 250 g of crystallized sodium carbonate at room temperature and filter it through a pleated filter into a larger glass flask. Then pass a slow stream of pure carbon dioxide washed with $NaHCO_3$ solution through the soda solution, causing sodium hydrogen carbonate to precipitate. Cooling the solution and shaking it occasionally accelerates crystallization. After about 2 h, the precipitated salt is sucked off on a glass filter slide and washed out with ice-cold water containing CO_2 until no more Cl^- and SO_4^{2-} ions can be detected in the wash water. The salt is then dried at 105 °C, transferred to a spacious platinum crucible, the crucible and its contents weighed and then heated to 260–270 °C.

The contents of the crucible should be stirred from time to time with a platinum wire. After about an hour, the crucible is allowed to cool in the desiccator, which should be freshly filled with calcium chloride, and weighed. Continue heating and weighing until the weight is constant. The pure soda obtained in this way must be stored in a well-sealed container.

To prepare the titer, three samples of approximately 0.2 g each of the purest, anhydrous sodium carbonate are accurately weighed out of a sealable weighing jar into three Erlenmeyer flasks made of borosilicate glass with a capacity of 300–400 ml. Extreme caution is required when weighing, as the anhydrous salt is easily dusty and also readily attracts water. Dissolve each sample in about 100 ml of water, add two to three drops of methyl orange solution (do not use too much indicator solution!) and titrate with the acid to be adjusted while continuously swirling the flask until the indicator just turns orange, i.e. has a slightly stronger orange hue than an adjacent reference solution prepared from 125 ml of water and the same number of drops of methyl orange solution. As the solution is saturated with CO_2 and therefore slightly more acidic, the equivalence point has not yet been reached. The solution is therefore heated to boiling for 2–3 min to drive off the CO_2, cooled and the cooled and now yellow solution is titrated until the color change (gold tone) begins. The use of the contrast indicator methyl orange indigo carmine (see p. 95) is also recommended because the end point is easier to recognize [30]. The titer is calculated from the volume of acid consumed in the usual way (see p. 51). It is the **use titer** of the acid when using methyl orange as an indicator.

The usage titer contains the titration error (see p. 98). This is used to obtain the **corrected titer** in the following way: Dissolve an approximately equivalent amount of common salt in the same volume of distilled water that the titrated solution occupies when the indicator turnover begins, add the same volume of indicator solution and titrate with the acid to be adjusted to ensure that the color is the same as the titrated solution. At room temperature, the correction for 100 ml is approximately 0.1 ml acid ($c(eq) = 0.1$ mol/l). This volume is subtracted from the volume of the acid solution used for the first titration and the corrected titer is calculated from the value obtained. The color change is much more sensitive if methyl red is used as an indicator. This is due to the different color change ranges of the two indicators. With a final volume of 100 ml, 0.75 ml of acid ($c(eq) = 0.1$ mol/l) is theoretically required for a pH change from 4.4 to 3.1 (changeover area of methyl orange), while only 0.04 ml of acid is required for a pH change from 6.2 to 4.4 (changeover area of methyl red). Kolthoff [89] recommends adjusting the acid against rosolic acid (turnover range 6.9–8.0) or phenol red (6.4–8.2) in the heat. A titration error does not occur here so that no reference solution is required.

If you want to determine the titer for the use of phenolphthalein as an indicator, add one to two drops of phenolphthalein solution to the Na_2CO_3 solution, titrate at room temperature until just decolorized, heat the solution to boiling for 5 min and titrate the red-colored solution again until the red color disappears. These operations are carefully repeated until no more pink coloration occurs even after boiling for 10 min. When using phenolphthalein as an indicator, complete boiling of the CO_2 is essential. If this is not done, the acid consumption will be too low, as phenolphthalein already changes color in the weak alkaline range.

All color changes during the titration should be viewed and evaluated against a white background (tile, porcelain plate, paper):

1 ml acid, $c(eq) = 0.1$ mol/l, corresponds to 5.2994 mg Na_2CO_3.

Adjustment with sodium oxalate: The purest anhydrous sodium oxalate (see p. 158) is heated to 330–350 °C. It decomposes according to

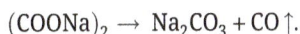

$$(COONa)_2 \rightarrow Na_2CO_3 + CO\uparrow.$$

The resulting sodium carbonate is titrated with the acid to be adjusted as described. The decomposition of the precisely weighed sample (approx. 0.3 g) is carried out in a platinum crucible. After about half an hour, the conversion is quantitatively complete. This method described by Sörensen [91] makes the properties of the primary standard sodium oxalate and can also be used for acid-base titration:

1 ml acid, $c(eq) = 0.1$ mol/l, corresponds to 6.700 mg $Na_2C_2O_4$.

3.1.4.2 Adjustment of alkaline solutions

The solutions used for alkalimetric determinations, usually sodium hydroxide or potassium hydroxide solution with a concentration of 0.2 or 0.1 mol/l, should be as carbonate-free as possible, as the presence of carbonate significantly influences the color change of methyl orange and especially of phenolphthalein, which changes color in the weakly alkaline region. In the latter case, a carbonate-containing alkali consumes less acid than a carbonate-free alkali. This results in two requirements for the preparation and use of adjusted alkaline solutions:

- The lye should be as low in carbonate as possible from the outset, or even better, carbonate-free.
- The lye must be stored in such a way that it cannot attract carbon dioxide from the surrounding air.

The second requirement can be met by storing the standard solution in a spacious storage bottle made of borosilicate glass or plastic, which is connected to a titration apparatus (see Section 2.1, Fig. 2.16). The rubber hand blower is connected with a soda lime tube in between. To prevent water from overdistilling from the storage vessel into the soda lime tube when standing, it is closed off from the storage vessel by a pinch tap on the rubber hose of the blower.

For most practical purposes, a sufficiently low-carbonate sodium hydroxide solution with a concentration of 0.2 mol/l is obtained as follows: weigh 9–10 g of the purest sodium hydroxide in cookie form raw on a top pan balance, rinse it quickly three times in a row with deionized water in a porcelain dish to remove the adhering crust of sodium carbonate, immediately transfer the sodium hydroxide to the clean storage bottle (borosilicate glass or plastic), through which a carbon dioxide-free air stream has been passed for about 2 h, and top up to the desired volume with freshly boiled, cooled deionized water. Finally, the clean and dry burette is placed on the bottle. When all the sodium hydroxide has dissolved, shake the bottle well, wait for the temperature to equalize, fill the burette using the rubber hand blower and determine the exact concentration by titration. A conductometric titration (see Section 4.3) can be used to prove that such an alkali is practically carbonate-free.

A completely carbonate-free alkali can be produced from metallic sodium dissolved in absolute ethyl alcohol. The resulting sodium alcoholate is hydrolyzed with previously boiled deionized water added in portions, the alcohol is completely boiled and the solution is suitably diluted with boiled deionized water. All operations must be carried out with CO_2-free air (see [93]).

The method first recommended by Sörensen [94] for the production of carbonate-free sodium hydroxide solution by siphoning off and diluting highly concentrated lye produced from solid sodium hydroxide and water in equal proportions, in which sodium carbonate is practically insoluble, can only be used if the preparation is carried out in plastic containers. The concentrated sodium hydroxide solution also attacks chemically resistant glass by dissolving out silicon and aluminum oxide; the solution can therefore be easily contaminated by sodium silicate and aluminate. Potassium hydroxide solution cannot be produced in this way, as potassium carbonate is too soluble in the concentrated potassium hydroxide solution.

The absence of carbonates is guaranteed by the use of barite lye under the precautionary measures mentioned. A barite lye with an approximate concentration of 0.1 mol/l (based on equivalents) is obtained as follows: About 20 g of crystallized barium hydroxide is brought into solution with vigorous shaking. When the solution, which is cloudy due to barium carbonate, has clarified after standing for some time, the solution is carefully lifted into a bottle filled with CO_2-free air (see above) and immediately sealed by placing the clean and dry burette on top.

The adjustment of the alkaline solutions is best carried out by titration with an acid of corresponding equivalent concentration, the titer of which has been determined according to one of the rules listed in the previous section. Methyl orange, methyl red, methyl orange-indigocarmine, phenolphthalein, etc. can be used as indicators. The use titres found when using methyl orange and phenolphthalein must not differ by more than 0.1%. In addition, when using phenolphthalein, the alkali must consume the same volume of acid standard solution if it is titrated directly at room temperature and if it is boiled shortly before reaching the end point, cooled again and then titrated to the end. If this is not the case, it contains too much carbonate. The adjustment of a barium hydroxide solution is only adjusted using phenolphthalein as an indicator.

You can also adjust the alkali directly with a suitable primary standard such as oxalic acid dihydrate, benzoic acid, potassium hydrogen phthalate or amidosulfonic acid.

Example: For adjustment with crystallized oxalic acid, 0.3–0.4 g of the purest oxalic acid dihydrate (see p. 99) is weighed precisely, dissolved in approximately 200 ml of boiled deionized water and titrated using phenolphthalein with the alkali to be adjusted at the approximate equivalent concentration $c(eq) = 0.2$ mol/l until a pink color appears:

$$1 \text{ ml lye, } c(eq) = 0.2 \text{ mol/l, corresponds to } 12.607 \text{ mg } (COOH)_2 \cdot 2 \text{ H}_2O.$$

Always use boiled water to dilute the solutions for all settings of the standard solutions in acidimetry and alkalimetry.

3.1.4.3 Determination of strong and weak bases

Strong and weak bases are always titrated with strong acids. The indicator is selected according to the aspects specified in Section 3.1.3. The concentration of the alkali to be titrated should correspond approximately to the acid used for the titration. If the content of solid hydroxides is to be determined, the absorption of water vapor and carbon dioxide from the ambient air must be avoided as far as possible during weighing-in. Weighing is therefore carried out in a weighing jar. Concentrated ammonia solution is also always weighed out in a sealed weighing jar. The jar is then opened under water to avoid NH_3 losses during dilution. Diluted ammonia solution is measured with a pipette. It is advisable to squeeze the solution into the pipette, not to aspirate it. The determination can be carried out either by direct titration or by back titration.

3.1.4.4 Determination of the total alkali content of technical sodium hydroxide

Procedure: About 5 g of the substance is weighed in a sealed weighing jar, dissolved in water and made up to 1 l in a volumetric flask:
1. **Direct titration:** 50 ml of the solution is titrated in the cold with sulfuric acid ($c(H_2SO_4) = 0.1$ mol/l) against methyl orange as an indicator. The proportion of the alkali content present as carbonate is also determined.
2. **Back titration:** 50 ml of the solution is mixed with $V_1 = 30$ ml of sulfuric acid ($c(H_2SO_4) = 0.1$ mol/l). The carbon dioxide displaced by the excess sulfuric acid is completely driven off by gently boiling the solution. Phenolphthalein is then added as an indicator and the still hot solution is titrated back with sodium hydroxide solution ($c(NaOH) = 0.2$ mol/l) until the pink coloration begins (consumption V_2 in ml). The difference $V_1 - V_2$ gives the volume of sulfuric acid required to neutralize the total alkali content (carbonate + hydroxide):

1 ml sulfuric acid, $c(H_2SO_4) = 0.1$ mol/l, corresponds to 6.198 mg Na_2O. The analysis result is given as a percentage by mass.

3.1.4.5 Determination of carbonates as well as hydroxides and carbonates side by side

All carbonates react with acids according to the equation:

$$Na_2CO_3 + H_2SO_4 \rightarrow Na_2SO_4 + H_2O + CO_2 \uparrow.$$

In solution, the base CO_3^{2-} reacts with the acid according to $CO_3^{2-} + 2\,H_3O^+ \rightarrow 3\,H_2O + CO_2 \uparrow$. Carbonates can therefore be titrated directly with acids in the same way as hydroxides. The determination is carried out using methyl orange as an indicator in the cold. If working with more dilute acid solutions, e.g. with an equivalent concentration of $c = 0.1$ mol/l, it is better to first titrate the carbonate solution with the acid in the cold until the indicator changes color, then briefly boil the solution to drive off the carbon dioxide and finish titrating the cooled solution with two drops of methyl orange solution. This procedure has already been discussed in the titration of acids with sodium carbonate (see p. 99).

The carbonates can also be titrated in the boiling heat using phenolphthalein as an indicator. Proceed as described in the previous example for the titration of alkali metal hydroxides. The carbonates can therefore be determined by direct and back titration. Back titration is used to determine carbonates that are not soluble in water. If the hydroxide and carbonate content are to be determined side by side in a sodium hydroxide solution containing carbonate, which has absorbed CO_2 through prolonged standing, for example, the method described by Winkler [95] is used. First, the total alkali content (carbonate + hydroxide) of the lye is determined by acidimetric titration in the cold using methyl orange as an indicator. In the second sample, the carbonate ions are determined by adding an excess of neutral barium chloride solution as sparingly soluble barium carbonate according to

$$Na_2CO_3 + BaCl_2 \rightarrow 2\,NaCl + BaCO_3 \downarrow$$

precipitated. The hydroxide remaining in the solution is then titrated with oxalic acid standard solution ($c(C_2H_2O_4) = 0.05$ mol/l) using phenolphthalein as an indicator.

The mass of alkali metal hydroxide present is calculated from the consumption of oxalic acid solution, and the mass of alkali metal carbonate is calculated from the difference between the total alkali and hydroxide content.

Procedure: Dilute 25 ml of the sodium hydroxide solution to be tested ($c \approx 2$ mol/l) to 500 ml with boiled water in a volumetric flask. After adding methyl orange, 25 ml of this solution is titrated with hydrochloric acid ($c(HCl) = 0.1$ mol/l) at room temperature (V_1 in ml). A second sample of 25 ml of the solution is added to 50 ml of barium chloride solution ($c(BaCl_2) = 0.05$ mol/l), to which a few drops of phenolphthalein solution were previously added and neutralized with sodium hydroxide solution. After a waiting time of about 10 min, the solution clouded by precipitated barium carbonate is slowly titrated with the oxalic acid solution while constantly swirling until the indicator is decolored (V_2 in ml). Oxalic acid is used because hydrochloric or sulfuric acid would partially convert the barium carbonate even if added very carefully:

   25 ml of the diluted lye contains $4.000 \cdot V_2$ mg NaOH and $5.299 \cdot (V_1 - V_2)$ mg Na_2CO_3.

3.1.4.6 Determination of carbonate and hydrogen carbonate side by side

According to Winkler [95], the total alkali content of the solution to be tested is determined by titration with an acid of known content (see p. 104). Then determine the volume of a carbonate-free (!) sodium hydroxide solution of known content required to remove the hydrogen carbonate according to

$$HCO_3^- + OH^- \rightleftharpoons CO_3^{2-} + H_2O$$

quantitatively into the carbonate.

The mass of the hydrogen carbonate is obtained directly from the volume of the sodium hydroxide solution.

Procedure: In an aliquot of the sample solution, determine the total alkali content (carbonate + hydrogen carbonate) by titration with hydrochloric acid (c(HCl) = 0.1 mol/l) at room temperature using methyl orange (consumption: V_1 in ml). A second aliquot of the sample solution is mixed with a measured excess of sodium hydroxide solution (c(NaOH) = 0.1 mol/l) (volume: V_2 in ml) to convert the hydrogen carbonate into carbonate. Then titrate the unused sodium hydroxide solution next to the carbonate in the manner described in the above procedure, after adding neutral barium chloride solution with oxalic acid solution and phenolphthalein as an indicator (consumption: V_3 in ml). The difference $V_2 - V_3$ is the volume of NaOH solution required to convert the hydrogen carbonate, from which the mass of the hydrogen carbonate is obtained:

1 ml sodium hydroxide solution, c(NaOH) = 0.1 mol/l, corresponds to 6.1017 mg HCO_3^- or 8.4007 mg NaHCO$_3$.

The difference $V_1 - (V_2 - V_3)$ is the volume of the acid solution required to convert the carbonate:

1 ml acid solution, c(eq) = 0.1 mol/l, corresponds to 3.0005 mg CO_3^{2-} or 5.2994 mg Na$_2$CO$_3$.

If the sodium hydroxide solution used for titration is not carbonate-free, its carbonate content must be determined in a separate determination and its titer corrected accordingly.

3.1.4.7 Determination of borax

Just as carbonates of alkali metals can be titrated, the alkali metal salts of other weak acids, e.g. arsenic acid, hydrocyanic acid, telluric acid and boric acid, can also be determined by direct titration with mineral acids using methyl orange or, better, methyl red or methyl orange-indigocarmine. This type of titration used to be called **displacement titration** because the weak acid is displaced from its salts by the strong mineral acid. According to Brønsted's theory, however, displacement titrations are also acid-base titrations. The anion formed when the salt is dissolved is a base that is weaker than the base OH$^-$, but can still be titrated directly with an acid. As an example, the determination of borax, Na$_2$[B$_4$O$_5$(OH)$_4$] · 8 H$_2$O, will be described. The titration is carried out according to

$$[B_4O_5(OH)_4]^{2-} + 2\ H_3O^+ + H_2O \rightarrow 4\ H_3BO_3.$$

The weak boric acid is formed from the tetraborate (pK_{a1} = 9.24); the pH value of its solution at the concentration $c(H_3BO_3)$ = 0.2 mol/l is 4.97 (see p. 84). The titration is therefore complete when the pH falls below 5, which is indicated by the change in methyl red from yellow to red.

> **Procedure:** Accurately weigh about 6–7 g of borax, dissolve in boiled water in a volumetric flask and make up to 250 ml. Add a few drops of methyl red to 50 ml of this solution and titrate with hydrochloric acid ($c(HCl)$ = 0.2 mol/l) at room temperature. It is advisable to use a reference solution containing sodium chloride, boric acid and methyl red in approximately the same concentration as the sample solution in boiled water:
>
> 1 ml hydrochloric acid, $c(HCl)$ = 0.2 mol/l, corresponds to 38.137 mg $Na_2B_4O_7 \cdot 10\ H_2O$ or 20.122 mg $Na_2B_4O_7$ or 6.198 mg Na_2O or 4.598 mg Na.

3.1.4.8 Determination of nitrogen according to Kjeldahl

The determination of ammonia in ammonium salts, nitric acid in nitrates and the nitrogen content of organic compounds is based on the fact that the ammonium ion is expelled as gaseous ammonia by adding excess caustic soda:

$$NH_4^+ + OH^- \rightleftharpoons NH_3 \uparrow\ + H_2O.$$

After absorption in a measured volume of an acid solution of known concentration, which is provided in excess, the ammonia is determined by back titration of the unreacted acid.

If the substance to be determined is not present in the form of an ammonium salt, it must be converted into such a salt by suitable operations. Organic nitrogen-containing compounds, especially amino compounds, are destroyed according to Kjeldahl [96] by heating with concentrated sulfuric acid. The carbon is oxidized to carbon dioxide, while the previously organically bound nitrogen is quantitatively converted to ammonium sulfate.

The reduction of nitrates can be carried out in both acidic and alkaline solutions. According to Ulsch (1891), nitrate in boiling sulfuric acid solution is quantitatively converted into ammonium by reduction with *ferrum reductum* into ammonium without the formation of intermediate products. In alkaline solution, the reduction is best carried out with the alloy specified by Devarda (1894). It consists of 50% copper, 45% aluminum and 5% zinc. As it is very brittle, it is easy to pulverize.

The destruction of organic nitrogen-containing compounds according to Kjeldahl is greatly facilitated by the addition of dehydrating agents such as phosphorus(V) oxide or potassium sulfate. The presence of a catalyst such as mercury(II) oxide, metallic mercury, anhydrous copper(II) sulfate or platinum(IV) chloride also has an accelerating effect.

The conversion of the bound nitrogen of organic nitro and cyano compounds into ammonium sulfate is only quantitatively successful if the organic substance is destroyed in the presence of phenol-sulfuric acid. Otherwise the nitrogen escapes, at

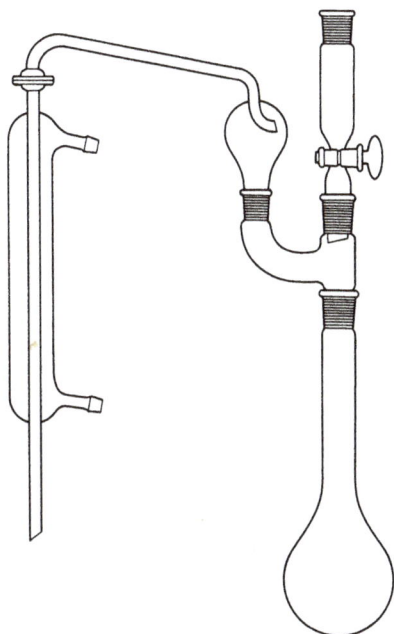

Fig. 3.6: Kjeldahl distillation apparatus.

least partially, in the form of nonbasic volatile compounds. Pyridine and quinoline compounds cannot be determined using the Kjeldahl method.

The apparatus shown in Fig. 3.6 is used for all determinations. The mostly pear-shaped long-necked flask (Kjeldahl flask) with a capacity of about 500 ml has a dropping funnel and a drop catcher, which effectively prevents liquid droplets from splashing out of the flask into the subsequent parts of the apparatus. The drip catcher is connected to a Liebig condenser. The distillate is collected in an absorption vessel.

The following examples are suitable for the ammonia distillation method and show its importance.

3.1.4.9 Determination of the nitrogen content of saltpetre

Procedure: Dissolve 10 g of the substance to be analyzed in water to make one liter. Use 50 or 25 ml of this solution, which is not filtered, for each of the following determinations.

Determination of the ammonium content
Pour 50 ml of the initial solution into the Kjeldahl flask, dilute it with about 200 ml of water and carefully connect the flask to the rest of the apparatus after throwing in a few glass beads with a roughened surface to ensure a uniform boiling process. The absorption vessel is filled with 50 ml sulfuric acid of concentration $c(\frac{1}{2} H_2SO_4) = 0.2$ mol/l and about 150 ml water. Then approximately 30 ml sodium hydroxide solution, $c(NaOH) = 0.2$ mol/l, is added to the flask through the dropping funnel. The dropping funnel is closed and the contents of the flask are heated to a lively boil for about 30 min. After completion of the distillation, the excess sulfuric acid not converted by the excess ammonia is

measured back by titration of the cooled distillate with sodium hydroxide solution, $c(NaOH) = 0.2$ mol/l, using methyl red or methyl orange as an indicator:

1 ml H_2SO_4, $c(\frac{1}{2} H_2SO_4) = 0.2$ mol/l, corresponds to 3.406 mg NH_3 or 3.608 mg NH_4^+ or 2.801 mg N.

Determination of the nitrate content

After reduction in acidic solution: 25 ml of the starting solution is mixed in a Kjeldahl flask with 5 g *ferrum reductum* and 10 ml of a sulfuric acid prepared by mixing one part by volume of concentrated acid and two parts by volume of water. A funnel filled with water and melted at the bottom is suspended in the neck of the flask to provide cooling. The flask is now slowly heated with a low flame. The liquid should only begin to boil after about 5 min; it is boiled for another 20 min. Finally, allow the contents of the flask to cool, rinse the cooling funnel carefully and dilute the solution with about 100 ml of water. The flask is then connected to the other parts of the distillation apparatus. About 30 ml of a sodium hydroxide solution with a concentration of about 2 mol/l is added through the dropping funnel. Finally, proceed as described above.

After reduction in alkaline solution: In a Kjeldahl flask, 25 ml of the starting solution is mixed with approximately 2 g of finely powdered Devarda's alloy and diluted with water to approximately 100 ml. The flask is then connected to the distillation apparatus. Now add 50–60 ml of a sodium hydroxide solution containing about 2 mol/l through the dropping funnel. Reduction is promoted by gentle heating. The actual distillation is only started after 1 h of gentle heating.

In both cases, the absorption vessel contains 50 ml of sulfuric acid, $c(\frac{1}{2} H_2SO_4) = 0.2$ mol/l. The experiment results in the sum of the nitrate and ammonia nitrogen. In both cases, it is necessary to carry out a blank test to determine how much ammonia is formed by the reducing agent used:

1 ml sulfuric acid, $c(\frac{1}{2} H_2SO_4) = 0.2$ mol/l, corresponds to 12.401 mg NO_3^-.

3.1.4.10 Determination of the nitrogen content of hard coal

Procedure: Approximately 0.75 g of the finely powdered coal is accurately weighed into the Kjeldahl flask and 10 g of anhydrous potassium sulfate and 1–2 g of dehydrated copper(II) sulfate are added. After adding 10–12 ml of concentrated sulfuric acid, the flask is loosely closed with a funnel suspended in its opening, which has been filled with water after melting the outlet pipe and thus acts as a cooler, and heated slowly and carefully over a wire mesh until the sulfuric acid has almost boiled, until its initially brown-black contents have become completely clear and colorless. In most cases, this operation takes 2–3 h. After cooling, the flask is connected to the distillation apparatus. After adding first 100 ml of water and then adding 80 ml of a sodium hydroxide solution containing about 6 mol/l through the dropping funnel, distillation can begin. The receiver contains 10 ml of sulfuric acid, $c(\frac{1}{2} H_2SO_4) = 0.2$ mol/l, which is back-titrated with sodium hydroxide solution with a concentration of 0.2 mol/l. It is taken from a micro-burette with a nominal capacity of 5 ml.

Instead of copper(II) sulfate, 0.1 g of mercury(II) oxide can be used. However, after decomposition, a few milliliters of a concentrated sodium sulfide solution must be added in addition to the sodium hydroxide solution in order to precipitate the mercury as sulfide and thus prevent the formation of mercury-ammonia compounds. Distillation can then begin.

3.1.4.11 Determination of the total nitrogen content of a garden fertilizer

The fertilizer contains urea, potassium nitrate and ammonium phosphate.

Procedure: About 1 g of the substance is weighed into the Kjeldahl flask and dissolved in 15-ml phenolsulfuric acid. The phenolsulfuric acid is prepared by mixing a cold solution of 20 g phosphorus(V) oxide in concentrated sulfuric acid and a cold solution of 4 g phenol in a little concentrated sulfuric acid and making up the mixture to 100 ml with concentrated sulfuric acid. After dissolving, add 1–2 g of sodium thiosulfate and, after it has decomposed, 10 ml of concentrated sulfuric acid and a drop of mercury. Then heat carefully and proceed as described.

The addition of phenolsulfuric acid leads to the formation of nitrophenol, which is reduced to aminophenol by the decomposition products of sodium thiosulfate.

3.1.4.12 Anhydrous titrations

A number of weakly basic substances, including some pharmaceuticals, can be determined titrimetrically without Kjeldahl digestion by switching from water to another solvent. If an analyte, e.g. a tertiary amine, is very weakly basic and its pK_b value is too close to the pK_b value of the water (15.74), then it is no longer possible to determine this analyte by titration in aqueous solution, as the water as a base competes with the analyte for the protons originating from the titrant substance. In such cases, titration is therefore carried out in anhydrous acetic acid. As a rule, a solution of perchloric acid (0.1 mol/l) in anhydrous acetic acid is used as the titrant substance. It should be noted at this point that such a mixture can easily ignite. A little acetic anhydride is added to convert the remaining water. A positive effect is that basic analytes that are difficult to dissolve in water can often be dissolved much better in acetic acid.

The reactions that take place during anhydrous titration are explained below. A tertiary amine analyte is at least partially protonated when dissolved in anhydrous acetic acid, producing acetate:

$$R_3N + CH_3COOH \rightleftharpoons R_3NH^+ + CH_3COO^-$$

The titer substance perchloric acid is practically completely dissociated in anhydrous acetic acid, whereby the acetic acid is protonated to the acetacidium ion:

$$HClO_4 + CH_3COOH \rightleftharpoons ClO_4^- + CH_3COOH_2^+$$

As a protic solvent, acetic acid undergoes autoprotolysis in the same way as water:

$$2\,CH_3COOH \rightleftharpoons CH_3COOH_2^+ + CH_3COO^-$$

In accordance with this equilibrium, which in pure acetic acid is practically completely with the acetic acid molecules, the acetacidium of the standard solution reacts with the acetate in the sample, whereby this is successively reproduced by the reaction of the tertiary amine as the titration progresses. At the equivalence point, the pH value decreases abruptly, which is indicated with a pH indicator or potentiometrically with a glass membrane electrode.

Crystal violet, Sudan III/IV (caution: carcinogen class III) or naphtholbenzein (change from brown-yellow to green) are frequently used as indicators in anhydrous titration with perchloric acid [96a]:

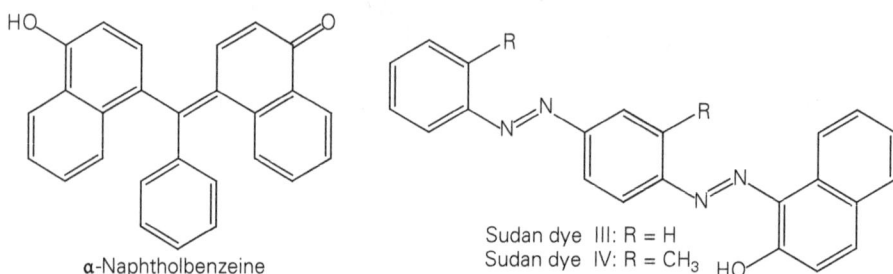

α-Naphtholbenzeine

Sudan dye III: R = H
Sudan dye IV: R = CH₃

The very frequently used triphenylmethane dye crystal violet shows a change from violet to green (polymethine dye: cyanine), after further addition of standard solution the color changes to yellow (polyene dye). In some titrations, the yellow color is specifically titrated (overtitrated):

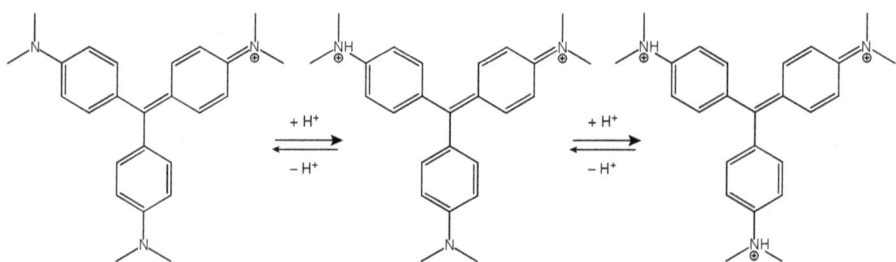

Violet	Green	Yellow
Polymethine dye	Polymethine dye	Polyene dye
Cyanine	Cyanine	

Numerous anions such as carboxylates (e.g. citrates, tartrates and acetates), sulfates, fluorides, nitrates or phosphates can also be determined as weak bases using anhydrous titration with perchloric acid in glacial acetic acid.

Anions such as chloride or bromide can also be determined anhydrous. An excess of mercury(II) acetate, a nondissociated compound, can be added to the analyte, which then exchanges an equivalent amount of acetate ions for chloride or bromide. These acetate ions are titrated with perchloric acid. Due to the use of mercury(II) acetate, the process should no longer be used if possible.

In many cases, however, chlorides or bromides can also be determined in a mixture of glacial acetic acid or formic acid and acetic anhydride anhydrous with perchloric acid. Following one theory the chloride or bromide ions react with acetic anhydride to form acetate ions and acetyl chloride or acetyl bromide. The acetate formed can then be titrated again with perchloric acid.

3.1.4.13 Determination of nicotinamide

Nicotinamide (vitamin B_3) can be titrated anhydrous with perchloric acid with protonation on the pyridine nitrogen.

Procedure: Accurately weigh approximately 0.250 g of the sample substance and dissolve in 20 ml anhydrous acetic acid. If necessary, heat the solution. After adding 5 ml acetic anhydride, stir for some time. Titration is carried out using perchloric acid (0.1 mol/l), as the standard solution and some crystal violet solution as an indicator. The equivalence point is indicated when it changes to greenish blue:

1 ml perchloric acid, c = 0.1 mol/l, corresponds to 12.21 mg $C_6H_6N_2O$ [8a].

3.1.4.14 Determination of fenbendazole

The chemotherapeutic agent used in veterinary medicine against worm infections is determined according to Ph. Eur. in an anhydrous environment with perchloric acid by protonation of the guanidine structure.

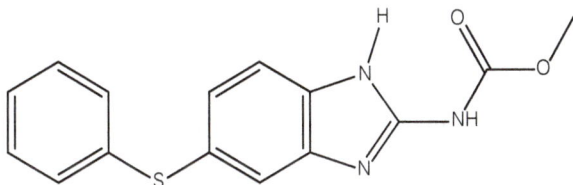

Procedure: Accurately weigh approximately 0.200 g of the sample substance and dissolve in 30 ml anhydrous acetic acid. If necessary, this must be heated. After cooling, titrate with perchloric acid (0.1 mol/l). The equivalence point is determined potentiometrically (see Section 4.4):

1 ml perchloric acid, c = 0.1 mol/l, corresponds to 29.94 mg $C_{15}H_{13}N_3O_2S$ [8a].

3.1.4.15 Determination of pantoprazole sodium salt

The proton pump inhibitor pantoprazole sodium salt (drug to reduce gastric acidity) is determined according to Ph. Eur. in an anhydrous environment by protonation on the deprotonated benzimidazol and on pyridine nitrogen with perchloric acid (2 equiv):

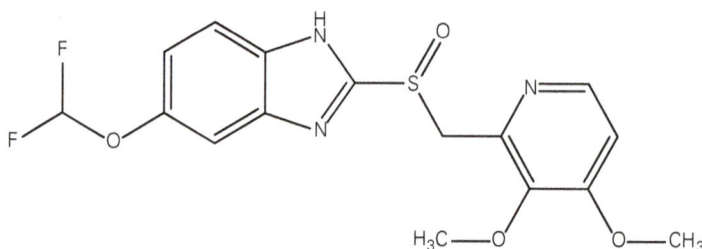

Procedure: Accurately weigh about 0.200 g of the sample substance and dissolve in 80 ml anhydrous acetic acid. Stir for at least 10 min after adding 5 ml acetic anhydride. Titration is carried out with perchloric acid 0.1 mol/l. The equivalence point is determined potentiometrically (see Section 4.4):
  1 ml perchloric acid, c = 0.1 mol/l, corresponds to 20.27 mg $C_{16}H_{14}F_2N_3NaO_4S$ [8a].

3.1.4.16 Determination of diltiazem hydrochloride according to Ph. Eur

The calcium-channel blocker diltiazem hydrochloride is quantified in the European Pharmacopoeia using the chloride determination method described above.

Procedure: About 0.400 g substance, dissolved in a mixture of 2 ml anhydrous formic acid and 60 ml acetic anhydride, is titrated with perchloric acid (0.1 mol/l). The end point is determined using potentiometry [8a]:
  1 ml perchloric acid, c = 0.1 mol/l, corresponds to 45.1 mg $C_{22}H_{27}ClN_2O_4S$.

The use of acetic anhydride as a solvent in anhydrous titration opens up further possibilities for titration:

–   Acetic anhydride can bind water in analytes, which would interfere in larger quantities.
–   Acetic anhydride is more lipophilic than acetic acid and is therefore important for the solubility of individual analytes.
–   In substance mixtures of primary or secondary amines with tertiary amines, the former can be masked by acetic anhydride, as these can be acetylated and are therefore no longer sufficiently basic for titration with perchloric acid.

3.1.4.17 Determination of formaldehyde using the sulfite method

Formaldehyde reacts analogously to other carbonyl compounds with hydrogen sulfite ions to form an addition compound:

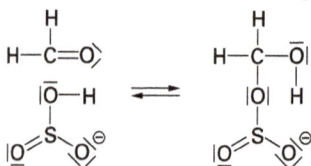

Even in very dilute solutions, the reaction equilibrium is almost completely on the side of the addition compound. The determination method is based on the fact that sulfite ions in aqueous solution undergo an acid-base reaction with the solvent water and that hydrogen sulfite and hydroxide are formed in the amount of substance corresponding to the further reaction of the hydrogen sulfite with the formaldehyde:

$$SO_3^{2-} + H_2O \rightleftharpoons HSO_3^- + OH^-.$$

This reaction takes place until practically all of the formaldehyde has been converted with hydrogen sulfite and thus an equimolar amount of hydroxide is present in the solution. This can be determined by titration with hydrochloric acid. The additional consumption of hydrochloric acid due to the excess sulfite must be determined and taken into account in the evaluation. Phenolphthalein is suitable as an indicator, but Poethke recommends, however, a mixture of phenolphthalein and α-naphtholphthalein for a sharper indication of the end point [28].

Procedure: Weigh approximately 3 g of formaldehyde solution accurately and add it to 50 ml of freshly prepared sodium sulfite solution (1 mol/l). The sodium sulfite should be practically free of hydrogen sulfite, hydroxide or carbonate. After adding the indicator (phenolphthalein or a mixture of phenolphthalein/α-naphtholphthalein), titrate slowly with hydrochloric acid. The target concentration of the standard solution depends on the content range of the formaldehyde. For concentrated solutions, 1 mol/l hydrochloric acid is used; for the determination of low formaldehyde contents, a hydrochloric acid (0.1 mol/l) is used. The consumption of hydrochloric acid is x ml. There is therefore an excess of (50 − x) ml sulfite solution. In a blank test, this volume of sulfite solution is diluted to, e.g. approximately 100 ml, added with indicator and titrated with hydrochloric acid. The volume used in this process is subtracted as a blank value during the evaluation. If the formaldehyde solution is acidic, it must be neutralized before analysis:

1 ml of standard solution corresponds to 30.03 mg formaldehyde (at 1 mol/l target concentration) or 3.003 mg (at 0.1 mol/l) [28].

3.1.4.18 Determination of strong and weak acids

Strong and weak acids are titrated with solutions of strong bases – NaOH, KOH and $Ba(OH)_2$. The following also applies here: The concentration of the acid to be titrated should correspond approximately to the concentration of the alkali used for the titration. Care must be taken to select the correct indicator (see p. 96 for aspects). The titration of a strong acid is described as a first example.

3.1.4.19 Determination of sulfuric acid

The reaction proceeds according to the equation:

$$H_2SO_4 + 2\,NaOH \longrightarrow Na_2SO_4 + 2\,H_2O.$$

Sulfuric acid is also strongly protolyzed in the second stage. The acidity constant of the hydrogen sulfate ion is $K_{a2} = 1.20 \cdot 10^{-2}$ at 25 °C ($pK_{a2} = 1.92$).

> **Procedure:** Dilute the analysis solution to approximately 100 ml with freshly boiled, CO_2-free water, add a few drops of indicator solution and titrate with sodium hydroxide solution (0.1 mol/l) at room temperature while continuously swirling the titration beaker until the indicator changes (observation against a reference solution). Suitable indicators for the end point include methyl orange, methyl red, bromothymol blue or bromocresol green-methyl red mixed indicator:
>   1 ml NaOH solution, $c(NaOH) = 0.1$ mol/l, corresponds to 4.9037 mg H_2SO_4.

Other strong acids, such as HCl, HBr, HI, $HClO_4$ and HNO_3, can be titrated in the same way.

Concentrated and fuming acids are weighed, not pipetted, to avoid evaporation losses. As an example the determining of the content of **fuming sulfuric acid** (oleum) is used to illustrate this procedure. For weighing, you can use a weighing burette (see p. 33) or a simple thin-walled glass sphere that runs out into a long capillary. A sphere should be able to hold a maximum of 2 g of acid. It is carefully dried and weighed. Then heat it carefully and dip the tip of the capillary into the fuming acid. As the sphere cools, the acid rises through the capillary. The tip is cleaned and melted to close it, and the sphere is weighed again. It is then smashed in a thick-walled stopper bottle containing a little water by shaking vigorously, making sure that the capillary is also completely smashed. Dilute with CO_2-free water, transfer the acid solution to a 250 ml volumetric flask, rinse the bottle and funnel with water and leave to stand for 1–2 h to equalize the temperature before filling up to the mark with CO_2-free water. If using a weighing pipette, place about 100 ml of boiled water in the 250 ml volumetric flask, weigh out a few grams of the acid with the weighing pipette, pour about 1–2 g of it into the flask via a funnel and weigh the pipette back. Fill the flask as described. Now take 20 or 25 ml samples with a pipette and titrate according to the instructions above.

Weaker acids, such as oxalic acid or acetic acid, are titrated using phenolphthalein as an indicator. Particular attention must be paid here to the absence of carbonate. Titration is therefore carried out with alkali lyes in the heat or with baryta lye in the cold. It is often advisable to add excess, carbonate-free alkali to the solution of the weak acid and to retitrate the excess alkali with adjusted mineral acid. The titration of acetic acid is described as an example of the determination of a weak acid.

3.1.4.20 Determination of acetic acid
The conversion is carried out according to the equation:

$$CH_3COOH + NaOH \longrightarrow CH_3COONa + H_2O.$$

Due to the low acid strength of acetic acid ($K_a = 1.78 \cdot 10^{-5}$ at 25 °C, $pK_a = 4.75$), the equivalence point is given by the pH value of the sodium acetate solution. It is 8.72

when titrating 100 ml of acetic acid with a concentration of 0.1 mol/l with sodium hydroxide solution of the same concentration.

Procedure: Dilute the analysis solution to approximately 100 ml with freshly boiled, CO_2-free water, add a few drops of phenolphthalein solution and titrate with sodium hydroxide solution (0.1 mol/l) at room temperature while continuously swirling the titration beaker until a pink coloration appears, which must persist for at least 60 s:
  1 ml NaOH solution, $c(NaOH) = 0.1$ mol/l, corresponds to 6.0053 mg CH_3COOH.

The same rule can be used to titrate acids with $K_a \geq 10^{-5}$, such as hydrofluoric acid ($K_a = 7.24 \cdot 10^{-4}$) in plastic beakers, formic acid ($K_a = 1.77 \cdot 10^{-4}$), propionic acid ($K_a = 1.34 \cdot 10^{-5}$), oxalic acid ($K_{a2} = 6.17 \cdot 10^{-5}$) and tartaric acid ($K_{a2} = 4.55 \cdot 10^{-5}$).

To determine the content of **glacial acetic acid** approximately 2 g is accurately weighed into a dry conical flask of 50 ml capacity with a ground-glass stopper, mixed with approximately 20 ml water and transferred to a 250-ml volumetric flask. The weighing flask is rinsed several times. After topping up to the mark with CO_2-free water, take aliquots of 20 or 25 ml and titrate as described.

Wine vinegar usually contains 4–5% acetic acid. To determine **the acetic acid in wine vinegar** weigh out about 20 g of wine vinegar as described, dilute to 100 ml in the volumetric flask and titrate a sample of 25 ml diluted with the same volume of water after adding phenolphthalein with sodium hydroxide solution (0.1 mol/l) as described. The acetic acid content is calculated in gram acetic acid per 100 g wine vinegar (mass fraction in percent).

3.1.4.21 Determination of boric acid

Boric acid, H_3BO_3, has such a low acidity constant with $K_{a1} = 5.75 \cdot 10^{-10}$ that it does not change the color of methyl orange. Even in the presence of phenolphthalein, boric acid cannot be titrated with alkaline solutions, as the indicator changes color long before the equivalence point is reached. The addition of polyhydric alcohols, such as mannitol or glycerol or glucose or fructose, increases the acidity constant of boric acid by complex formation to $K_{a1} = 3.6 \cdot 10^{-7}$ (mannitol) so that it can be titrated like a monobasic medium-strength acid (against phenolphthalein).

If there is an excess of a polyalcohol which must contain two neighboring hydroxyl groups, esterification of the OH groups of the boric acid with two molecules of the alcohol takes place. As boron strives for a coordination number of 4, a further bond to the oxygen atom of a neighboring alcoholic hydroxyl group occurs. This loosens the bond to the hydrogen and facilitates its cleavage as H^+. A boundary structure of the boric acid complex can be described as follows:

$$
\left[
\begin{array}{c}
\mathrm{HC-O} \quad \overset{(-)}{\underset{\mathrm{B}}{\diagdown}} \quad \mathrm{O-CH} \\
\mathrm{HC-O} \diagup \quad \diagdown \mathrm{O-CH}
\end{array}
\right]^{-} \mathrm{H}^{+}
$$

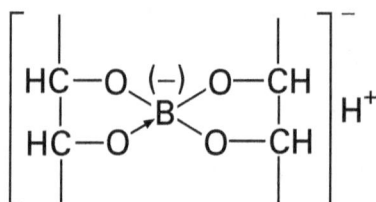

If the alcohol is present in the lower amount, a compound is formed with a ratio of boric acid: polyalcohol = 1:1, which also behaves like a monobasic acid, but whose acidic character is not as pronounced as that of the (1:2) compound. It is noteworthy that the reaction proceeds at an astonishingly high rate for an esterification. The titration of boric acid is predominantly carried out in the presence of an excess of glycerol against phenolphthalein (pH ≈ 8). The complex compounds with fructose, mannitol, dulcite or sorbitol, however, are much stronger acids. With a sufficiently high concentration of one of these polyalcohols, the acidic character of the complex compound is strengthened to such an extent that titration to pH ≈ 6 is possible (methyl red or bromocresol purple as an indicator). This behavior is important for the determination of boric acid in addition to protolyzing salts or weak acids, which are also detected in a titration against phenolphthalein. However, iron and aluminum must always be separated [97].

The success of the boric acid titration is dependent on the following conditions:
1. The alcohol used must react neutrally. Since glycerol usually reacts acidically, it must first be carefully neutralized against phenolphthalein.
2. The polyalcohol must be added in sufficient excess so that the boric acid is quantitatively converted into the complex compound.
3. The titration must be carried out with careful exclusion of CO_2.

Procedure: A volume of 20 ml glycerol is precisely neutralized with sodium hydroxide solution (0.1 mol/l) after adding two to three drops of phenolphthalein solution and 5 ml water (weak pink color). After adding 20 ml of analysis solution, titrate with the standard solution at room temperature without further dilution while continuously inverting the titration beaker until the pink coloration persists for at least 60 s.

If mannitol is used, dilute 20 ml of the analysis solution to 100 ml with CO_2-free water, add 3.6 g mannitol and two to three drops of indicator solution and titrate as described:

1 ml NaOH solution, $c(NaOH) = 0.1$ mol/l, corresponds to 6.1832 mg H_3BO_3 or 3.4809 mg B_2O_3.

Another example is the determination of the boric acid content of an alkali metal borate.

Procedure: Approximately 1.5 g of the carbonate-free borate is accurately weighed and dissolved in boiled water, and the solution is made up to 100 ml. Titrate an aliquot of 25 ml with hydrochloric acid, $c(HCl) = 0.2$ mol/l, according to the procedure given on p. 105 to determine the alkali content of the borate. Then a second sample of the borate solution, also 25 ml, is exactly neutralized by adding the volume of hydrochloric acid determined in the first titration, a few drops of phenolphthalein and 50 ml of glycerol neutralized against phenolphthalein (see above) are added and titrated with carbonate-free sodium hydroxide solution, c(NaOH) = 0.2 mol/l, until the color remains pink. Then add another 10 ml glycerol and, if the solution decolors, titrate again to pink coloration. If the color remains when glycerol is added again, the end point has been reached.

If the borate sample contains carbonate, the solution is neutralized with hydrochloric acid against methyl red as described and the CO_2 is expelled by brief boiling under reflux (so that the boric acid, which is volatile in water vapor, does not escape) and by slowly passing through a CO_2-free air stream. After the solution has cooled down, the reflux condenser is rinsed out and the boric acid is titrated:

1 ml sodium hydroxide solution, c(NaOH) = 0.2 mol/l, corresponds to 6.9618 mg B_2O_3.

3.1.4.22 Determination of magnesium by back titration

Just as salts of weak acids can be determined by acidimetric titrations, salts of weak bases can also be titrated alkalimetrically. By adding strong bases to the solutions of the mineral acid salts of metals that form poorly soluble hydroxides, these are precipitated, as shown in the following equation:

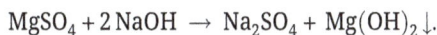

$$MgSO_4 + 2\,NaOH \rightarrow Na_2SO_4 + Mg(OH)_2 \downarrow.$$

For example, the nitrates, chlorides and sulfates of cobalt, nickel, manganese and copper. The solutions are first neutralized against dimethyl yellow, methyl orange or methyl orange-indigo carmine, a measured excess of NaOH solution is added, the hydroxide precipitate is filtered off and the excess sodium hydroxide solution is titrated back with hydrochloric acid in an aliquot of the filtrate. The procedure is explained using the example of determining the magnesium content of a magnesium chloride solution.

Procedure: The weakly acidic sample solution, whose magnesium concentration should be about 0.5 mol/l, must not contain any ammonium salts. Neutralize 25 ml of the solution precisely in a 250-ml volumetric flask by adding sodium hydroxide solution, c(NaOH) = 0.2 mol/l, drop by drop using dimethyl yellow or methyl orange. The flask is then filled up to the mark with boiled water. Add 100 ml of the NaOH solution to 100 ml of the solution in a second volumetric flask, make up to 250 ml and shake well. The solution is filtered through a dry filter and the precipitate is not washed out. Discard the first 50 ml of the filtrate. Then titrate back the excess NaOH in 100 ml of the filtrate with hydrochloric acid, c(HCl) = 0.2 mol/l, using dimethyl yellow or methyl orange:

1 ml sodium hydroxide solution, c(NaOH) = 0.2 mol/l, corresponds to 2.431 mg Mg or 9.521 mg $MgCl_2$.

3.1.4.23 Precipitation process with pH change display

Magnesium or calcium can be determined directly by acid-base titration. A standard solution with potassium palmitate is suitable for this as the titrant substance. Magnesium palmitate, for example, precipitates during the titration:

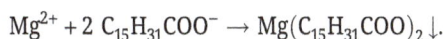

$$Mg^{2+} + 2\,C_{15}H_{31}COO^- \rightarrow Mg(C_{15}H_{31}COO)_2 \downarrow.$$

After reaching the equivalence point, palmitate ions are further added and partially protonated by the water, releasing hydroxide:

$$C_{15}H_{31}COO^- + H_2O \rightarrow C_{15}H_{31}COOH + OH^-.$$

Phenolphthalein is used as an indicator. The titration end point is considered to be reached when a pink coloration occurs [24a].

3.1.4.24 Determination of magnesium by direct titration

Procedure:
Standard solution: Dissolve 15 g KOH in 100 ml ethanol while heating. In another vessel, combine 25.6 g palmitic acid ($C_{15}H_{31}COOH$), 0.1 g phenolphthalein, 500 ml propanol and 300 ml water and titrate slowly with the KOH solution until a clear pink-colored liquid is obtained. Make up to the mark with water in the 1-l volumetric flask to obtain a 0.1 mol/l solution. Adjust against $MgCl_2$ or $CaCl_2$. The solution absorbs CO_2 from the air and should therefore be used up very soon.

Procedure: Pipette 20 ml of the sample solution into the titration flask, neutralize and dilute to approximately 100 ml with water. After adding a little phenolphthalein, titrate with 0.1 mol/l standard solution until a pink coloration appears (V_1). A blank titration is carried out (V_2). The difference $V_1 - V_2$ gives the volume of the 0.1 mol/l standard solution consumed during the conversion of the analyte:
   1 ml of this corresponds to 1.215 mg Mg [24a].

3.1.4.25 Determination of ammonium salts

Ammonium salts of strong acids, such as NH_4Cl and $(NH_4)_2SO_4$, cannot be titrated with sodium hydroxide solution, even in the 1 mol/l concentration range, as the end point cannot be identified with sufficient precision. NH_4^+ is a too weak acid (pK_a = 9.25) and the corresponding base NH_3 is therefore a too strong base (pK_b = 4.75), meaning that phenolphthalein cannot be used as an indicator. In the presence of excess formaldehyde, however, titration with caustic soda is possible. The NH_3 released when caustic soda is added condenses in a rapid reaction with formaldehyde to form the very weak base hexamethylenetetraamine (urotropine, pK_b = 9.1) to

$$4\,NH_4^+ + 6\,CH_2O \rightleftharpoons (CH_2)_6N_4 + 6\,H_2O + 4\,H^+,$$

which no longer acts on phenolphthalein.

Procedure: The sample solution, which should contain about 0.1–0.2 g of ammonium salt and react neutrally, is mixed with 10 ml of an approximately 35% formaldehyde solution previously neutralized against phenolphthalein and, after dilution to 100 ml with boiled water and addition of a few drops of phenolphthalein solution, titrated with carbonate-free sodium hydroxide solution, $c(NaOH)$ = 0.1 mol/l, at room temperature until a stable pink color appears:
   1 ml sodium hydroxide solution, $c(NaOH)$ = 0.1 mol/l, corresponds to 1.7031 mg NH_3 or 1.8039 mg NH_4^+.

3.1.4.26 Formol titration (titration according to Sörensen)

Amino acids can also be titrated with aqueous sodium hydroxide against phenolphthalein (or potentiometric indication) using the same principle. Amino acids are
present as zwitterions and therefore cannot be titrated in an aqueous medium despite
pK_a values of approximately 2–3. The addition of formaldehyde leads to imine formation with the primary amino group (more detailed investigations have shown, however, that more complex compounds than just the formation of formimines occur)
and thus to a significant decrease in basicity so that a zwitterion is no longer formed
and the carboxylic acid can be titrated classically in the aqueous against phenolphthalein or potentiometrically.

3.1.4.27 Determination of glycine

Glycine: R = H

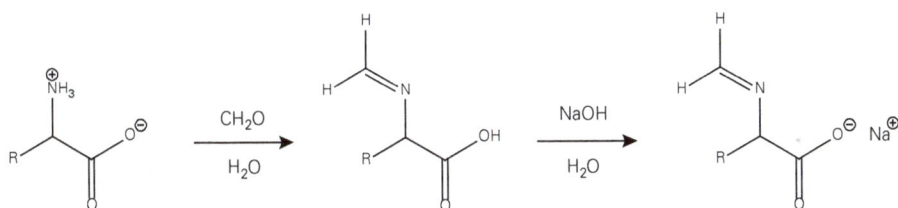

Procedure: Dissolve 70 mg glycine in 50 ml water, add 5 ml formalin (previously neutralized against
phenolphthalein) and 0.4 ml phenolphthalein solution and then titrate with sodium hydroxide solution
(0.1 mol/l):

  1 ml sodium hydroxide solution, $c = 0.1$ mol/l, corresponds to 7.507 mg $C_2H_5NO_2$

3.1.4.28 Determination of cationic acids by displacement titration

Salts of organic amines usually have pK_a values of around 8–10. This means that they
cannot be titrated in water with sodium hydroxide solution. This problem was previously solved by using two-phase titration, in which the cationic acids were titrated in
a mixture of ethanol and chloroform with aqueous sodium hydroxide solution against
phenolphthalein. In this case, the free base formed during the titration is constantly
shaken out into the chloroform, thus shifting the equilibrium to the right so that the
cationic acid is completely deprotonated. Due to the use of chloroform with its carci-
nogenic and environmentally toxic potential, this titration is no longer up to date (e.g.
DAB 7 determination of morphine hydrochloride).

According to pharmacopoeias such as Ph. Eur. [8a] or USP, cationic acids are
therefore titrated today by displacement titration in ethanol with aqueous sodium hydroxide solution for potentiometric indications. A small amount of hydrochloric acid
(0.01 M) is added to the ethanol solvent, which facilitates the dissolution of the salt

and protonates small amounts of free base. During the titration, the hydrochloric acid solution is first titrated back (first equivalence point) and any acidic impurities present in the solvent are also titrated, followed by the titration of the cationic acid (second equivalence point). The difference between the two equivalence points therefore provides the consumption for titration of the cationic acid.

3.1.4.29 Determination of imipramine hydrochloride according to Ph. Eur

The tricyclic antidepressant imipramine hydrochloride (pK_a = 9.59) is determined in the European Pharmacopoeia as cationic acid using 1 equiv of sodium hydroxide solution.

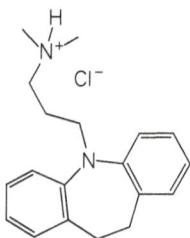

Procedure: About 0.250 g substance, dissolved in 50 ml ethanol 96%, and 5.0 ml hydrochloric acid (0.01 mol/l) are added and titrated with sodium hydroxide solution (0.1 mol/ml). The volume added between the two inflection points determined using potentiometry (see Section 4.4) is read off [8a]: 1 ml sodium hydroxide solution, c = 0.1 mol/l, corresponds to 31.69 mg $C_{19}H_{25}ClN_2$.

3.1.4.30 Titration of weak acids in anhydrous medium

Another method for determining weak acids such as phenols, sulfonamides, ammonium ions and imides with pK_a values between approximately 8 and 11 is titration in an anhydrous medium. Tetrabutylammonium hydroxide (TBAH), ethanolic potassium or sodium hydroxide solution, lithium methanolate or other alcoholates are used as standard solutions. The analytes can be dissolved in solvents such as dimethyl sulfoxide (DMSO), dimethyl formamide, pyridine or alcohols and then titrated with the corresponding standard solution against thymolphthalein, thymol blue or potentiometric indication.

3.1.4.31 Preparation of 0.1 M tetrabutylammonium hydroxide solution

TBAH solution is prepared by dissolving tetrabutylammonium iodide and silver oxide in methanol and then diluting with toluene:

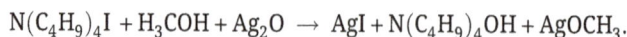

$$N(C_4H_9)_4I + H_3COH + Ag_2O \longrightarrow AgI + N(C_4H_9)_4OH + AgOCH_3.$$

Procedure: About 40 g tetrabutylammonium iodide is dissolved in 90-ml anhydrous methanol. After adding 20 g of finely powdered silver oxide, shake vigorously for 1 h. A few milliliters of the mixture are centrifuged and the identity test for iodide is carried out with the supernatant liquid. If the reaction is positive, a further 2 g of silver oxide is added to the mixture and shaken for 30 min. This procedure is repeated until the supernatant liquid contains no longer any iodide. The mixture is filtered through a narrow-pored glass sinter crucible and the vessel and filter are rinsed thrice with 50 ml toluene each. The washing liquids are combined with the filtrate and diluted with toluene to 1000.0 ml. CO_2-free nitrogen is introduced into the solution for 5 min. Instead of toluene, the standard solution can also be prepared in 2-propanol (= isopropanol) [8a].

3.1.4.32 Adjustment of 0.1 M tetrabutylammonium hydroxide standard solution

Procedure: A volume of 10 ml dimethylformamide is titrated with the TBAH solution with the addition of 0.05 ml of a solution of thymol blue (3 g/l) in methanol until a pure blue coloration is obtained. Approximately 0.200 g of benzoic acid urtiter, which has been accurately weighed beforehand, is immediately added to this solution. Shake until the substance is dissolved and titrate with the TBAH solution until a new blue coloration is obtained. During the titration, the solution must be protected from CO_2 in the air. The factor is calculated from the titration volume of the second titration. The factor must be determined immediately before use:

1 ml tetrabutylammonium hydroxide solution, c = 0.1 mol/l, corresponds to 12.21 mg $C_7H_6O_2$.

3.1.4.33 Determination of hydrochlorothiazide according to Ph. Eur. 9.0

The thiazide diuretic hydrochlorothiazide was determined in the European Pharmacopoeia 9.0 by anhydrous titration as a weak acid with TBAH. The two weakly acidic sulfonamides (pK_{a1} 8.6–9.5 and pK_{a2} 9.9–10.4) are deprotonated in the process:

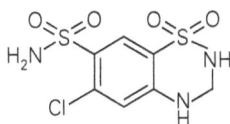

Procedure: About 0.120 g substance, dissolved in 50 ml DMSO, is titrated with TBAH solution (0.1 mol/l) in 2-propanol to the second inflection point. The end point is determined using potentiometry. A blank titration is carried out [8a]:

1 ml tetrabutylammonium hydroxide solution, c = 0.1 mol/l, corresponds to 14.88 mg $C_7H_8ClN_3O_4S_2$.

3.1.4.34 Determination of rutoside according to Ph. Eur. 11.0

The flavonoid rutoside is determined in the European Pharmacopoeia by anhydrous titration with TBAH as a weak acid. Two of the four phenolic hydroxy groups are detected during the titration:

$$O-C_6H_{10}O_4-O-C_6H_{11}O_4$$

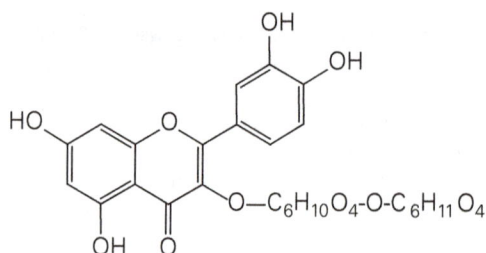

Procedure: About 0.200 g substance, dissolved in 20 ml dimethylformamide, is titrated with TBAH solution (0.1 mol/l). The end point is determined using potentiometry [8a]:
1 ml tetrabutylammonium hydroxide solution, c = 0.1 mol/l, corresponds to 30.53 mg $C_{27}H_{30}O_{16}$.

3.1.4.35 Determination of glycerol according to Ph. Eur. 11.0

The trivalent alcohol glycerol can be determined after reaction with sodium periodate in a Malaprade cleavage (diol cleavage) by titration with sodium hydroxide solution. By oxidative cleavage with sodium periodate, vicinal alcohols are first cleaved to the aldehydes. In the case of glycerol, formaldehyde and 2-hydroxyacetaldehyde are obtained first. The latter is split again to formic acid and formaldehyde. The formic acid can then be titrated with sodium hydroxide solution. Before titration, the excess sodium periodate must be destroyed with an excess of ethylene glycol. The same procedure can also be used to determine other sugar alcohols such as sorbitol or mannitol; these two sugar alcohols each produce four molecules of formic acid and two molecules of formaldehyde. Malaprade cleavage can also be evaluated iodometrically as part of redox titrations. However, this method is now considered obsolete due to its complexity and the use of a standard solution of arsenious acid (see Comment to Ph. Eur.):

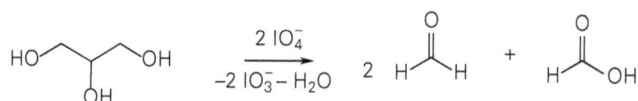

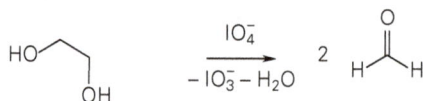

Procedure: Mix 75 mg of substance thoroughly with 45 ml of water. Add 25.0 ml of a mixture of 1 part by volume of sulfuric acid (0.1 mol/l) and 20 parts by volume of sodium periodate solution (0.1 mol/l) to the mixture and leave to stand for 15 min under light protection. After adding 5.0 ml of a solution of ethylene glycol (500 g/l), this mixture is allowed to stand for 20 min under light protection and titrated with sodium hydroxide solution (0.1 mol/l) with the addition of 0.5 ml phenolphthalein solution (0.1% in 80% ethanol). A blank titration is carried out [8a]:

1 ml sodium hydroxide solution, $c = 0.1$ mol/l, corresponds to 9.21 mg $C_3H_8O_3$.

3.1.4.36 Determination of phosphoric acid

An example for the determination of polybasic acids is the titration of phosphoric acid.

The protolysis of polybasic acids takes place in stages. If, for example, the tribasic orthophosphoric acid (H_3PO_4) is titrated with alkali, dihydrogen phosphate, hydrogen phosphate and phosphate are formed successively:

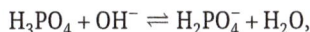

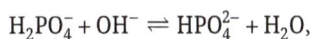

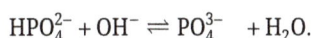

$$H_3PO_4 + OH^- \rightleftharpoons H_2PO_4^- + H_2O,$$

$$H_2PO_4^- + OH^- \rightleftharpoons HPO_4^{2-} + H_2O,$$

$$HPO_4^{2-} + OH^- \rightleftharpoons PO_4^{3-} + H_2O.$$

Each of these stages has its characteristic acidity constant. They are $K_{a1} = 6.92 \cdot 10^{-3}$ ($pK_{a1} = 2.16$), $K_{a2} = 6.17 \cdot 10^{-8}$ ($pK_{a2} = 7.21$) and $K_{a3} = 1.97 \cdot 10^{-13}$ ($pK_{a3} = 12.32$). The hydrogen ion concentration in the solution of one of the phosphates is obtained as the geometric mean of the respective acid constants or the pH value as the arithmetic mean of the respective pK_a values. The solution of the dihydrogen phosphate has an H_3O^+ ion concentration of $2.16 \cdot 10^{-5}$ mol/l, that of the hydrogen phosphate of $1.15 \cdot 10^{-10}$ mol/l and phosphate of about $10^{-13.5}$ mol/l. It is therefore possible to titrate orthophosphoric acid and other polybasic acids in stages if the acid constants of the individual stages are far enough apart (approx. 10^4). The indicators must be selected so that their transition point coincides as closely as possible with the pH value in the solution of the desired salt.

Procedure:

First-stage titration: Titrate the sample solution diluted to approximately 100 ml with water at room temperature with sodium hydroxide solution, 0.1 mol/l, against methyl orange until a strong orange coloration or, better still, against dimethyl yellow to a pure yellow color or even more precisely against bromophenol blue (conversion interval 3.0–4.6) to pH = 4.5 while inverting the titration beaker. Always titrate to the same color with a reference solution containing NaH_2PO_4 at a concentration of 0.05 mol/l and the same indicator concentration.

Second-stage titration: To achieve pH = 9.7, use thymolphthalein as an indicator (transition interval 9.3–10.5) and titrate until a weak blue coloration is obtained. If phenolphthalein is to be used, the reaction of the hydrogen phosphate with water must be suppressed by saturating the titration solution with sodium chloride. The error in both cases is about 1%.

> Third-stage titration: Direct titration is not possible due to the small acid constant. However, by adding calcium chloride in a suitable concentration, calcium phosphate can be precipitated and the acidity determined according to
>
> $$2\,H_2PO_4^- + 3\,Ca^{2+} + 4\,H_2O \rightarrow Ca_3(PO_4)_2 + 4\,H_3O^+.$$
>
> H_3O^+ ions can be titrated against phenolphthalein.
>
> **Procedure according to Kolthoff [90]:**
> The solution neutralized to dimethyl yellow is mixed with 30 ml of a neutral (!) 40% calcium chloride solution, heated to boiling point and cooled to 14 °C. After adding phenolphthalein, titrate with carbonate-free alkali while swirling vigorously until pink coloration. The flask is then sealed and the solution, whose color slowly disappears, is titrated to completion after standing for 2 h at 14 °C. The error is 1–2%:
>
> 1 ml NaOH solution, $c(NaOH) = 0.1$ mol/l, corresponds to 9.7995 mg H_3PO_4 (titration of the first stage) or 4.8998 mg H_3PO_4 (joint titration of the first two stages).

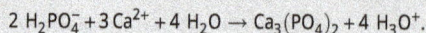

3.1.4.37 Determination after ion exchange

A series of determinations are analytically cumbersome and difficult; e.g. the determination of alkali metal ions or of nitrate, perchlorate, acetate and other anions in salt solutions can be carried out easily and with sufficient accuracy using ion exchangers [98]. The ion exchangers used for analytical purposes are synthetic resins that are produced by polymerization and contain groups with exchangeable ions. A distinction is made between cation exchangers (symbol: RH), which exchange H^+ for metal cations or vice versa, and anion exchangers (symbol: ROH), which exchange OH^- for anions in a stoichiometric manner. The process can be represented schematically by the following equations:

$$RH + Me^+ \rightleftharpoons RMe + H^+,$$

$$ROH + A^- \rightleftharpoons RA + OH^-.$$

The determination of a salt is based on an alkalimetric or acidimetric titration.

A cation exchanger consists of a polymeric resin skeleton with anionic groups and ionogenically bound cations, while an anion exchanger is a polymeric cation with active anions.

The schematic structure of a cation exchanger can be seen in the formula diagram in Fig. 3.7, which shows a widely used exchanger. It is produced by copolymerization of styrene with a smaller proportion of divinylbenzene and subsequent sulfonation.

In addition to these strongly acidic cation exchangers with $-SO_3H$ groups, weakly acidic ones with $-COOH$ groups are also commercially available, e.g. copolymers of methacrylic acid, $CH_2=C(CH_3)-COOH$ and glycol bis-methacrylate, $CH_2=C(CH_3)-COOCH_2-CH_2OOC-C(CH_3)=CH_2$.

Anion exchangers are also cross-linked, highly polymeric substances whose basic character is given by the incorporation of amino, substituted amino or quaternary

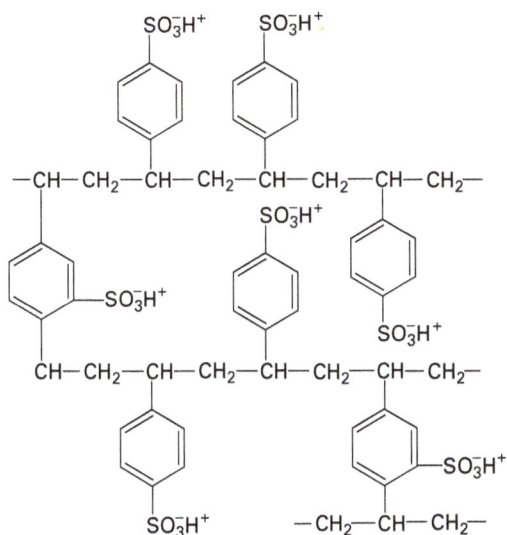

Fig. 3.7: Cation exchanger (H⁺ form).

ammonium groups. The polymers with the latter groups are strongly basic anion exchangers. A frequently used exchanger of this type is the copolymer of styrene with
some divinylbenzene, into which the group $-CH_2Cl$ is subsequently introduced by
chloromethylation and reacted with a base such as trimethylamine. A picture of the
formula is shown in Fig. 3.8.

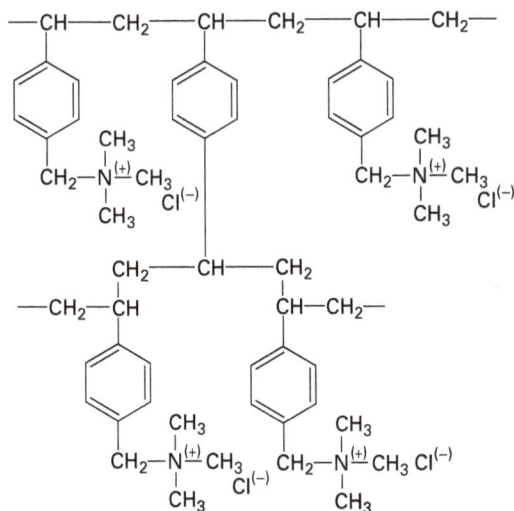

Fig. 3.8: Anion exchanger (Cl⁻ form).

For volumetric analysis purposes only the strongly acidic or strongly basic ion exchangers should be used. The cation exchange takes place at the sulfonic acid groups

$$R(SO_3^-)_n(H^+)_n + nMe^+ \rightarrow R(SO_3^-)_n(Me^+)_n + nH^+,$$

and the anion exchange at the quaternary ammonium groups

$$R(N^+(CH_3)_3)_n(OH^-)_n + nA^- \rightarrow R(N^+(CH_3)_3)_n(A^-)_n + nOH^-.$$

Trade names for ion exchangers are e.g. Amberlite (Rohm and Haas, USA), Duolite (Diamont Alkali, USA), Dowex (Dow Chemical Co., USA) and Lewatit (Merck, Sigma-Aldrich, Fluka, Riedel-de Haën).

The exchange is carried out in glass columns into which the resin is poured (Fig. 3.9). The dimensions of the column depend on the amount of substance of the ions to be exchanged and the number of analyses to be carried out without regenerating the exchanger. Columns about 20 cm long and 2 cm in diameter are suitable for working with concentrations in the 0.1 mol/l range. The column tube has a spherical extension at the top and is fitted with a drain cock at the lower end. Above the tap is a frit plate (G2) or a sieve plate made of glass with a layer of glass wool, which carries the resin grains. The column is filled with resin to two-third to three-fourth of its length. To prevent the inclusion of air bubbles, the resin is poured into the column filled with water.

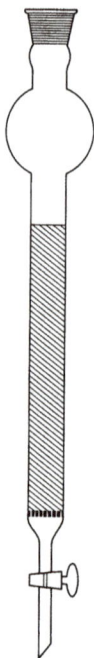

Fig. 3.9: Ion exchange column.

Ion exchangers for volumetric analysis purposes are available in Na^+ or Cl^- form. It is therefore necessary to convert them into the acidic or basic form. The conversion to the H^+ form is carried out with hydrochloric acid at a concentration of 3 mol/l, and sodium hydroxide solution (c = 1–2 mol/l) is used to convert the anion exchanger to the OH^- form. The acid or alkali must run slowly through the resin so that the exchange can take place. The flow rate is adjusted using the drain tap. Any initial brown coloration of the solution due to organic resin components is irrelevant. The end of regeneration is determined by checking for the absence of a reaction characteristic of the ion to be replaced. The last traces of the ion do not have to be removed. The "loaded" resin is then washed with water until the reaction of the effluent is neutral. The exchangers are kept under water at all times to prevent the column from running dry. Before using the exchanger column, check the reaction of the water in the outlet. In the event of an acidic or alkaline reaction, the resin must again be washed with water until neutral.

Anion exchangers made from polystyrene resins have an exchange capacity of about 1 mmol/ml wet resin based on equivalents. The capacity of cation exchangers is about twice as high. A column with the dimensions given above contains about 50 ml of resin. Assuming that 20 ml of a neutral solution with a concentration of c = 0.1 mol/l is used for each analysis, an anion exchanger must be regenerated after 18–20 determinations and a cation exchanger after 35–40 determinations if the capacity of 70–80% is utilized. This is done by adding 1 l of sodium hydroxide solution (c = 1–2 mol/l) or hydrochloric acid (c = 3 mol/l), which is allowed to run through the column at a rate of 10–12 drops/min. Then rinse with water until the reaction is neutral.

Determination of phosphates, nitrates and perchlorates of alkali metals.
The diluted sample solution is added to the column from which the water has been drained to the resin surface. Any salt solution adhering to the walls is rinsed off with a little water. Rinse with 150–200 ml of water in small portions and wait for the solution to sink to the resin surface each time. The flow rate should be 5–10 ml/min. Finally, rinse with about 50 ml of water. The resulting alkaline solution is titrated with an adjusted acid solution in the usual way. The time required for a determination is about 30 min.

3.1.4.38 Acid value, saponification value and ester value in the analysis of fatty pharmaceuticals and foods (characterization of fatty oils)

The acid value of sample substances containing fat is a measure of the content of free acids present and thus of the partial decomposition of fats that has already taken place as a result of saponification reactions. It is given in milligrams of potassium hydroxide per gram of sample substance. After neutralization of the solvent mixture not yet mixed with the sample substance, the sample solution is quickly titrated with potassium hydroxide solution or sodium hydroxide solution. Phenolphthalein is used as an indicator.

The saponification value includes the consumption of hydroxide through saponification of the previously undecomposed fats. It is also given in milligrams of potassium hydroxide per gram of sample substance. After saponification of the fats with an excess of potassium hydroxide in a boiling ethanolic solution, retitration is carried out with hydrochloric acid. Phenolphthalein is again used as an indicator.

The ester value (EV) *is* the difference between the saponification value (SV) and the acid value (AV) and thus serves as a measure of the content of undecomposed fats:

$$EV = SV - AV.$$

The three fat indices are an important part of the purity tests of fatty oils in the pharmacopoeias (e.g. Ph. Eur.). The terms "ester number," "acid number" and "saponification number" are also commonly used.

3.1.4.39 Determination of the acid value (AV)

Procedure: Twenty-five milliliters of ethanol (96%) and 25 ml petroleum ether are mixed, and after adding phenolphthalein they are neutralized with the standard solution (potassium hydroxide solution (0.1 mol/l) or sodium hydroxide solution (0.1 mol/l)). Approximately 10.00 g of sample substance is weighed accurately and dissolved in the solvent mixture. If heating is necessary, maintain a temperature of 90 °C during dissolution and titration. Titrate rapidly until the pink coloration persists for at least 15 s [8a]:

1 ml of standard solution, c = 0.1 mol/l, corresponds to an acid number of 5.610 mg KOH per g of sample substance.

3.1.4.40 Determination of the saponification value (SV)

Procedure: First, a suitable sample weight is determined, e.g. approximately 200 times the reciprocal value of the expected saponification number, as a mass in grams. Boil the sample with 25.0 ml ethanolic potassium hydroxide solution (0.5 mol/l) for 30 min under reflux. After addition of phenolphthalein the still hot solution is titrated with hydrochloric acid 0.5 mol/l. The European Pharmacopoeia refers to the necessity of a blank test "under identical conditions:"

1 ml hydrochloric acid, c = 0.5 mol/l, corresponds to a saponification value of 28.05 mg KOH per g sample [8a].

3.1.4.41 Determination of NH-acid compounds

NH-acidic compounds such as imines (e.g. barbiturates and hydantoins) or sulfonamides with pK_a values between 5 and 9 can, in many cases, be titrated directly with aqueous sodium hydroxide solution after silver complexes or silver precipitates have been formed with an excess of silver nitrate and an equivalent amount of protons has been released. A solution of silver nitrate in pyridine is often added so that an equivalent amount of pyridinium ions is formed, which is then titrated. The method, which can also be used for terminal alkynes (see below), is known as argentoacidimetry (silver-pyridine method according to Dutrieux).

Determination of phenobarbital according to Ph. Eur. 4.0

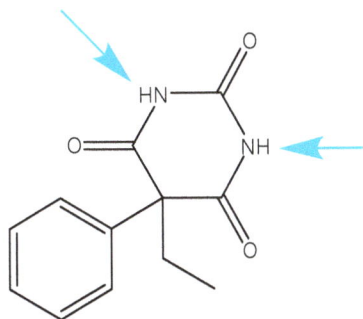

Like all barbiturates, the antiepileptic phenobarbital can be determined argentoacidimetrically. By forming a silver complex, both acidic protons (pK_a 7.36 and 12.14) are detected (in contrast to titration as a weak acid of barbiturates).

Procedure: About 0.100 g substance, dissolved in 5 ml pyridine, is titrated with ethanolic sodium hydroxide solution (0.1 mol/l) after the addition of 0.5 ml thymolphthalein solution and 10 ml silver nitrate-pyridine (85 g/l) until pure blue coloration. A blank titration is carried out [8a]:
   1 ml ethanolic sodium hydroxide solution, $c = 0.1$ mol/l, corresponds to 11.61 mg $C_{12}H_{12}N_2O_3$.

3.1.4.42 Determination of ethinylestradiol

Very weak organic acids can be determined titrimetrically using special techniques. Ethinylestradiol, for example, can act as an acid with its terminal ethinyl group in the presence of an excess of silver ions and release a proton to water, forming oxonium, which can then be titrated with sodium hydroxide solution [98a]:

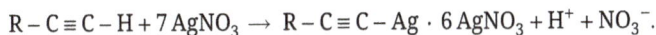

$$R-C\equiv C-H + 7\,AgNO_3 \rightarrow R-C\equiv C-Ag \cdot 6\,AgNO_3 + H^+ + NO_3^-.$$

Procedure: Accurately weigh approximately 0.200 g of the sample substance and dissolve in 40 ml tetrahydrofuran. After adding 5 ml silver nitrate solution (100 g/l), titrate with sodium hydroxide solution (0.1 mol/l). The equivalence point is determined potentiometrically (see Section 4.4). A blank titration is carried out. The electrode is rinsed with acetone after each titration:
   1 ml sodium hydroxide solution, $c = 0.1$ mol/l, corresponds to 29.64 mg $C_{20}H_{24}O_2$ [8a].

3.1.4.43 Determination of salicylic acid

Salicylic acid in pharmaceuticals is determined alkalimetrically, although a bromato-metric-iodometric method exists (see Section 3.3.6).

Procedure: Accurately weigh approximately 0.12 g of sample substance and dissolve in 30 ml ethanol 96%. After adding 20 ml water and 0.1 ml phenol red solution (1 g/l) as indicators, titrate with sodium hydroxide solution, c = 0.1 mol/l:

   1 ml sodium hydroxide solution, c = 0.1 mol/l, corresponds to 13.81 mg $C_7H_6O_3$[8a].

3.1.4.44 Determination of acetylsalicylic acid

The painkiller acetylsalicylic acid is determined in many pharmacopoeias by saponification titration. With an excess of sodium hydroxide solution, acetylsalicylic acid is saponified to sodium acetate and sodium salicylate. The excess sodium hydroxide solution is then retitrated with hydrochloric acid against phenolphthalein (consumption 2 equiv).

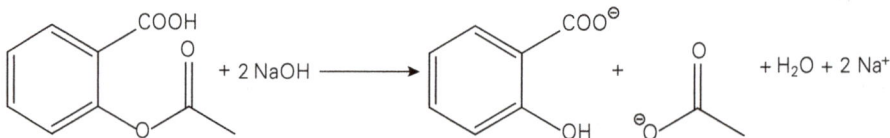

Procedure: About 1 g substance is dissolved in 10 ml ethanol 96% R in an Erlenmeyer flask with ground glass stopper. After adding 50 ml sodium hydroxide solution, c = 0.5 mol/l, close the flask and leave to stand for 1 h. After adding 0.2 ml phenolphthalein solution (0.1% in 80% ethanol), titrate this solution with hydrochloric acid (0.5 mol/l). A blank titration is carried out:

   1 ml sodium hydroxide solution, c = 0.5 mol/l, corresponds to 45.04 mg $C_9H_8O_4$.

3.2 Precipitation titrations

In precipitation analyses, the substance to be determined is converted into a sparingly soluble compound of defined composition during the titration. The end point of the titration is reached when the precipitant contained in the standard solution has been added in an equivalent quantity.

   In gravimetry an extraordinarily large number of precipitation reactions are used to carry out quantitative determinations. However, only a few of these reactions are suitable as a basis for precipitation titrations. As a result, only a relatively small number of useful analytical precipitation methods are available. The main reason for this is the lack of generally applicable and reliable methods for recognizing the titration end point. Also, some precipitation reactions – especially in dilute solutions – do not proceed with the speed required for titration. In addition, coprecipitation phenomena can occur which influence the stoichiometry of the reaction.

Among the useful and frequently used precipitation titrations are the methods for the determination of halides and pseudohalides with silver nitrate standard solution (argentometric titrations) and silver with a sodium chloride solution.

3.2.1 Theoretical foundations

3.2.1.1 Solute equilibrium

After precipitation has taken place, there is an equilibrium between the precipitate (soil body) and the liquid above it. This liquid is a **saturated solution**; it contains the precipitated substance in the highest possible concentration, the **saturation concentration** which depends on the temperature. The poorly soluble compounds formed during precipitation titrations are usually salts made up of ions. The components of the ion lattice, the cations and anions, pass directly from the ground body into the solution during the dissolving process. Conversely, they form the ion lattice of the crystal directly during the precipitation process. As strong electrolytes, salts are completely dissociated into their ions in the solution, i.e. there are no undissociated molecules in the solution apart from the ions.

If, for example, a silver nitrate solution is mixed with a sodium chloride solution, solid silver chloride precipitates:

$$Ag^+ + Cl^- \rightleftharpoons AgCl_{solid}.$$

The process is reversible, even if the equilibrium is far to the right. This means that solid silver chloride dissolves in water, but only to a very small extent. In the process, silver and chloride ions pass from the surface of the crystals into the solvent in equivalent numbers. When equilibrium is reached, the number of ion pairs leaving the crystal lattice in a unit of time is equal to the number of ion pairs deposited on the surface of the crystals from the solution. A **dynamic equilibrium** exists. If the temperature remains constant, the number of ion pairs is constant.

3.2.1.2 Solubility product and solubility

The application of the law of mass action to the solution equilibrium of AgCl results in

$$c(Ag^+) \cdot c(Cl^-) = K_{sp}.$$

The ground body, AgCl, **does not** appear in this equation as it is a pure solid substance. K_{sp} is called the **solubility product** of a salt; in this case, AgCl. If the silver or chloride ion concentration in the solution is increased, it is **supersaturated**. Solid silver chloride then precipitates until the solubility product is reached again.

The solubility product is therefore a measure of the solubility of the compound in question.

In a solution of pure silver chloride, the concentrations of silver and chloride ions are equal. The following therefore applies

$$c^2(\text{Ag}^+) = c^2(\text{Cl}^-) = K_{sp}.$$

If we use the assumed value $c(\text{AgCl})$ to denote the total concentration of silver chloride in the solution, then $c(\text{AgCl}) = c(\text{Ag}^+) = c(\text{Cl}^-)$ and the following applies:

$$c(\text{AgCl})^2 = K_{sp}$$

and

$$c(\text{AgCl}) = \sqrt{K_{sp}}.$$

The saturation concentration in pure water is also known as the **solubility S**. For silver chloride, it is equal to the square root of the solubility product. Since in the present case $K_{sp} = 10^{-10}$ mol^2/l^2, the concentration of the saturated silver chloride solution is $c(\text{AgCl}) = 10^{-5}$ mol/l. It contains 10^{-5} mol/l AgCl or 10^{-5} mol/l Ag$^+$ and 10^{-5} mol/l Cl$^-$.

By adding excess silver or chloride ions to the saturated solution, the solubility of the silver chloride can be further reduced. It can be calculated as follows:

In a silver nitrate solution containing 10^{-4} mol/l silver ions, the chloride ion concentration is

$$c(\text{Cl}^-) = \frac{K_{sp}}{c(\text{Ag}^+)} = \frac{10^{-10}\ \text{mol}^2/\text{l}^2}{10^{-4}\ \text{mol}/\text{l}} = 10^{-6}\ \text{mol}/\text{l}.$$

As the dissolved silver chloride is practically completely dissociated, $c(\text{AgCl}_{dissolved})$ = $c(\text{Cl}^-)$ must be set so that the solubility is 10^{-6} mol/l.

These considerations apply to electrolytes of composition AB (ratio cation A:anion B = 1:1). In principle, however, similar considerations can be made for all other precipitation processes. In the most general case, the formation of a compound of the composition $A_m B_n$ applies:

$$A_m B_n \rightleftharpoons m\,\text{A}^{n+} + n\,\text{B}^{m-},$$

$$c(\text{A}^{n+})^m \cdot c(\text{B}^{m-})^n = K_{sp}.$$

Since $c(A_m B_n) = \dfrac{1}{m} c(\text{A}^{n+}) = \dfrac{1}{n} c(\text{B}^{m-}) = S$ is

$$K_{sp} = (mS)^m \cdot (nS)^n,$$

$$K_{sp} = m^m L^m \cdot n^n L^n,$$

$$K_{sp} = m^m \cdot n^n \cdot L^{m+n}.$$

and

$$S = \sqrt[m+n]{\frac{K_{sp}}{m^m \cdot n^n}}.$$

The special case with $m = n = 1$, which we learned about with silver chloride, leads to

$$S = \sqrt{K_{sp}}.$$

The following applies in all cases:
- The deposition of a precipitate always begins when the solubility product of the ion types involved is exceeded.
- The solubility of a poorly soluble compound can be further reduced by an excess of precipitating agent. The maximum solubility is at the equivalence point.

Exceptions from this rule can always be observed when the excess reagent can form a soluble complex compound with the precipitate. For example, silver chloride is considerably soluble in excess hydrochloric acid. Chlorocomplexes of silver are formed, e.g. $H[AgCl_2]$.

These considerations show that all direct precipitation titrations have a common source of error. It consists in the fact that absolutely insoluble compounds do not exist and that the solubility of the precipitate is greatest at the equivalence point, the determination of which is the aim of the volumetric analysis.

3.2.2 Titration curves

The change in the ion concentration of the ion type to be determined that occurs during precipitation is represented by the titration curve. The negative decadic logarithm of the numerical value of the ion concentration, the **ion exponent** (pIon value in analogy to the pH value), is plotted against the volume of added standard solution.

Let us consider the case of titrating 100 ml of a dilute silver nitrate solution with a relatively concentrated sodium chloride solution. For simplification, it can be assumed that the volume of the titrated solution does not change during the titration. The temperature should also remain constant during the titration process. The initial concentration of silver ions is $c(Ag^+) = 10^{-1}$ mol/l, and the standard solution has the concentration $c(Cl^-) = 10$ mol/l. As a result of the precipitation of silver chloride, when the standard solution is added, the silver ion concentration decreases. If 0.9 ml NaCl solution has been added, 9 mmol Ag^+ ions have precipitated as AgCl or 90% of the silver ions present. The Ag^+ concentration is now 10^{-2} mol/l. After adding a further 0.09 ml NaCl solution (0.99 ml in total), 9.9 mmol silver ions have precipitated. In relation to the initial concentration, 1% of the silver ions are still in solution. The Ag^+ concentration is now 10^{-3} mol/l. If the amount of chloride is equivalent to the amount of silver that has just been added, the silver ion concentration in the solution resulting from the solubility product is $c(Ag^+) = 10^{-5}$ mol/l. Once the equivalence point is exceeded, further AgCl precipitates, as the considerations in the previous section have shown. If only 0.1% of the equivalent amount of NaCl is added in excess, the Ag^+ concentration is still

$$c(Ag^+) = \frac{K_{sp}}{c(Cl^-)} = \frac{10^{-10}\,\text{mol}^2/\text{l}^2}{10^{-4}\,\text{mol}/\text{l}} = 10^{-6}\,\text{mol}/\text{l}.$$

With an excess of 1%, the Cl^- concentration is 10^{-3} mol/l and the Ag^+ concentration is 10^{-7} mol/l. A 10% excess of NaCl lowers the silver ion concentration to 10^{-8} mol/l, a 100% excess to 10^{-9} mol/l.

Figure 3.10 shows the titration curve. The pAg value is plotted on the ordinate and the degree of titration τ (see p. 87) on the abscissa.

It can be seen that the equivalence point, i.e. the desired end point of the titration, is identical to the inflection point of the titration curve. It is the point at which the relative change in the silver ion concentration reaches its greatest value during the course of the titration.

Figure 3.11 shows the titration curve that results for a tenfold diluted silver nitrate solution with the same sodium chloride solution, $c(Ag^+) = 10^{-2}$ mol/l. The jump in the curve has become considerably smaller.

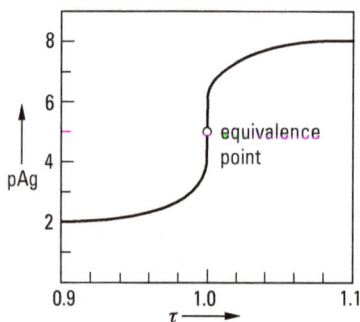

Fig. 3.10: Titration of $AgNO_3$ solution (0.1 mol/l) with NaCl solution (10 mol/l) (pAg = $-$lg $c(Ag^+)$, τ degree of titration).

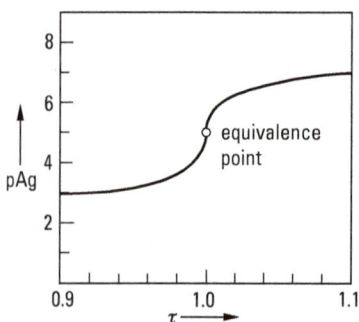

Fig. 3.11: Titration of $AgNO_3$ solution (0.01 mol/l) with NaCl solution (10 mol/l) (pAg = $-$lg $c(Ag^+)$, τ degree of titration).

If, on the other hand, a silver nitrate solution of concentration $c(Ag^+) = 10^{-2}$ mol/l is titrated with a sodium iodide solution of concentration $c(NaI) = 10$ mol/l, the titration curve shown in Fig. 3.12 is obtained. The jump in the curve is significantly larger. This is due to the considerably lower solubility of silver iodide compared to silver chloride. The solubility product of AgI is about 10^{-16} mol^2/l^2 so that an equilibrium concentration of silver ions of 10^{-8} mol/l is reached at the equivalence point.

The examples of the titration curves show that the result of a titrimetric precipitation analysis becomes more accurate and the the following conditions are fulfilled:

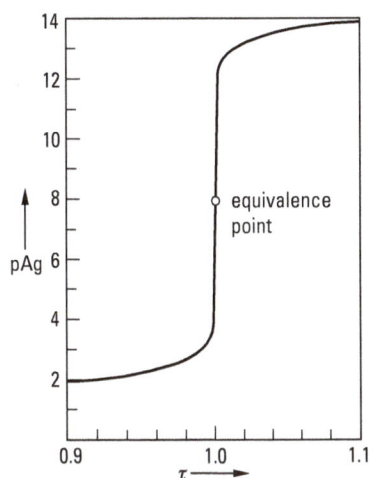

Fig. 3.12: Titration of AgNO$_3$ solution (0.01 mol/l) with NaI solution (10 mol/l) (pAg = $-$lg $c(Ag^+)$, τ degree of titration.

- The solubility product of the precipitated compound should be as small as possible.
- The initial concentration of the ion type to be titrated must be high enough.
- The same applies here:
- The practically recognizable end point of the titration must be as close as possible to the inflection point of the titration curve.

In most cases, fulfilling the last requirement presents particular difficulties. Therefore, it is precisely the lack of generally applicable and reliable methods for determining the end point that is responsible for the fact that only so few of the numerous gravimetrically exploitable precipitation processes could also be made the basis of measurement-analytical precipitation methods.

3.2.3 Methods of end point determination

The oldest and simplest method for detecting the titration end point does not use the addition of an indicator. The titration is carried out until a further addition of standard solution in the solution clarified by vigorous agitation and sedimentation of the precipitate no longer causes turbidity. However, this method of determining the end point is quite laborious and time-consuming. In addition, it can lead to incorrect results in all cases where the precipitate does not settle quickly. This is because the reaction products often remain in the solution in colloidal distribution or in the form of very fine suspensions before their final precipitation. For example, the method described by Gay-Lussac in 1828 for the titrimetric determination of sulfate using a barium salt solution has not been able to establish itself. The method also proposed by Gay-Lussac in 1832 for the volumetric analysis of silver [99] with an adjusted sodium chloride solution is one of the most accurate methods of volumetric analysis. The reason is that the silver chloride initially precipitated in colloidal distribution during the titration at the equivalence point, at which all the ions stabilizing the AgCl hydrosol are consumed, flocculates completely. In this case, titration is carried out until the so-called clear point is reached.

As a reversal of a precipitation method that works without the addition of an indicator, is the cyanide determination described by Liebig in 1851 [100]. The cyanide solution is titrated with an adjusted silver nitrate solution. As long as there is an excess of cyanide ions in the sample solution, complex dicyanoargentate ions are formed with the added silver ions $[Ag(CN)_2]^-$ so that the end point of the titration is recognized by the occurrence and not by the end of precipitate formation (see p. 146).

Another method for determining the end point uses an indicator which changes the color of the solution at the titration end point. The indicator can either form a colored, soluble compound with the ions that disappear during the titration or with the new ions added by the standard solution, which disappears or forms when the equivalence point is reached. If possible, the concentration of the newly added ions should therefore be high enough at the equivalence point for the indicator to form a colored compound. Or, conversely, the concentration of the disappearing ion must already be so low at the equivalence point that the indicator is no longer able to continue forming the colored compound. In both cases, a color change of the solution occurs at the end point of the titration. A practical example is the use of iron(III) ions as an indicator in the titration of silver ions with an adjusted potassium thiocyanate solution according to Volhard [101]. The concentration of thiocyanate in a saturated solution of silver thiocyanate is not sufficient to produce the dark red iron thiocyanate with the iron(III) ions. Only a small excess of SCN^- ions will color the solution slightly pink.

A third method useful for precipitation analyses for end point detection uses indicators that form a colored sparingly soluble precipitate with the standard solution as soon as the reactive ions in the sample solution have precipitated as a sparingly soluble precipitate when the equivalence point is reached, i.e. as soon as there is a possibility of a small excess of precipitant occurring. An example of this is the use of chromate ions as an indicator in the titration of halide ions according to Mohr [102]. As soon as the chloride ions have precipitated as silver chloride during the titration of a chloride solution, even a small excess of silver ions forms reddish-brown silver chromate, which is difficult to dissolve. The most important prerequisite for the usability of such an indicator is that in the saturated solution of the precipitate (e.g. AgCl) formed during the titration, the concentration of those ions (here Ag^+) that form the precipitate (Ag_2CrO_4) used for end point detection with the indicator ions (CrO_4^{2-}) is not sufficient to exceed its solubility product. Otherwise, the end point would be displayed before the equivalence point is reached, i.e. the changeover would occur too early.

In such cases, a fourth method of end point detection can lead to the goal: the spotting method. In this method, a drop is taken from the sample solution after each addition of standard solution and placed on a suitable surface, e.g. on a porcelain plate or on a piece of filter paper, together with a drop of the indicator solution. The end point detection thus takes place outside the solution to be titrated (external indicator). As an example, the zinc determination according to Schaffner (1858) is cited as an example. The zinc salt solution is titrated with a sodium sulfide solution of known concentration. The spot indicator is a cobalt salt solution, which reacts on the spotted plate with the precipitation of black cobalt sulfide as soon as a small excess of sulfide ions is present. However, the drop taken must not contain any zinc sulfide; otherwise this will react with the indicator and the end point will appear too early. A spot indicator can also be used to determine iron(II) ions with potassium dichromate (see p. 177). All spot methods are cumbersome and less precise. Titration methods with direct end point detection are therefore preferred.

A fifth method for determining the titration end point was given in 1923 by Fajans [103]. He describes the use of adsorption indicators in argentometry [104]. This method utilizes the occurrence of adsorption effects, which otherwise often represent a source of error in precipitation processes, in that foreign components (unused reagent of the standard solution) are entrained (occlusion inside the precipitate particles or adsorption on their surface). The most common adsorption indicators are eosin and fluorescein. The mode of action of the indicators is as follows: If a bromide solution is titrated with silver nitrate solution in the presence of eosin, colloidal particles of silver bromide are formed in the pink-colored solution, the surface of which adsorbs the bromide ions still in the solution, causing it to become negatively charged. Once the equivalence point has been exceeded, a small excess of silver ions is present instead of the bromide ions, which are now adsorbed by the colloidal silver bromide particles. The particles are now positively charged and are able to accumulate the anions of the dye. This deforms the electron shells of the dye anions, which appears as a color change. As soon as the equivalence point is exceeded, the precipitate and colloid solution turn red. The coloration disappears as soon as the solution again contains an excess of bromide ions and returns when the silver ions predominate; the phenomenon is reversible as long as colloidal silver bromide particles are still present in the solution. An adsorption indicator is only suitable if it is only strongly adsorbed in the immediate vicinity of the equivalence point and does not color the precipitate long before the end point is reached. The latter is the case when eosin is used to titrate chloride. Fluorescein is better suited for this purpose. The presence of higher electrolyte concentrations can interfere by favoring the flocculation of the silver halide sol. However, this can often be counteracted by using a protective colloid. For further adsorption indicators, see [82–84].

3.2.4 Determination of silver and argentometric determinations

3.2.4.1 Preparation of the customized solutions

The most important methods of precipitation analysis are based on the poor solubility of silver halides and silver thiocyanate. They enable the determination of silver with halide and thiocyanate solutions and of soluble halides and thiocyanates with silver nitrate solutions. The latter methods are summarized under the term "argentometry" together. To carry out the titrations, standard solutions of silver nitrate, sodium chloride and ammonium or potassium thiocyanate of the respective concentration $c(eq) = 0.1$ mol/l are required.

Silver nitrate solution: Either the purest metallic silver (fine silver), which is commercially available in the form of sheet or wire, or chemically pure silver nitrate is used. According to Richards and Wells (1908), metallic silver can also be produced by reducing silver nitrate with ammonium formate. The ammonium formate solution is prepared by introducing ammonia into freshly distilled formic acid. The silver precipitate is washed free of ammonia and melted in a hydrogen stream. About 10.7868 g of fine silver is accurately weighed and dissolved in 100 ml of the purest chloride-free nitric acid ($\rho = 1.20$ g/ml). The solution is heated to boiling point to destroy the nitrous acid and remove the nitrogen oxides, transferred to a liter flask after cooling and made up to the mark with deionized water. The concentration of nitric acid is approximately 0.5 mol/l. No special adjustment is required.

If for the determination according to Mohr a neutral silver nitrate solution 16.9873 g of pure silver nitrate, dried at 150 °C to constant mass, is accurately weighed, dissolved in water and the solution is diluted to 1 l. The silver nitrate used must not contain any metallic silver; its solution must react neutrally. It can be obtained pure by recrystallization from water containing weak nitric acid. A special titer preparation is not necessary here. However, it is advisable to check the titer of the silver nitrate solution using the purest sodium chloride (see below). The silver nitrate solution is stored in a brown bottle to protect it from direct exposure to sunlight.

Sodium chloride solution: The purest sodium chloride is used to prepare the solution, which can be prepared as follows: gaseous hydrogen chloride is introduced into a saturated solution of the purest commercially available salt and cooled externally with ice water. The sodium chloride precipitates, is separated using a glass filter and washed out several times with a small amount of ice water. The salt is then predried at 110 °C, finely powdered and finally heated in an electric oven at around 500 °C until constant mass is achieved. If a gas flame is used to heat the crucible, the combustion gases must be prevented from reaching the crucible contents. The salt must be free of bromide, iodide and sulfate and must not contain any potassium or alkaline earth metals.

To prepare the standard solution, dissolve about 5.85 g of the purest sodium chloride in 1 l of water and adjust the solution with the approximate concentration of 0.1 mol/l with a silver nitrate standard solution of the same concentration or with fine silver. The adjustment is carried out according to the same procedure that is to be used later with the solution (according to Gay-Lussac or according to Fajans), and if possible under the same conditions. This results in an empirical titer that eliminates the error caused by the method to a certain extent (e.g. by taking into account the solubility of the silver chloride).

Ammonium thiocyanate solution: Ammonium thiocyanate is hygroscopic and decomposes if you try to dry it at high temperatures. Therefore, only a solution with an approximate concentration of 0.1 mol/l is prepared by dissolving 8–9 g (theoretical 7.612 g) of the salt, which should be as dry as possible, in 1 l of water. The salt used must be free of chloride. The test for this is carried out according to Kolthoff [89]: 200 mg of thiocyanate is dissolved in 25 ml of water, 15 ml of sulfuric acid and $c(H_2SO_4)$ = 2 mol/l, and then potassium permanganate solution is added until the reddish-brown color remains (brown from the precipitated manganese dioxide). Then heat to boiling under the fume hood for 10–15 min until the hydrogencyanide has volatilized and the volume is about 10–15 ml. The manganese dioxide is reduced with perhydrol. After cooling, no more than a weak opalescence may occur with silver nitrate.

The thiocyanate solution is adjusted according to Volhard (see p. 140 ff.) with a silver nitrate solution with a concentration of 0.1 mol/l by adding 20 ml of boiled nitric acid, $c(HNO_3)$ = 2 mol/l, and 2–3 ml of the nitric acid ammonium iron(II) sulfate indicator solution to 25 ml of the silver nitrate solution. The solution is diluted to approximately 100 ml and titrate slowly with the thiocyanate solution, swirling constantly, until a faint red color just remains.

3.2.4.2 Determination of silver according to Gay-Lussac

The method is mainly used in coin laboratories to determine the silver content of alloys due to its high accuracy. The principle has already been described (see p. 131). A sodium chloride solution adjusted under the conditions of the subsequent titrations against fine silver or a silver nitrate solution of known concentration is used and thus avoids the procedural error of about 0.1%, which is caused by the solubility of the silver chloride (K_{sp} = 1.12 · 10^{-10} mol^2/l^2, corresponding to S = 1.43 mg/l) and the excess chloride solution required to achieve more complete precipitation.

For the analysis of silver alloys the interfering influence of other metals must be taken into account. Such metals, which form easily soluble nitrates and chlorides, do not interfere. Mercury must be removed before the determination by remelting the alloy in an electric furnace. Lead may only be present in traces. Antimony and bismuth are kept in solution by adding tartaric acid. If the alloy contains more than one-sixth of its mass in gold, it is no longer completely soluble in nitric acid. In this case, it is melted together with a weighed portion of the purest silver.

Procedure: Dissolve the alloy in 10 ml chloride-free nitric acid (ρ = 1.2 g/ml) and briefly boil the solution to drive off the nitrogen oxides. The cooled solution is made up to 100 ml – if necessary after filtering off sparingly soluble meta-tinic acid – and titrated as follows:

Dilute 25 ml of the weakly acidic silver nitrate solution with 50 ml of water in a glass bottle of about 200 ml. Add sodium chloride solution with a concentration of 0.1 mol/l in portions of 1 ml each, later 0.5 ml each, close the bottle after each addition and shake vigorously. As soon as the addition of half a milliliter no longer causes turbidity in the clear solution above the precipitate, the first, orienting experiment is finished. Remove another 25 ml of solution, dilute with 50 ml of water and now add 1 ml less of the saline solution at a time than in the previous experiment. The liquid is shaken again until the precipitate has settled. After the precipitate has settled, 0.5 ml of a sodium chloride solution with a concentration of 0.01 mol/l, which was prepared by diluting the solution containing 0.1 mol/l, is added in portions from a microburette in such a way that the solution flows down the glass wall. As long as not all of the silver chloride has precipitated, a clearly visible cloudiness can be observed on the surface of the liquid, which is particularly easy to recognize when the shaking flask is viewed in reflected light. The liquid is shaken again and the addition of the sodium chloride solution is continued until a further 0.5 ml no longer causes opalescence. The last addition of reagent is not taken into account when reading the burette. At least two control determinations are necessary to ensure the result. A precision of 0.05% can be achieved with this method if the same temperature is maintained when setting and using the standard solutions:

1 ml sodium chloride solution, $c(NaCl)$ = 0.1 mol/l, corresponds to 10.787 mg Ag or 16.987 mg $AgNO_3$. For silver alloys, the result must be given in milligram per gram.

The silver determination according to Gay-Lussac is more precise if a potassium bromide solution is used instead of the sodium chloride solution as AgBr (K_{sp} = 4.8 · 10^{-13} mol²/l²) is less soluble than AgCl (K_{sp} = 1.1 · 10^{-10} mol²/l²). Chloride-free potassium bromide is obtained by carefully melting pure potassium bromate in a platinum dish. The bromate decomposes releasing oxygen. The melting cake is powdered and the potassium bromide is heated to 500 °C in an electric furnace until the mass is constant.

3.2.4.3 Determinations according to Volhard

Determination of silver: Less cumbersome and less complicated than Gay-Lussac's method, which is in itself extremely accurate, is the method given by Volhard (1874). It is based on the precipitation of the sparingly soluble silver thiocyanate (K_{sp} = 6.84 · 10^{-13} mol²/l²):

$$Ag^+ + SCN^- \rightarrow AgSCN \downarrow.$$

An excess of thiocyanate ions is detected with the aid of an iron(III) salt solution (see p. 136):

$$Fe^{3+} + 3\,SCN^- \rightarrow Fe(SCN)_3.$$

The indicator solution is a cold saturated solution of ammonium iron(III) sulfate to which boiled nitric acid is added until the brown coloration disappears. Use 2 ml of this solution for every 100 ml of the solution to be titrated.

The titration is carried out in a cold nitric acid solution. The acid concentration should be approximately 0.4 mol/l. Slight fluctuations in the hydrogen ion concentration have no influence. However, the nitric acid must not contain nitrous acid because this forms red nitrosyl thiocyanate with thiocyanate. If a silver alloy is being analyzed, its nitric acid solution (see p. 138) must be boiled out. The presence of other metal ions does not interfere if they form easily soluble, dissociated thiocyanates and do not have a strong inherent color. Mercury forms a less soluble thiocyanate that is not dissociated in solution and must therefore be removed before analysis. If the copper content of an alloy is below 70%, the interference is negligible.

Procedure: As described on p. 139 f. for adjusting the NH_4SCN standard solution, add 20 ml boiled (HNO_2 free) nitric acid, $c(HNO_3)$ = 2 mol/l, to the silver-containing sample solution, dilute to about 100 ml with water and titrate slowly with ammonium thiocyanate solution, swirling continuously, until a permanent weak red coloration appears:
  1 ml ammonium thiocyanate solution, $c(NH_4SCN)$ = 0.1 mol/l, corresponds to 10.7868 mg Ag.

It is important to maintain the same working conditions as for the titration of the thiocyanate standard solution, as silver ions are adsorbed on the freshly precipitated silver thiocyanate, which means that a constant excess of thiocyanate solution is required under the same experimental conditions before the red color of the iron(III) thiocyanate becomes visible.

Determination of thiocyanate and copper: An excess of silver nitrate solution is added to the thiocyanate solution and the excess silver ions are titrated back as described. Direct titration is not possible because the precipitating silver thiocyanate adsorbs iron(III) thiocyanate so that the decolorization to which titration would have to be performed cannot be observed exactly:
  1 ml silver nitrate solution, $c(AgNO_3)$ = 0.1 mol/l, corresponds to 5.808 mg SCN^-.

Argentometric thiocyanate titration can also be used to determine copper. Copper(II) ions are reduced to copper(I) ions by sulfurous acid:

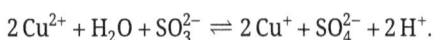

$$2\,Cu^{2+} + H_2O + SO_3^{2-} \rightleftharpoons 2\,Cu^+ + SO_4^{2-} + 2\,H^+.$$

The copper(I) ions precipitate as sparingly soluble copper(I) thiocyanate after the addition of excess thiocyanate solution:

$$Cu^+ + SCN^- \rightleftharpoons CuSCN \downarrow.$$

In the filtrate of the white precipitate (with a tinge of violet), the excess thiocyanate cannot be retitrated directly with silver nitrate for the reason mentioned. Instead, silver nitrate solution must be added in excess and the excess silver titrated back with ammonium thiocyanate solution.

Silver, mercury, chloride, bromide, iodide and cyanide ions must not be present. In silver-containing copper ores the sum of the copper and silver is first determined according to the method describedearlier and then the silver alone according to Gay-Lussac.

Procedure: Add an excess of freshly prepared sulfurous acid (approx. 30 ml) and 100 ml ammonium thiocyanate solution (0.1 mol/l) to 50 ml of the neutral or very slightly sulfurous copper(II) sulfate solution. Heat the solution to boiling point to drive off the excess sulfur dioxide, allow to cool, transfer the solution and precipitate quantitatively to a 250-ml volumetric flask and make up to the mark with water. The solution is shaken well and filtered through a dry filter. Discard the first 25 ml of the solution and collect the remainder in a dry beaker. Pipette 50 ml of the solution into a beaker, add 30 ml silver nitrate solution (0.1 mol/l), 20 ml nitric acid, 2 mol/l and 2 ml indicator solution. The excess silver ions are then titrated back with ammonium thiocyanate solution (0.1 mol/l).

Let V_1 be the volume of thiocyanate solution added for copper precipitation (exactly 0.1 mol/l), V_2 the volume of silver nitrate solution and V_3 the volume of thiocyanate solution for back titration; then V_x is the volume of thiocyanate solution used to precipitate the copper:

$$V_x = V_1 - 5 \, (V_2 - V_3).$$

1 ml ammonium thiocyanate solution, $c(NH_4SCN) = 0.1$ mol/l, corresponds to 6.3546 mg Cu.

Determination of halides and cyanide: Volhard's method for silver determination is particularly useful because it allows the halide content of acidic solutions to be determined as a so-called residual method. Excess silver nitrate solution is added to the halide solution and the excess silver ions are retitrated with thiocyanate solution.

Chlorides: The chloride determination cannot be carried out according to the simple rule valid for bromides. The excess silver can only be titrated with thiocyanate once the precipitated silver chloride has been separated. The conversion of the indicator would otherwise be blurred because the precipitated silver chloride

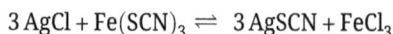

$$3\,AgCl + Fe(SCN)_3 \rightleftharpoons 3\,AgSCN + FeCl_3$$

would convert with iron thiocyanate into the less soluble silver thiocyanate until the equilibrium state in the solution between chloride and thiocyanate ions has been reached (1 SCN^- ion to 164 Cl^- ions). The red color initially achieved would fade permanently, resulting in too high a consumption of thiocyanate solution and thus too low a consumption of silver nitrate for chloride precipitation. The solubility of silver bromide and silver thiocyanate, on the other hand, is the same.

Procedure:

1. Chlorides: 25 ml of the chloride sample solution is made up to the mark with silver nitrate solution (0.1 mol/l) in a 100-ml volumetric flask. Shake the solution well for a few minutes, filter through a dry filter, discard the first 20 ml and collect the remaining filtrate in a dry beaker. Pipette off 50 ml and determine the excess silver in this solution as described.

    However, as freshly precipitated silver chloride adsorbs silver ions, you always use a little too much silver nitrate. In practice, it has been found that 0.7% must be subtracted from the value found in order to obtain the correct value.

    According to Caldwell and Moyer [105], the correction is not necessary if nitrobenzene (1 ml per 0.05 g chloride) and 1–4 ml silver nitrate solution are added in excess to the nitric acid sample solution. The solution is then shaken for 30–40 s to agglomerate the precipitate into large flocs. The hydrophobic nitrobenzene covers the precipitate so that the indicator can be added without filtering and the excess silver can be retitrated with thiocyanate solution as described.

    Alary et al. [106] replace the toxic and environmentally hazardous nitrobenzene with polyvinylpyrrolidone (PVP). To 100 ml of the sample solution with 0.2–5 mg chloride, add 1 ml 65% nitric acid, 3 ml indicator solution (dissolve 470 g $Fe(NO_3)_3 \cdot 9\,H_2O$ in 800 ml water, add 20 ml 65% nitric acid and make up to 1 l), 5 ml PVP solution (dissolve 35 g PVP in approx. 700 ml water) and then 2 or 5 ml $AgNO_3$ solution (0.05 mol/l). The PVP forms a stable layer on the AgCl particles as soon as the precipitate forms. No turbidity occurs so that the end point is easier to recognize.

2. Bromides: Add 20 ml boiled nitric acid (2 mol/l), 2 ml indicator solution and 50 ml silver nitrate solution (0.1 mol/l) to 25 ml sample solution. The excess silver is then determined by titration with ammonium thiocyanate solution (0.1 mol/l).

3. Iodides: The determination is carried out as described for the bromides. It gives excellent results if the indicator solution is only added after the iodide ions have already precipitated due to excess silver nitrate and the solution has been shaken vigorously for 5 min. Otherwise the iodide ions will reduce the iron(III):

    $$2\,Fe^{3+} + 2\,I^- \rightleftharpoons I_2 + 2\,Fe^{2+}.$$

4. Cyanides: Determination is carried out according to the instructions given for chlorides. Silver cyanide also adsorbs silver ions, and so a correction of −0.7% must also be applied to the mass determined:

1 ml silver nitrate solution, $c(AgNO_3)$ = 0.1 mol/l, corresponds to 3.5453 mg Cl⁻ or 7.9904 mg Br⁻ or 12.690 mg I⁻ or 2.6018 mg CN⁻.

3.2.4.4 Determinations according to Mohr

In the determinations according to Mohr the halide ions are titrated directly with silver nitrate solution. Potassium chromate is used as an indicator. The end point is recognized by the fact that a small excess of silver ions leads to the precipitation of reddish-brown silver chromate (see p. 137 f.):

$$2\,Ag^+ + CrO_4^{2-} \rightarrow Ag_2CrO_4 \downarrow.$$

The pH value during titration must be between 6.5 and 10.5. In acidic solutions according to the equilibria

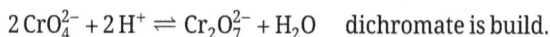

$$2\,CrO_4^{2-} + 2\,H^+ \rightleftharpoons Cr_2O_7^{2-} + H_2O \quad \text{dichromate is build.}$$

The resulting dichromate ions do not form a sufficiently sparingly soluble silver dichromate ($K_{sp} = 2 \cdot 10^{-7}$ mol^3/l^3). Acidic solutions must therefore be blunted with sodium hydrogen carbonate or borax. If the pH values are too high, the determination cannot be carried out because silver hydroxide and possibly also silver carbonate will precipitate. Phosphate, arsenate, sulfite and fluoride ions interfere with the titration.

The most favorable indicator concentration of $c(K_2CrO_4) = 5 \cdot 10^{-3}$ mol/l is obtained by adding 2 ml of a neutral 5% potassium chromate solution per 100 ml of sample solution. Since the solubility of silver chromate increases sharply with increasing temperature, titration should only be carried out at room temperature.

If possible, all determinations must be carried out under the same conditions with regard to the halide and chromate ion concentration as for the adjustment of the standard solution so that the excess of silver ions required for the recognizable reddish-brown coloration is always the same.

This applies, in particular, to the determination of iodides. As the following considerations show, due to the large difference in solubility between silver iodide (K_{sp} (AgI) $= 10^{-16}$ mol^2/l^2) and silver chromate ($K_{sp}(Ag_2CrO_4) = 2 \cdot 10^{-12}$ mol^3/l^3), a noticeable excess of silver ions is required to exceed the solubility product of silver chromate. Applying the law of mass action to the two equilibria,

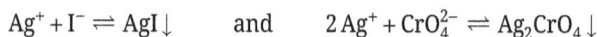

$$Ag^+ + I^- \rightleftharpoons AgI \downarrow \qquad \text{and} \qquad 2\,Ag^+ + CrO_4^{2-} \rightleftharpoons Ag_2CrO_4 \downarrow$$

shows that silver chromate only precipitates when the iodide ion concentration becomes smaller than the concentration given by the relationship

$$\frac{c(I^-)}{\sqrt{c(CrO_4^{2-})}} = \frac{K_{sp}(AgI)}{\sqrt{K_{sp}(Ag_2CrO_4)}}$$

given value $c(I^-) = 7.1 \cdot 10^{-11} \cdot \sqrt{c(CrO_4^{2-})}$ mol/l. At the usual indicator concentration of $c(K_2CrO_4) = 5 \cdot 10^{-3}$ mol/l, precipitation therefore begins at $c(I^-) = 7.1 \cdot 10^{-11} \cdot \sqrt{5 \cdot 10^{-3}} = 5 \cdot 10^{-12}$ mol/l. This corresponds to a silver ion concentration of $c(Ag^+) = 2 \cdot 10^{-5}$ mol/l, which is three powers of 10 greater than at the equivalence point, where it is only $c(Ag^+) = \sqrt{K_{sp}(AgI)} = 10^{-8}$ mol/l. The excess silver required for practical recognition of the end point is of course even greater. It is therefore essential to use a silver nitrate solution that has been adjusted to potassium iodide under the conditions of the subsequent titration.

The method is mainly used to determine chlorides and bromides. It provides good results even with relatively dilute solutions. The hazardous substance classification of chromate should be noted.

Determination of chloride in sodium chloride solution

Procedure: To 25 ml of the neutral sodium chloride solution, which should contain approximately 0.1 mol/l NaCl, add 2 ml of neutral 5% potassium chromate solution. The clear yellow solution is slowly titrated with silver nitrate solution (0.1 mol/l) with constant shaking until the reddish-brown coloration that occurs after each new addition of reagent no longer disappears, but persists for several minutes: 1 ml silver nitrate solution, $c(AgNO_3) = 0.1$ mol/l, corresponds to 3.5453 mg Cl.

Determination of chloride in drinking water and waste water: The pH value of the water must be between 6.5 and 10.5 according to the Mohr titration conditions. If the water is strongly colored or contains hydrogen sulfide, boil 1 l of it for 5 min with a little potassium permanganate solution. The still red liquid is decolorized with 30% hydrogen peroxide solution, made up to 1 l again after cooling and filtered (the first parts of the filtrate are discarded).

Procedure: Add 2 ml of 5% potassium chromate solution to each of two 100 ml samples of water and titrate with silver nitrate solution (0.01 mol/l) from a microburette until the first faint reddish-brown coloration. Then dilute to 150 ml with water. The solutions decolorize again. One sample is now used as a reference solution and the other is titrated to the end until the permanent change in color is visible. In this way, you always have the same final volume and can therefore deduct the same correction for the necessary excess of standard solution, namely 0.6 ml, from the consumption.

3.2.4.5 Determinations according to Fajans

The end point detection with adsorption indicators has already been discussed on p. 137 f. Eosin is well suited for the determination of bromide, iodide and thiocyanate, but not for chloride, as it is already adsorbed by silver chloride at the start of the titration. Fluorescein, on the other hand, is useful in this case. Silver can be determined using the basic dye rhodamine 6G.

As the photochemical decomposition of the silver halides is greatly accelerated by the dyes mentioned, the titrations described below should be carried out quickly and not in direct sunlight.

Determination of bromide, iodide and thiocyanate: The titration is carried out in a weak acetic acid solution with eosin as indicator (1% aqueous solution of the sodium salt; add two drops to each 10 ml halide solution, $c(Hal) \approx 0.1$ mol/l). Titrate while shaking vigorously until the precipitate suddenly takes on a distinct red ($c(Hal) \approx 0.1$ mol/l) or pink ($c(Hal) \approx 0.01$ mol/l) color. If the halide ion concentration is only 0.001

mol/l, the silver halide no longer flocculates, but the color of the solution changes sharply from pink to purple at the equivalence point.

Determination of chloride: The neutral solution, which should contain approximately 0.1 mol/l chloride, is mixed with fluorescein as an indicator (0.2% aqueous solution of the sodium salt; add two drops to every 10 ml of sample solution) and titrated until the precipitate suddenly turns bright red. According to Kolthoff [89], the silver chloride can be kept colloidal in solution by adding 5 ml of 2% chlorine-free dextrin solution to every 10 ml of sample solution; the color of the liquid then changes sharply to pink at the equivalence point. Polyvalent ions can make it difficult to recognize the end point due to their coagulating effect.

Determination of silver: Silver is titrated in acetic acid solution with potassium bromide solution and rhodamine 6G as an indicator until the silver bromide suddenly turns blue-violet. The error of the determination is approximately 0.1%.

3.2.4.6 Determination of cyanide according to Liebig

In contrast to the precipitation analysis methods discussed so far, in the cyanide determination according to Liebig (1851), the precipitate is only formed at the end of the titration, when the equivalence point has just been exceeded. It is also a simple example of the use of a complex formation reaction in volumetric analysis (see Section 3.4).

If a weakly alkaline alkali metal cyanide solution is mixed dropwise with silver nitrate solution, the appearance of a white precipitate of silver cyanide is observed at the drip point, but this disappears again immediately on shaking, as the excess cyanide forms the soluble dicyanoargentate(I) complex with the silver cyanide:

$$AgCN + CN^- \rightleftharpoons [Ag(CN)_2]^-.$$

However, if practically all the cyanide ions have been bound in this way during the titration, the first excess drop of silver nitrate solution produces a permanent turbidity of silver cyanide:

$$[Ag(CN)_2]^- + Ag^+ \rightleftharpoons 2\,AgCN \downarrow,$$

which indicates the end of the titration. The calculation is based on the fact that 1 mol Ag^+ corresponds to 2 mol CN^-.

The titration should be carried out slowly and with constant shaking, especially toward the end, as the silver cyanide primarily precipitated at the drop-in point only slowly goes into solution when there is only a small excess of cyanide. The solution should be slightly alkaline (pH < 13) and must not contain any ammonium salts, as the presence of ammonia prevents the precipitation of silver cyanide. The presence of chloride, bromide, iodide and thiocyanate ions has no interfering influence. As an example, the analysis of technical potassium cyanide is described as an example.

Procedure: Several samples of about 0.3 g potassium cyanide are weighed out precisely, dissolved in 100 ml water each and, after adding 2 ml potassium hydroxide solution, $c(KOH) = 2$ mol/l, to each, titrated slowly with silver nitrate solution, $c(AgNO_3) = 0.1$ mol/l, while shaking until a permanent turbidity just recognizable occurs. To make it easier to recognize the end point, place the titration vessel on a dark surface (e.g. black paper):

1 ml silver nitrate solution, $c(AgNO_3) = 0.1$ mol/l, corresponds to 5.2036 mg CN^- or 5.4052 mg HCN or 13.024 mg KCN. Convert the analysis result into mass fraction KCN in percent. If the technical product contains sodium cyanidethe numerical value may exceed 100%.

3.2.4.7 Argentometric titrations of organic ammonium salts

Hydrochlorides or hydrobromides of organic ammonium salts can also be titrated analogously by argentometry. As an example, the titrations of the salts levomethadone HCl (strong analgesic) and ipratropium bromide (therapeutic agent for chronic respiratory diseases) from Ph. Eur. are listed [8a].

Determination of levomethadone HCl according to Ph. Eur.

Procedure: About 0.300 g substance dissolved in a mixture of 40 ml water and 5 ml acetic acid is titrated with silver nitrate solution (0.1 mol/l). The end point is determined by potentiometry using a silver electrode as indicator electrode [8a]:

1 ml silver nitrate solution, 0.1 mol/l, corresponds to 34.59 mg $C_{21}H_{28}ClNO$.

Determination of ipratropium bromide according to Ph. Eur.

Procedure: About 0.350 g substance dissolved in 50 ml water is titrated with silver nitrate solution (0.1 mol/l) after adding 3 ml diluted nitric acid (12.5%). The end point is determined by potentiometry using a silver electrode as indicator electrode [8a]:

1 ml silver nitrate solution, 0.1 mol/l, corresponds to 41.24 mg $C_{20}H_{30}BrNO_3$.

3.2.4.8 Determination according to Budde

NH-acid compounds such as barbiturates or sulfonamides can be determined by direct titration with silver nitrate. No indicators are required for this titration, as soluble silver complexes with the analyte are present before the equivalence point and silver salts precipitate with the analyte after the equivalence point is exceeded.

Determination of cyclobarbital calcium according to DAB 7 – GDR: Barbiturates that are substituted at only one N or no N can be titrated directly with silver nitrate according to Budde. For barbiturates unsubstituted on N, a soluble 1:1 silver/barbiturate complex is formed before the equivalence point; for barbiturates substituted on N, a 2:1 complex is formed. In the described procedure, the calcium is precipitated as carbonate before titration [28]:

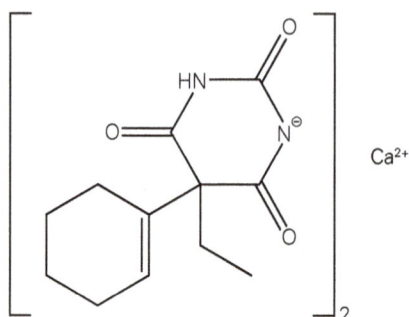

Procedure: About 0.300 g dried substance is mixed with 5.0 ml sodium carbonate solution (20.0 g/100 ml) and 20.0 ml water in a 100-ml Erlenmeyer flask. The mixture is heated in a water bath at 55 °C for 15 min with repeated swirling and then filtered. The Erlenmeyer flask and the filtrate are washed twice with 10.0 ml each of the mixture of 10.0 ml ethanol and 10.0 ml water. The combined filtrates are titrated with silver nitrate solution (0.1 mol/l) until a turbidity arises which persists for 5 min:
   1 ml silver nitrate solution (0.1 mol/l) corresponds to 25.53 mg $C_{24}H_{30}CaN_4O_6$ [8b].

3.2.4.9 Argentometric determination of organically bound chlorine or bromine

Numerous organic halogen compounds can be titrated argentometrically after mineralization or digestion. In some cases this can already be done with aqueous sodium hydroxide solution (cf. titration of lindane or cyclophosphamide); in most cases a carbonate melt or reductive digestion with elemental sodium is required.

Titration of lindane according to Ph. Eur.: The insecticide lindane (1,2,3,4,5,6-hexachlorocyclohexane) can be titrated argentometrically after reaction with ethanolic potassium hydroxide solution to 1,3,5-trichlorobenzene according to Volhard [8a].

Procedure: Add 10 ml ethanol (96%) to 0.200 g substance. Heat the mixture on a water bath until completely dissolved, cool, add 20 ml ethanolic potassium hydroxide solution (0.5 mol/l) and leave to stand for 10 min with frequent shaking. After adding 50 ml water, 20 ml dilute nitric acid (12.5%), 25.0 ml silver nitrate solution (0.1 mol/l) and 5 ml ammonium iron(III) sulfate solution (100 g/l), titrate the solution with ammonium thiocyanate solution (0.1 mol/l) until the color changes to reddish yellow. A blank titration is carried out:
   1 ml silver nitrate solution, $c = 0.1$ mol/l, corresponds to 9.694 mg $C_6H_6Cl_6$.

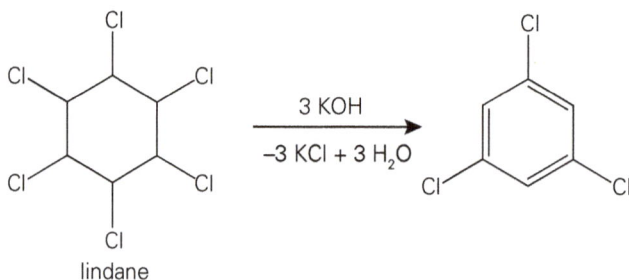

lindane

Titration of cyclophosphamide according to Ph. Eur.: The cytostatic drug cyclo-phosphamide is determined according to the European Pharmacopoeia after alkaline digestion and subsequent argentometric titration according to Volhard. In the case of the aliphatic alkyl halide, quantitative hydrolysis is already possible with aqueous sodium hydroxide solution [8a].

Procedure: Dissolve 0.100 g of substance in 50 ml of a solution of sodium hydroxide (1 g/l) in ethylene glycol. The solution is heated to reflux for 30 min. After cooling, the cooler is washed with 25 ml of water. After adding 75 ml 2-propanol, 15 ml dilute nitric acid (12%), 10.0 ml silver nitrate solution (0.1 mol/l) and 2.0 ml ammonium iron(III) sulfate solution (100 g/l), the mixture is titrated with ammonium thiocyanate solution (0.1 mol/l):
  1 ml silver nitrate solution, $c = 0.1$ mol/l, corresponds to 13.05 mg $C_7H_{15}Cl_2N_2O_2P$.

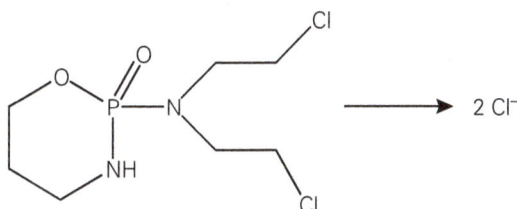

3.3 Oxidation and reduction titrations

3.3.1 Theoretical foundations

3.3.1.1 Oxidation and reduction

In the historical development of chemistry, the terms "oxidation" and "reduction" have often undergone a significant change in meaning. Initially, oxidation was understood as the chemical conversion of a substance with oxygen (lat. oxygenium) and reduction as the return (lat. reducere) of the oxidized substance to its original state, i.e. the removal of oxygen (Lavoisier, 1777). Every combustion process is therefore an oxidation, e.g.

$$2\,H_2 + O_2 \rightarrow 2\,H_2O,$$
$$2\,CO + O_2 \rightarrow 2\,CO_2$$

and the process

$$PbO + H_2 \rightarrow Pb + H_2O$$

a reduction.

In the first half of the nineteenth century, both terms were expanded. At one point, the process of hydrogen removal was also referred to as oxidation, e.g.

$$2\,NH_3 + \tfrac{3}{2}\,O_2 \rightarrow N_2 + 3\,H_2O$$

and that of the combination with hydrogen as a reduction, e.g.

$$N_2 + 3\,H_2 \rightarrow 2\,NH_3.$$

On the other hand, the terms have also been extended to chemical reactions in which oxygen or hydrogen are not absorbed or released in elementary form, but are transferred from other compounds, e.g.

$$2\,HBr + Cl_2 \rightarrow Br_2 + 2\,HCl$$

(oxidation of HBr, reduction of Cl_2),

$$CO + PbO \rightarrow CO_2 + Pb$$

(oxidation of CO, reduction of PbO).

In 1860, the term "oxidation" was expanded again. Oxidation should be understood as a chemical reaction that leads to an increase in the oxidation state (see p. 39) of the element in question, e.g.

$$2\,Na + Cl_2 \rightarrow 2\,NaCl.$$

("Combustion" of sodium in a chlorine atmosphere, oxidation of "zero-valent" to "monovalent" sodium; in today's terms: increase in the ionic valence of sodium from 0 to 1+) or

$$4\,FeO + O_2 \rightarrow 2\,Fe_2O_3.$$

(Oxidation of "bivalent" iron to "trivalent"; increase in the ionic valence of the iron from 2+ to 3+).

For the first time, Kossel and Lewis in 1916, the current definition of the terms "oxidation" and "reduction" is based on the electron theory of chemical bonding.

According to this theory, all oxidation and reduction reactions are based on the transfer of electrons. There is formal agreement with the theory of acid-base reactions by Brønsted (see Section 3.1.1), according to which protons are transferred.

Oxidation is the **release of electrons**, **reduction** is the **acceptance of electrons**. The following reactions are oxidation processes:

$$Fe^{2+} \rightarrow Fe^{3+} + e^-,$$

$$2\,I^- \rightarrow I_2 + 2\,e^-,$$

$$Na \rightarrow Na^+ + e^-.$$

Examples of reduction processes may be mentioned:

$$Fe^{3+} + e^- \rightarrow Fe^{2+},$$

$$Sn^{4+} + 2e^- \rightarrow Sn^{2+},$$

$$Cl_2 + 2e^- \rightarrow 2Cl^-.$$

If a substance – as in the examples given – occurs in two forms that differ only in the number of electrons, these two forms form a **corresponding redox pair (redox system)**. The oxidized form A_{ox} can be converted into the reduced form A_{red} by accepting electrons (reduction) and this can be converted into the oxidized form by removing electrons (oxidation):

$$A_{ox} + ze^- \rightleftharpoons A_{red},$$

where z is the number of electrons involved in the redox process.

If a substance is oxidized by releasing electrons, another substance must inevitably be reduced at the same time by accepting electrons, as free electrons do not occur under normal chemical reaction conditions. Every oxidation of one substance therefore also means the reduction of another. This coupling is expressed by the term **redox reaction**.

Two corresponding redox pairs must always take part in every redox reaction:

$$A_{red} + B_{ox} \rightleftharpoons A_{ox} + B_{red}.$$

The particles A_{red} have a reducing effect by transferring electrons to the particles B_{ox}. A_{red} is therefore called a **reducing agent**. Reducing agents are electron donors. They are oxidized by the release of electrons. The particles B_{ox} oxidize A_{red} by accepting electrons. B_{ox} is therefore an **oxidizing agent**, an electron acceptor. During the reaction, it changes into the reduced form.

A redox reaction is a transfer of electrons from a reducing agent to an oxidizing agent.

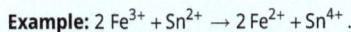

Example: $2\,Fe^{3+} + Sn^{2+} \rightarrow 2\,Fe^{2+} + Sn^{4+}$.
In an acidic solution, Fe^{3+} ions are reduced by Sn^{2+} ions or – in another formulation – Sn^{2+} ions are oxidized by Fe^{3+} ions. The corresponding redox pairs involved are Fe^{3+}/Fe^{2+} and Sn^{4+}/Sn^{2+}.

Example: $Fe + Cu^{2+} \rightarrow Fe^{2+} + Cu$.
This process takes place when an iron nail is immersed in a copper salt solution. Iron reduces the copper ions. The redox pairs involved are Fe^{2+}/Fe and Cu^{2+}/Cu.

Substances such as iron(II) ions, which can act as oxidizing agents by accepting electrons ($Fe^{2+} + 2e^- \rightarrow Fe$), but can also release electrons ($Fe^{2+} \rightarrow Fe^{3+} + e^-$), are referred to as **redoxamphoteric substances**.

3.3.1.2 Oxidation number

For the quantitative description of redox reactions, it is useful to designate the oxidation state of the element atoms involved by numbers. For reactions between ions, the ionic charge number with a positive or negative sign, the **ionic valence** can be used. However, ionic bonding is only a borderline case of chemical bonding, and by no means all compounds are made up of ions. The transfer of electrons is therefore not so easy to recognize in many redox processes.

For example, when hydrogen is oxidized by oxygen, covalent, nonpolar bonds are broken in the element molecules and strongly polar O–H bonds are formed in the water molecule:

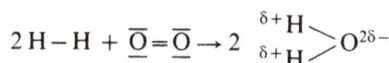

$$2\,H-H + \underline{\overline{O}}=\underline{\overline{O}} \rightarrow 2\;\begin{array}{c}{}^{\delta+}H\\{}^{\delta+}H\end{array}\!\!\!>O^{2\delta-}$$

This process is also to be understood as a partial electron transfer. In the newly formed bonds, the binding electrons are more strongly attracted to the more electronegative bonding partner. In such cases, the polarities of the relevant covalent bonds must be considered. Another example is the formation of gaseous hydrogen chloride from the elements:

$$H-H + |\underline{\overline{Cl}}-\underline{\overline{Cl}}| \rightarrow 2\,{}^{\delta+}H-\underline{\overline{Cl}}|^{\delta-}$$

Instead of ionic valence, the term **oxidation number** has therefore been introduced for an elementary atom. The oxidation number of an elementary atom in an atomic compound is the charge that the atom would have if the electrons of all bonds on this atom were assigned to the more electronegative bonding partner.

A compound is formally divided into positively and negatively charged components and the oxidation number of the respective atom is indicated by a Roman numeral, which is written at the top right next to or above the element symbol. Numbers without a sign indicate positive oxidation numbers; negative oxidation numbers are indicated by a minus sign in **front of** the number.

Example: $\quad\quad\quad\quad\quad\quad\quad\quad C^{IV}O_2^{-II}\quad N^{-III}H_3^{I}\quad H_2^{\pm 0}.$

The algebraic sum of the oxidation numbers of all atoms present is equal to zero in an uncharged molecule; in a molecular ion it is equal to the ionic charge, e.g.

CO_2	C:	IV	SO_4^{2-}	S:	VI
	O:	$2\cdot(-II)$		O:	$4\cdot(-II)$
Sum:		0	Sum:		-2

To determine the oxidation numbers the following rules can be applied:

1. Atoms in molecules of elements and in elementary metals have an oxidation number of zero. They have no excess charge and do not differ in their electronegativity:

$$\overset{\pm 0}{N} \equiv \overset{\pm 0}{N} \qquad \overset{\pm 0}{Cu}.$$

2. For atomic ions, the oxidation number is equal to the ionic charge:

$$Na^+,\ Mg^{2+},\ Fe^{3+},\ O^{2-}.$$

3. In molecules and molecular ions consisting of several elements, the oxidation number of an atom is equal to the formal charge that would be assigned to this atom if the molecule in which it is incorporated consisted entirely of atomic ions.

To determine the oxidation numbers, proceed as follows: Write down the line formula (Lewis formula) of the compound with all (bonding and nonbonding) valence electrons. Then formally separate all covalent bonds, assigning the bonding electrons to the more electronegative partner. A criterion for the relative electronegativity are the coefficients of Pauling [107]. In the case of covalent bonds between similar atoms, the bonding electrons are assigned equally. The difference between the number of valence electrons of the atom under consideration in the uncharged state and the number of valence electrons that the atom receives after the formal bond separation is its oxidation number. To check this, the electron balance is established.

Example: Phosphoric acid, H_3PO_4:

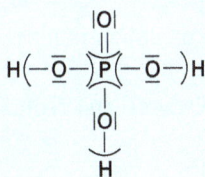

Oxidation numbers: Hydrogen I, Oxygen -II, Phosphorus V
Balance: $3 \cdot I + 4 \cdot (-II) + V = 0$.

Example: Oxalic acid, $H_2C_2O_4$

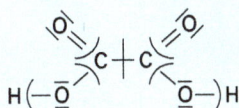

Oxidation numbers: Hydrogen I, Oxygen-II, Carbon III
Balance: $2 \cdot I + 4 \cdot (-II) + 2 \cdot III = 0$.

Example: Potassium permanganate, $KMnO_4$

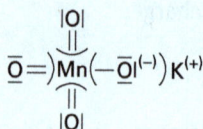

$$\bar{O}=\!\!\!\big)\!Mn\!\big(\!-\overline{O}l^{(-)}\big)K^{(+)}$$

with $|O|$ above and $|O|$ below the Mn.

Oxidation numbers: Potassium I, Oxygen –II, Manganese VII
Balance: $I + 4 \cdot (-II) + VII = 0$.

Example: Hydrogen peroxide, H_2O_2

$$H\!\big(\!-\overline{O}+\overline{O}-\!\big)H$$

Oxidation numbers: Hydrogen I, Oxygen –I
Balance: $2 \cdot I + 2 \cdot (-I) = 0$.

As the rules and examples show, the oxidation numbers – apart from the atomic ions – have no concrete physical meaning, but only a formal meaning. They are a tool for recognizing redox processes and the complete or partial electron transitions associated with them. With the help of the oxidation numbers, the definition of a redox reaction can be formulated in a new and more comprehensive way:

- Any chemical reaction in which there is any change in the oxidation numbers of the elements involved is a redox reaction.
- Oxidation means increasing the oxidation number of an element, and reduction means decreasing the oxidation number.

3.3.1.3 Redox potential

The ability of a substance to act as an oxidizing or reducing agent essentially depends on its electron affinity or its ionization energy.

If a zinc rod is immersed in a copper(II) sulfate solution, it becomes coated with metallic copper:

$$Zn + Cu^{2+} \rightarrow Cu + Zn^{2+}.$$

During the process, zinc is oxidized and copper is reduced. It can be broken down into two subprocesses:

$$Zn \rightleftharpoons Zn^{2+} + 2\ e^-,$$

$$Cu^{2+} + 2\ e^- \rightleftharpoons Cu.$$

In principle, a zinc rod immersed in a zinc salt solution reacts according to the first equation. However, the process very soon reaches equilibrium with its counter-process – indicated by the lower arrow in the equation. The fact that metallic zinc – even if only to a small extent – goes into solution as Zn^{2+} ions means that the zinc rod becomes negatively charged. If, on the other hand, a copper rod is immersed in a copper salt solution, a small amount of metallic copper is precipitated according to the second equation. This gives the copper rod a small positive charge compared to the solution. The combinations Zn (metal)/Zn^{2+} (dissolved) and Cu (metal)/Cu^{2+} (dissolved) are referred to as **half elements** (in the sense of galvanic elements), and their electrical potentials are given by the charge as **single potentials**. The combination of the half-elements creates the known galvanic elements in the case of the combination of the above half elements Zn/Zn^{2+} and Cu/Cu^{2+}, the **Daniell element**.

Even elementary hydrogen also assumes a certain individual potential in relation to the solution of its ions. If, for example, a platinum sheet (see p. 258) is immersed in an acid and rinsed with hydrogen gas, some of the gas dissolves in the platinum. This platinum rod saturated with hydrogen gas on the surface then behaves like a **hydrogen rod**.

The relative size of the individual potentials depends on the nature of the element in question and the concentration of its ions in the solution (see Section 4.4.1). It can be determined experimentally by combining the different half-elements one after the other with one and the same half-element as a reference element and reading the different voltages on the voltmeter. Table 3.7 shows a series of **standard potentials $E°$** is compiled, which were measured at 20 °C and $p = 1.103$ bar. The **standard hydrogen electrode** is used as the reference electrode which was arbitrarily assigned the potential zero.

The sequence of metallic elements arranged according to increasing (positive) individual potentials is called the **electrochemical voltage series**. It gives us a useful yardstick for assessing the bonding strength of the electrons in the outermost electron shells. Of the elements listed in Tab. 3.7, lithium, for example, releases valence electrons most easily, whereas gold releases them most heavily.

In the same way that the tendency of the elements to release or accept electrons, i.e. their oxidation or reduction capacity, can be compared with each other by setting up a voltage series, all other oxidizing and reducing agents can also be characterized by determining their individual potentials, their **redox potentials** are determined.

An electrode made of bare platinum immersed in a solution of an oxidizing or reducing agent assumes a certain potential. The greater the oxidizing capacity of the solution, the more positive or noble, the more reducing the solution, the more negative or ignoble is the potential of the electrode. The values read off the voltmeter for solutions of different oxidizing and reducing agents, using the standard hydrogen electrode as a reference system, are given in Table 3.8. The potentials provided by the

Tab. 3.7: Electrochemical series of some metals [6, 20].

Redox system	$E°$ (V)	Redox system	$E°$ (V)
$Li^+ + e^- \rightleftharpoons Li$	−2.96	$Co^{2+} + 2e^- \rightleftharpoons Co$	−0.28
$Ca^{2+} + 2e^- \rightleftharpoons Ca$	−2.76	$Pb^{2+} + 2e^- \rightleftharpoons Pb$	−0.13
$Mg^{2+} + 2e^- \rightleftharpoons Mg$	−2.38	$2H^+ + 2e^- \rightleftharpoons H_2$	±0.00
$Al^{3+} + 3e^- \rightleftharpoons Al$	−1.67	$Cu^{2+} + 2e^- \rightleftharpoons Cu$	+0.35
$Mn^{2+} + 2e^- \rightleftharpoons Mn$	−1.18	$Ag^+ + e^- \rightleftharpoons Ag$	+0.80
$Zn^{2+} + 2e^- \rightleftharpoons Zn$	−0.76	$Hg^{2+} + 2e^- \rightleftharpoons Hg$	+0.85
$Fe^{2+} + 2e^- \rightleftharpoons Fe$	−0.45	$Au^{3+} + 3e^- \rightleftharpoons Au$	+1.42

respective oxidation and reduction processes again depend on the concentration of the substances involved. The standard potentials contained in Tab. 3.8 $E°$ refer to solutions at 25 °C.

Read in the sense of the upper arrows, the reaction equations listed in Tab. 3.8 describe reduction processes, in the sense of the lower arrows oxidation processes. The reduction capacity of the titanium(III) ion is greater than that of the tin(II) ion, for example, and this in turn is more strongly reducing than the $[Fe(CN)_6]^{4-}$ ion. On the other hand, the permanganate ion in acidic solution is a stronger oxidizing agent than the bromate ion or the hypochlorite ion. The decreasing oxidizing capacity of the halogens is also expressed in the potential values in the series $Cl_2/Br_2/I_2$.

In summary, it can be said that the oxidation or or reduction capacity of each oxidizing or reducing agent can be characterized numerically by the magnitude of the electrical potential that an unassailable electrode immersed in its solution assumes compared to the standard hydrogen electrode.

Tab. 3.8: Electrochemical series for ions [6,20].

Redox system	$E°$ (V)
$S(solid) + H_2O + 2e^- \rightleftharpoons HS^- + OH^-$	−0.48
$Cr^3 + e^- \rightleftharpoons Cr^{2+}$	−0.44
$[Ti(OH)]^{3+} + H^+ + e^- \rightleftharpoons Ti^{3+} + H_2O$	−0.06
$Sn^{4+} + 2e^- \rightleftharpoons Sn^{2+}$	+0.15
$[Fe(CN)_6]^{3-} + e^- \rightleftharpoons [Fe(CN)_6]^{4-}$	+0.36
$I_3^- + 2e^- \rightleftharpoons 3I^-$	+0.54
$Fe^{3+} + e^- \rightleftharpoons Fe^{2+}$	+0.77
$ClO^- + H_2O + 2e^- \rightleftharpoons Cl^- + 2OH^-$	+0.88
$Br_2(solved) + 2e^- \rightleftharpoons 2Br^-$	+1.09
$IO_3^- + 6H^+ + 6e^- \rightleftharpoons I^- + 3H_2O$	+1.09
$Ce^{4+} + e^- \rightleftharpoons Ce^{3+}$	+1.71
$Cr_2O_7^{2-} + 14H^+ + 6e^- \rightleftharpoons 2Cr^{3+} + 7H_2O$	+1.36

Tab. 3.8 (continued)

Redox system	E° (V)
$Cl_2(\text{solved}) + 2\,e^- \rightleftharpoons 2\,Cl^-$	+1.36
$BrO_3^- + 6\,H^+ + 6\,e^- \rightleftharpoons Br^- + 3\,H_2O$	+1.44
$MnO_4^- + 8\,H^+ + 5\,e^- \rightleftharpoons Mn^{2+} + 4\,H_2O$	+1.51

The reduced form of a redox pair cannot donate electrons to any oxidized form of another redox pair or the oxidized form of a particular pair cannot accept electrons from any reduced form of other pairs. Whether the oxidized form of a redox pair can act as an oxidizing agent or whether the reduced form can act as a reducing agent depends on the experimental conditions. A substance is not an oxidizing agent or a reducing agent from the outset. It depends on the reactant. Furthermore, the behavior depends on the temperature and the concentrations. In many cases, redox equilibria occur.

3.3.2 Permanganometric determinations

In permanganometry, in which potassium permanganate solution is used as a standard solution, the high oxidation capacity of the MnO_4^- ion is utilized. However, the course of the redox reactions is very different depending on whether the reaction medium is acidic, neutral or alkaline.

The majority of the reactions used in permanganometry for titration purposes take place in acidic solution according to the equation:

$$MnO_4^- + 8\,H^+ + 5\,e^- \rightarrow Mn^{2+} + 4\,H_2O.$$

The permanganate ion, in which manganese has the oxidation number VII, is reduced to the manganese(II) ion with the participation of eight hydrogen ions, taking up five electrons supplied by the respective reducing agent. This produces four molecules of water from eight hydrogen ions.

In some cases, the titration is carried out with potassium permanganate in a neutral solution. This mainly applies to substances such as hydrazine which are only clearly oxidized by permanganate in solutions with a low hydrogen ion concentration.

In weakly acidic, neutral or weakly alkaline solutions, permanganate reacts according to

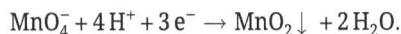

$$MnO_4^- + 4\,H^+ + 3\,e^- \rightarrow MnO_2\downarrow + 2\,H_2O.$$

Here, the manganese is reduced from oxidation number VII to manganese dioxide, in which manganese has oxidation number IV, with the participation of 4 H^+ ions and the absorption of only three electrons released by the reducing agent.

However, the actual process, the mechanism of the reactions that take place in the solutions in which permanganate is used as an oxidizing agent, is significantly more complex than the formulations indicate. Examples will show this later.

Another possibility is to work in strongly alkaline solutions. Here, the permanganate is reduced to manganate(VI):

$$MnO_4^- + e^- \rightarrow MnO_4^{2-}.$$

This reaction can also be used for volumetric analysis. It is used, for example, to determine phosphite, hypophosphite, methanol and formaldehyde (see Section 3.3.9) [124].

3.3 2.1 Preparation of the potassium permanganate solution

From the explanations on standard solutions and the oxidizing effect of permanganate, it follows that a potassium permanganate solution of concentration $c(\frac{1}{5} KMnO_4) = 1$ mol/l (formerly 1 normal $KMnO_4$ solution, 1 N) to be used for titrations in the acidic range must contain 158.034: 5 = 31.607 g $KMnO_4$ per liter. In practice, one usually works with 10-fold diluted solutions, $c(\frac{1}{5} KMnO_4) = 0.1$ mol/l.

Despite the purity of the commercially available potassium permanganate, it is not possible to prepare a precise measured solution by weighing out exactly 3.1607 g $KMnO_4$, dissolving it in water and filling it up to 1 l because such a solution does not exhibit titer constancy. Instead, weigh approximately the calculated portion of the salt, about 3.2 g, on a laboratory balance, dissolve it in a clean bottle to 1 l with deionized water and leave the solution to stand for about 1–2 weeks. Even with the most careful work, the titer slowly decreases in the first few days because traces of ammonium salts, dust particles and other organic impurities are gradually oxidized while permanganate is consumed. Instead of leaving the solution to stand for a long time, it can also be heated for an hour on a boiling water bath. This accelerates the oxidation processes. The solution is then filtered through a carefully cleaned glass filter crucible (not a paper filter!) into a likewise carefully cleaned storage bottle. If filtering is omitted, the titer of the solution will continue to decrease, as the manganese dioxide produced during the oxidation of the dust particles catalyzes the self-decomposition of the dissolved permanganate, according to the following scheme:

$$4\,KMnO_4 + 2\,H_2O \rightarrow 4\,MnO_2\downarrow + 4\,KOH + 3\,O_2\uparrow.$$

Store the storage bottle away from light.

The chemical effectiveness of the ready-to-use potassium permanganate solution still has to be determined precisely. This is done with the help of suitable primary standards. The most suitable substances are sodium oxalate, $Na_2(COO)_2$, and chemically pure iron.

Adjustment with sodium oxalate: Titration with sodium oxalate is based on the classic investigations by Sörensen in 1897–1906, which is by far the preferred method. In the following, the method is described in detail because it reveals the aspects that are important both for the titration of permanganate solutions and for titration in general.

Titration is based on the following redox equation:

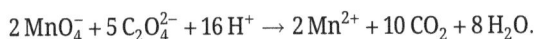

$$2\,MnO_4^- + 5\,C_2O_4^{2-} + 16\,H^+ \rightarrow 2\,Mn^{2+} + 10\,CO_2 + 8\,H_2O.$$

The oxidation of the oxalate ion to carbon dioxide takes place in a warm sulfuric acid solution within relatively wide limits of the hydrogen ion concentration without interfering side reactions according to the reaction equation given.

The advantages of sodium oxalate as a titer substance are: It can be easily obtained very pure by recrystallization according to its formulaic composition. It contains no water of crystallization. It is easy to dry. It is a neutral salt that does not absorb water, carbon dioxide or ammonia. It is therefore easy to weigh.

From the preparative preparation from soda and oxalic acid, the following impurities can be considered: moisture, sodium carbonate, sodium hydrogen oxalate, sodium sulfate or sodium chloride. The moisture can be easily removed by drying the salt in a drying oven at 230–250 °C. The sodium oxalate only begins to decompose above 330 °C ($Na_2C_2O_4 \rightarrow Na_2CO_3 + CO\uparrow$). An admixture of sodium carbonate or sodium hydrogen oxalate can be determined by titration with HCl or NaOH using phenolphthalein as an indicator and removed by recrystallization. Sulfates and chlorides can be detected by suitable precipitation reactions in an acidified solution of about 10 g of the salt. To determine the absence of organic impurities, 1 g of the salt is heated with 10 ml of pure concentrated sulfuric acid. The sulfuric acid must not turn brown or even black.

Procedure: Weigh three or four samples of approximately 0.15–0.2 g of dried sodium oxalate to ±0.1 mg. To do this, use an elongated weighing tube with a glass cap attached, which contains any mass of the substance to be weighed. After accurate weighing, open the tube, carefully insert the neck deep into the opening of a 400-ml titration beaker and carefully tap the tube held at an angle to allow the desired amount of substance to slide into the beaker. The weighing tube is then closed and weighed again. The difference between the two weighings gives the mass of the substance sample in the beaker. Care must be taken when pouring in the sample to ensure that the substance is not atomized. Each sample is dissolved in about 200 ml of water, and the solution is acidified with 10 ml of sulfuric acid (concentrated sulfuric acid diluted 1:4) and heated to 75–85 °C.

Titration is carried out by allowing the potassium permanganate solution set to the zero mark (for opaque solutions, take the reading at the upper edge of the meniscus!) to drip from the burette into the hot sodium oxalate solution while constantly swirling the titration beaker. Wait until the solution has decolorized before each new addition of permanganate. Initially, the oxidation of the oxalate ion only takes place slowly. The reaction equation only shows the initial and final stages. In reality, the reaction is much more complicated, with the manganese(II) ion playing a role as a catalyst (Skrabal, 1904). It is initially only present in trace amounts, but is formed in increasing amounts during the titration. After adding a few milliliters, the permanganate solution can be allowed to flow in somewhat faster. In order not to exceed the end point, it must be allowed to drip in again very slowly and carefully toward the end of the titration.

The end point can be recognized by the fact that the permanganate solution is no longer decolorized, but gives the solution a weak red-violet color. The color intensity of the permanganate ions can be recognized from the fact that – according to Kolthoff [89] – a solution with a concentration of 1–$2 \cdot 10^{-5}$ mol/l (based on equivalents) is still extremely weakly pink in color. Thanks to this circumstance and the fact that the manganese(II) ion appears completely colorless even in moderately diluted solutions, permanganometry does not require the addition of foreign indicators. To recognize the end point of the titration, the excess addition of a certain small volume of the permanganate solution is required. One drop, i.e. about 0.03 ml, is able to color 300 ml of colorless solution slightly pink. If 20–30 ml of permanganate solution were used, this would be an excess of about 0.1%. However, the same additional consumption of permanganate solution also occurs in later titrations so that it does not need to be taken into account in normal determinations. For very exact titrations, however, this excess must be taken into account, taking into account the volume of the solution provided as well as the associated titration.

The observation that the weak pink color of a titrated solution gradually disappears after some time can be explained not only by the addition of oxidizable dust particles from the air but also by the fact that the manganese(II) ions formed during the titration slowly reduce the permanganate ions.

The **calculation of the titer** of the potassium permanganate solution based on the titration results is carried out as follows:

Suppose we had titrated three samples of 0.2718, 0.1854 and 0.1922 g $Na_2C_2O_4$ with the $KMnO_4$ solution and used 40.15, 27.40 and 28.47 ml, respectively, up to the equivalence point. One milliliter of a permanganate solution containing exactly 0.1 mol/l equiv ($\frac{1}{5}$ $KMnO_4$) shows exactly 0.1 mmol equivalents of sodium oxalate ($\frac{1}{2}$ $Na_2C_2O_4$), i.e. 67,000 mg/mmol–0.1 mmol/ml = 6,700 mg/ml $Na_2C_2O_4$. The fractions 271.8 mg: 6.7 mg/ml = 40.55 ml, 185.4 mg: 6.7 mg/ml = 27.69 ml and 192.2 mg: 6.7 mg/ml = 28.69 ml therefore indicate the volume in milliliters of an exactly 0.1 mol/l $KMnO_4$ solution to which our three samples correspond. In reality, we did not use 40.55 ml, but only 40.15 ml or only 27.40 ml instead of 27.69 ml and only 28.47 ml instead of 28.69 ml of the standard solution. It is therefore slightly stronger than 0.1 mol/l. The fractions 40.55/40.15 = 1.011, 27.69/27.40 = 1.010 and 28.69/28.47 = 1.010, i.e. an average of 1.010, indicate the titer. One milliliter of the potassium permanganate solution used corresponds to 1.01 ml of a solution of the exact concentration 0.1 mol/l (the solution has the concentration $c(\frac{1}{5}$ $KMnO_4) = 0.101$ mol/l).

Oxalic acid itself, $H_2C_2O_4 \cdot 2 H_2O$, is less suitable for the adjustment of permanganate solutions because it is more difficult to obtain a preparation that corresponds exactly to the water content and because it easily attracts ammonia contained in the laboratory air and thus changes into ammonium oxalate trace by trace. However, the use of oxalic acid solution is often recommended because it can also be used for the determination of alkalis. The adjustment with oxalic acid is carried out in the same way as titration with sodium oxalate. The substance is used in an air-dry state:

1 ml $KMnO_4$ solution, $c(\frac{1}{5}$ $KMnO_4) = 0.1$ mol/l, corresponds to 1 ml oxalic acid solution, $c(\frac{1}{2}$ $H_2C_2O_4 \cdot 2 H_2O) = 0.1$ mol/l, or 6.303 mg $H_2C_2O_4 \cdot 2 H_2O$.

Adjustment with chemically pure iron: The method is very accurate if pure iron is available, which according to Mittasch (1928) is prepared by thermal decomposition of iron pentacarbonyl, $Fe(CO)_5$, and is also commercially available. The metallic iron is dissolved in dilute sulfuric acid in the absence of air to form iron(II) sulfate, and the solution is then titrated with potassium permanganate as described below.

Under no circumstances should so-called florist's wire be used for adjustment, as recommended in older textbooks. Flower wire can contain up to 0.3% foreign components such as carbon, silicon, phosphorus and sulfur, some of which dissolve in sulfuric acid to form oxidizable compounds. As a result, more permanganate solution is consumed than corresponds to the actual iron content of the wire so that the consumption can appear to be over 100% of the theoretical value. In addition, the carbide content of the wire is subject to considerable fluctuations.

The dissolution of the weighed iron samples in diluted sulfuric acid must be carried out in the absence of air to prevent oxidation of the iron to iron(III) ions. This is best done in a round-bottomed flask with a Bunsen valve (see Fig. 3.13). The titration is carried out as described in Section 3.3.2.2.

Adjustment with sodium thiosulfate solution according to Ph. Eur.: Add 2 g potassium iodide and 10 ml dilute sulfuric acid (98 g/l) to 20.0 ml of the potassium permanganate solution (0.02 mol/l). Titrate the mixture with sodium thiosulfate solution (0.1 mol/l). Toward the end of the titration, 1 ml starch solution (10 g/l) is added. The factor must be determined immediately before use [8a].

3.3.2.2 Determination of iron in sulfuric acid solution

The titration of iron(II) in a sulfuric acid solution is carried out according to the reaction equation:

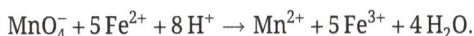

$$MnO_4^- + 5\,Fe^{2+} + 8\,H^+ \rightarrow Mn^{2+} + 5\,Fe^{3+} + 4\,H_2O.$$

With this titration, which delivers very accurate results, Margueritte founded permanganometry in 1846.

Titration of an iron(II) sulfate solution: Prepare enough of the solution so that the consumption of permanganate solution is 25–40 ml, add 10 ml of diluted sulfuric acid (1:4) and dilute to about 200 ml with air-free water. The titration can be carried out at room temperature or after heating the solution. The end point is reached when the solution remains a faint orange color for 1 min after the last addition of permanganate. The color results from the slightly yellowish color of the resulting iron(III) salt solution and the red-violet of the excess permanganate. By adding a little phosphoric acid, the iron(III) ions can be converted into colorless complex compounds so that the pink color of the permanganate ions is retained at the end of the titration. However, even without the addition of phosphoric acid, the end point of the titration is easy to determine:

   1 ml of a potassium permanganate solution, $c(\frac{1}{5}\,KMnO_4) = 0.1$ mol/l, corresponds to 5.5845 mg iron.

Titration of an iron(III) sulfate solution: If the iron is not present with oxidation number II from the outset, it must be quantitatively reduced to iron(II) before titration with potassium permanganate. Only reducing agents whose excess can be easily removed from the solution after the reduction are suitable for this purpose. Sulfurous acid, nascent hydrogen or tin(II) chloride are used, among others.

Reduction with sulfurous acid according to

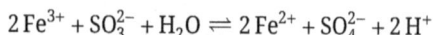

$$2\,Fe^{3+} + SO_3^{2-} + H_2O \rightleftharpoons 2\,Fe^{2+} + SO_4^{2-} + 2\,H^+$$

is incomplete in strongly acidic solutions according to the law of mass action. The iron(III) salt solution is therefore – if it is acidic – almost neutralized with soda solution, mixed with a freshly prepared solution of sulfurous acid in excess and heated to boiling for ¼ to ½ h while slowly passing through air-free carbon dioxide. After about 30–40 ml of water has been distilled from the flask, check whether the CO_2 stream passing through still contains SO_2. For this purpose, the outlet tube of the condenser is immersed in a bowl of weakly sulfuric acid water colored pink by a drop of the potassium permanganate solution. If decoloration no longer occurs, the solution is titrated directly in the flask after adding 10 ml of diluted sulfuric acid with permanganate solution.

Reduction with nascent hydrogen according to

$$Fe^{3+} + H \rightarrow Fe^{2+} + H^+$$

is produced in a sulfuric acid solution by adding pure metallic zinc or aluminum. The nascent hydrogen is produced when the metal is dissolved, e.g.

$$Zn + 2\,H^+ \rightarrow Zn^{2+} + 2\,H.$$

The following reaction also takes place:

$$2\,Fe^{3+} + Zn \rightarrow 2\,Fe^{2+} + Zn^{2+}.$$

A **reduction flask with a Bunsen valve** is used (see Fig. 3.13). The zinc is placed in the glass basket, which is attached to a glass rod that protrudes through the top of the ground joint cap in a gas-tight but movable manner. During the reduction, the glass basket, which consists of a spirally wound glass rod, is lowered almost to the bottom of the reduction flask; the solution is heated to around 70–80 °C. Its volume must not be unnecessarily large; otherwise the reduction will take too long. The lively hydrogen can leave the flask through the Bunsen valve, which consists of a glass tube widened in the middle to form a small sphere, the lower end of which protrudes through the ground-glass cap into the reduction flask. At the top is a small piece of rubber tubing slit open at the side, which in turn is sealed with a glass rod. The hydrogen can leave the flask through the side slit if there is excess pressure. However, if there is negative pressure in the flask, the tube is pressed tightly together and the slit is closed so that no air can enter. The purpose of the spherical expansion of the Bunsen valve is to allow any entrained droplets of liquid to settle and flow back into the reduction flask. The hydrogen gas escaping from the slit should not cause an acidic reaction on moist indicator paper if the Bunsen valve is constructed correctly and the reduction is conducted as prescribed. This is proof that no acidic solution is entrained from the flask.

After 1–2 h, the reduction is complete. The basket with the excess zinc or aluminum is pulled up, the ground joint cap is removed and the parts are carefully rinsed with boiled water. The still warm solution is then titrated. The metals used must of course be iron-free. If necessary, the permanganate consumption of a weighed and dissolved sample must be determined and taken into account as a correction.

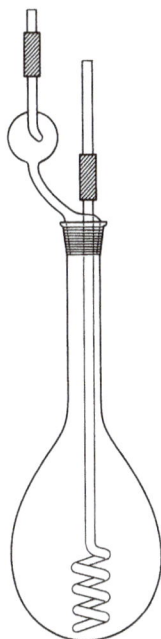

Fig. 3.13: Reducing flask with a Bunsen valve.

The reduction of the iron(III) ions can also be achieved with amalgamated zinc in the **Jones reductor** (H. C. Jones [108]) (see Fig. 3.14). A glass tube about 20 mm in diameter and 35–40 cm long contains a glass frit (G2) or a porcelain filter plate with fine holes above a drain tap and a cushion of glass wool on top. The column is filled with finely granulated, amalgamated, iron-free zinc, which is prepared by shaking the zinc granulate with a solution containing 2% of the zinc mass mercury(II) chloride or nitrate. Suck 50–100 ml of 2–3% warm sulfuric acid through the reducer. A drop of the permanganate solution must not be decolored by the sulfuric acid after passing through the reducer. Otherwise, repeat the treatment with the diluted sulfuric acid. The diluted iron(III) sulfate solution (approx. 100 ml) is sucked through the reducer in 1 min and collected in the titration vessel in an inert gas atmosphere (N_2 and CO_2). Then wash with 25–50 ml of 2–3% sulfuric acid followed by 100–150 ml of air-free water. The filtrate is then titrated immediately. Any iron(II) sulfate residues remaining in the reducer are rinsed out by washing again with water. The wash water is added to the sample solution and titrated. The reducer is kept filled with water and treated with diluted sulfuric acid as described above before each use. Instead with amalgamated zinc, the iron(III) solution can also be treated with cadmium or fine-grained silver (see [109]).

The reduction with **tin(II) chloride** is described in Section 3.3.2.3.

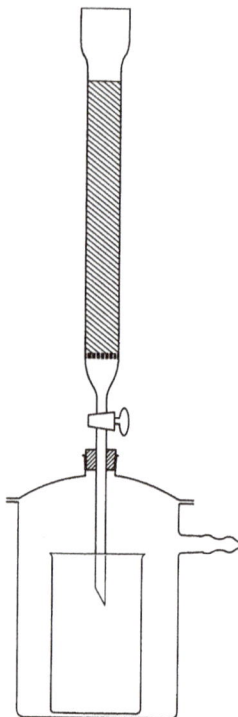

Fig. 3.14: Jones reductor.

Determination of iron(II) in addition to iron(III): Two titrations are carried out. One sample of the sulfuric acid solution is titrated directly; it provides the content of iron(II) ions. The second sample of the solution is reduced before titration; this gives the total iron content. The difference between the two values indicates the content of iron(III) ions.

3.3.2.3 Determination of iron in hydrochloric acid solution

The oxidation of the iron(II) ions in sulfuric acid solution proceeds exactly according to the reaction equation given so that this determination method provides very reliable values. However, if an analogous attempt is made to titrate iron(II) salts in the presence of hydrochloric acid, it can be observed that (1) the end point of the titration is much more difficult to recognize, (2) the solution smells distinctly of chlorine during the titration and (3) for the same quantity of iron(II) salt solution considerably more permanganate solution is used than in a sulfuric acid solution. The presence of hydrochloric acid therefore causes a different reaction process that deviates from the specified scheme, as some of the hydrochloric acid is oxidized by the permanganate to chlorine. A quantitative determination of hydrochloric acid iron(II) salt solutions by titration with potassium permanganate solution cannot be carried out without further precautions.

To clarify the reaction process note the following observations:

1. Dilute hydrochloric acid alone is not oxidized by permanganate solutions (Schäfer, 1954). Although oxidation should be possible according to the position of the redox potentials of the redox pairs, the process is obviously inhibited (see the oxidation of $C_2O_4^{2-}$ by MnO_4^-, p. 159).
2. The presence of iron ions must be held responsible for the development of chlorine and the resulting increased consumption of permanganate. However, since iron(III) salts also have no influence on the reaction of permanganate and hydrochloric acid, the error must be caused by the iron(II) ions.
3. Zimmermann [110] has made the observation that the reaction disturbance caused by hydrochloric acid can be reduced to a minimum if a sufficient excess of manganese(II) sulfate is added to the iron(II) salt solution before the titration begins.

The oxidation of the chloride is promoted by iron(II) ions, while the presence of manganese(II) ions in sufficiently high concentrations apparently largely prevents the reaction between permanganate and chloride. It must be concluded from this that the reaction process in the oxidation of iron(II) ions by permanganate is much more complicated than the reaction scheme indicates.

Zimmermann (1882) and Manchot (1902) were the first to attempt to clarify the induction of the oxidation of chloride ions to elemental chlorine in the presence of iron(II) ions. They assumed the intermediate formation of a labile higher oxidation state of iron. Manchot found that during the oxidation of Fe^{2+} to Fe^{3+} in the presence of an acceptor, i.e. a substance whose oxidation is induced by MnO_4^- of Fe^{2+}, such as hydrochloric acid or tartaric acid, as much permanganate is consumed as corresponds to the formation of an iron compound of oxidation number V. Manchot then assumed that, in general, when titrating iron(II) salt solutions with permanganate as an intermediate product, iron of this oxidation number occurs in a previously unknown short-lived compound.

Although the formation of iron with higher oxidation numbers (VI, IV) has also been made probable in other reactions, such as the catalytic decomposition of hydrogen peroxide (Bohnson and Robertson 1923, Bray and Gorin 1932) and the formation of compounds of iron with oxidation numbers IV to VI is possible (Scholder 1952, Klemm and Wahl 1954), there is no decisive proof that higher oxidation numbers of iron occur during the titration of iron salt solutions.

Another possibility is the intermediate formation of manganese with lower oxidation numbers, such as VI, V, IV and III, during the reduction of permanganate. Birch already suspected in 1909 that manganese(III) chloride is formed in the absence of manganese(II) ions. As manganese(III) salts are easily reduced to manganese(II) salts in acidic solutions, the resulting manganese(III) oxidizes the chloride ions to chlorine. Although little is known about the reaction process at the start of the titration and about the involvement of the various oxidation stages of the manganese, it is generally assumed today that the equilibrium is responsible for the redox reaction:

$$Mn^{3+} + e^- \rightleftharpoons Mn^{2+}$$

with $E° = 1.51$ V is of decisive importance (F. C. Kessler 1863, Skrabal 1904, Birch 1909 and Schleicher 1951, 1952, 1955).

The manganese(III) is either formed at the beginning of the reaction by reduction of MnO_4^- or it is formed later by the reaction of MnO_4^- with Mn^{2+} (cf. the acceleration of the redox reaction by the increasing Mn(II) concentration during the titration of oxalate (p. 159)).

In contrast to the reactions with MnO_4^-, the reactions with Mn^{3+} take place very quickly (Schleicher 1951). The redox potential of the Mn^{2+}/Mn^{3+} system is so high that, in addition to Fe^{2+} (Fe^{2+}/Fe^{3+}, $E° = +0.77$ V), chloride present in the solution (2 Cl^-/Cl_2, $E° = +1.36$ V) can also be oxidized. However, the oxidation of chloride ions during iron titration in hydrochloric acid solution is largely prevented if a mixture of manganese sulfate solution, phosphoric acid and sulfuric acid (**Reinhardt-Zimmermann solution**) is added to the solution. However, as numerous studies have shown, the oxidation of chloride ions can also be prevented more or less effectively in the presence of other salts such as sulfates, acetates, phosphates, fluorides and oxalates (Skrabal 1904, Barneby 1914, Ishibashi, Shigematsu and Shibata 1956, Somasundaram and Suryanarayana 1956). This also provides some confirmation of the occurrence of manganese(III) during titration with permanganate. Mn^{3+}, which tends to form complexes, forms complex compounds with the anions mentioned, which lowers the Mn^{3+} concentration and reduces the oxidation potential to such an extent (see Nernst's equation, p. 272 f.) that the oxidation of chloride ions is no longer possible (see Taube 1948).

The titration proceeds in the presence of Reinhardt-Zimmermann solution approximately as follows: The high Mn^{2+} concentration ensures the elimination of the reaction inhibition through the formation of Mn(III) and thus the rapid progress of the reaction:

$$Fe^{2+} + Mn^{3+} \rightarrow Fe^{3+} + Mn^{2+}.$$

At the same time, the high Mn^{2+} concentration lowers the oxidation potential of the Mn^{2+}/Mn^{3+} system to such an extent – to which the phosphoric acid also contributes through complex formation – that the oxidation of chloride can no longer take place. Furthermore, by forming a colorless iron(III) complex, the phosphoric acid not only makes it easier to identify the end point of the titration, which is made more difficult by deep yellow chloroacids of iron, e.g. $H_3[FeCl_6]$, in hydrochloric acid solution, but also causes the redox potential of Fe^{2+}/Fe^{3+} to decrease by lowering the Fe^{3+} concentration, which in turn promotes the oxidation of Fe^{2+}.

The processes briefly discussed here are described in detail in Laitinen [111]. On the subject of induction and catalysis in redox titrations, see also [9, 112] and Schleicher (1951–1958).

Titration of an iron(II) chloride solution: Add 10 ml Reinhardt-Zimmermann solution to the hydrochloric acid sample solution diluted to 100 ml and titrate while swirling vigorously.

According to Kolthoff and Smit (1924), the error in the determination can be kept below 0.2% if the iron concentration is 0.1 mol/l and the hydrochloric acid concentration does not exceed 1 mol/l. Reinhardt-Zimmermann solution should be added generously and the titration should not be carried out too quickly.

The Reinhardt-Zimmermann solution is composed as follows: A mixture prepared from 1 l of phosphoric acid (ρ = 1.3 g/ml), 600 ml of water and 400 ml of sulfuric acid (ρ = 1.84 g/ml) is added to a solution of 200 g of crystallized manganese(II) sulfate in 1 l of water.

Titration of an iron(III) chloride solution: The method based on Reinhardt [113] and Zimmermann has been widely used in metallurgical and smelting laboratories. It allows to titrate hydrochloric acid solutions of iron ores and iron alloys directly. Additions of cobalt, copper, lead, chromium and titanium as well as arsenic do not interfere; only antimony must not be present.

The reduction of the iron(III) ions is achieved by adding tin(II) chloride:

$$Sn^{2+} + 2\,Fe^{3+} \rightarrow Sn^{4+} + 2\,Fe^{2+},$$

which is only used in small excess. The excess is removed by adding a small amount of mercury(II) chloride:

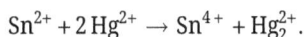

$$Sn^{2+} + 2\,Hg^{2+} \rightarrow Sn^{4+} + Hg_2^{2+}.$$

The precipitating mercury(I) chloride

$$Hg_2^{2+} + 2\,Cl^- \rightarrow Hg_2Cl_2 \downarrow$$

is practically not oxidized by permanganate. Nevertheless, only a small amount of mercury(I) chloride may be produced, as it may reduce the iron(III) chloride during the titration.

Procedure: Carefully add enough tin(II) chloride solution drop by drop to the boiling, strong hydrochloric iron(III) salt solution so that the liquid has just become completely colorless. After the solution has cooled, 10 ml of a clear, cold saturated sublimate solution is added in one pour, after which pure white, crystalline mercury(I) chloride must precipitate. If a precipitate colored gray by metallic mercury is formed (especially if the Sn^{2+} concentration is too high, the Hg^{2+} can be reduced to metallic Hg), do not continue working. After 2 min, dilute to 600–700 ml, add 10 ml Reinhardt-Zimmermann solution and titrate with potassium permanganate solution.

The tin(II) chloride solution is prepared as follows: by dissolving 120 g of pure tin in 500 ml of hydrochloric acid ($\rho = 1.124$ g/ml), a solution is obtained which is poured into a 4-l bottle already containing 1 l of hydrochloric acid of the same concentration and also 2 l of water. According to Kolthoff [89], the storage bottle should contain metallic tin.

Determination of iron(II) in addition to iron(III): The determination can also be carried out in hydrochloric acid solutions using Reinhardt-Zimmermann solution without any problems. One example is the titration of the dissolution of the types of magnetite.

3.3.2.4 Determination of uranium and phosphate

Uranyl salts, containing uranium with the oxidation number VI, can be reduced quantitatively to uranium(IV) in a sulfuric acid solution using aluminum sheet without fear of further reduction to uranium(III):

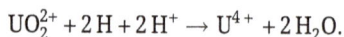

$$UO_2^{2+} + 2\,H + 2\,H^+ \rightarrow U^{4+} + 2\,H_2O.$$

If zinc, amalgamated zinc or cadmium is used as a reducing agent, however, partial reduction to uranium(III) always occurs. The solution reduced in the cuvette with Bunsen valve (see Fig. 3.13) is titrated with potassium permanganate.

The method also enables the permanganometric determination of phosphate, which is important in the analysis of artificial fertilizers. The method is based on the fact that phosphate ions are quantitatively precipitated from a weak acetic acid alkali phosphate solution containing NH_4^+ salts with the addition of excess uranyl(VI) acetate solution as sparingly soluble uranylammonium phosphate:

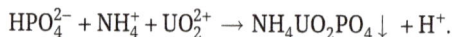

$$HPO_4^{2-} + NH_4^+ + UO_2^{2+} \rightarrow NH_4UO_2PO_4 \downarrow + H^+.$$

The precipitated $NH_4UO_2PO_4$ is filtered off, washed out and dissolved in sulfuric acid. The solution is reduced as described above and titrated with potassium permanganate solution.

3.3.2.5 Determination of oxalate

The titration of oxalic acid and oxalates has already been described in the methods for adjusting potassium permanganate solutions (see p. 158 ff.):

1 ml $KMnO_4$ solution, $c(\tfrac{1}{5}\,KMnO_4) = 0.1$ mol/l, corresponds to 4.5018 mg anhydrous $C_2H_2O_4$.

3.3.2.6 Determination of calcium

Calcium ions can be quantitatively precipitated from a weakly ammoniacal solution containing NH_4Cl in boiling heat with ammonium oxalate solution:

$$Ca^{2+} + C_2O_4^{2-} \rightarrow CaC_2O_4 \downarrow .$$

If the CaC_2O_4 is dissolved in sulfuric acid or hydrochloric acid after filtering and washing out, the released oxalic acid can be titrated with potassium permanganate solution. The method was frequently used in the mortar and cement industry and to determine the hardness of drinking and process water (see p. 237 f.).

> **Work specification:** A calcite consisting of $CaCO_3$ and admixtures of silicate gangue, solid alumina (Al_2O_3) and iron(III) oxide, Fe_2O_3, is to be analyzed. A portion of the finely ground material is weighed, placed in an Erlenmeyer flask via a funnel and dissolved in hydrochloric acid at boiling temperature. The insoluble gangue is filtered off. The cooled solution is filled up to the mark in a volumetric flask and mixed. An aliquot is taken for further analysis.
>
> After oxidation of the iron, this and the aluminum are removed by precipitation as hydroxides by adding ammonium carbonate-free ammonia solution at boiling heat. The precipitate is filtered off and washed out thoroughly. Calcium is determined in the filtrate by adding a little NH_4Cl solution to the weakly ammoniacal solution, bringing to the boil and adding a small excess of hot ammonium oxalate solution. The solution is left to stand for a few hours and then filtered, preferably through a membrane filter. The precipitate is first washed with water containing ammonium oxalate, then with pure water and finally rinsed in an Erlenmeyer flask. Dissolve in warm diluted sulfuric acid. If you are not working with membrane filters but with hardened paper filters, rinse the main mass of the precipitate from the filter and then dissolve the residues of adhering filling quantitatively by dripping hot diluted sulfuric acid into the same Erlenmeyer flask. If filter crucibles made of glass or porcelain are used for filtration, care must be taken to ensure that no precipitate or solution residues remain in the pores of the filter plates. Then dilute with hot water to about 300 ml and titrate with potassium permanganate solution:
>
> 1 ml $KMnO_4$ solution, $c(⅕ KMnO_4) = 0.1$ mol/l, corresponds to 2.004 mg Ca or 2.8040 mg CaO. The CaO content of the calcite is given as a percentage.

3.3.2.7 Determination of hydrogen peroxide

Hydrogen peroxide reacts in acidic solution with permanganate to

$$2\,MnO_4^- + 5\,H_2O_2 + 6\,H^+ \rightarrow 2\,Mn^{2+} + 5\,O_2\uparrow + 8\,H_2O.$$

The smooth reaction is also used for the gas volumetric determination of hydrogen peroxide.

The titration is carried out at room temperature after adding 30 ml of sulfuric acid (diluted 1:4) to the solution diluted to 200 ml. If hydrochloric acid is present, add 10 ml Reinhardt-Zimmermann solution (see p. 166 f.):

1 ml $KMnO_4$ solution, $c(⅕ KMnO_4) = 0.1$ mol/l, corresponds to 1.701 mg H_2O_2. The analysis result is calculated as a mass fraction in percent.

The reaction of MnO_4^- and H_2O_2 also shows an initial delay in the visible start of the reaction, an incubation period which can be attributed to the same causes as the phenomenon observed in the oxidation of oxalic acid with $KMnO_4$ (see p. 159).

Permanganometric H_2O_2 determination played a major role in the laboratories of bleaching plants. Peroxides, perborates and percarbonates can also be titrated in a similar way due to the hydrolytic cleavage of H_2O_2 in sulfuric acid solutions.

3.3.2.8 Determination of peroxodisulfate

Peroxodisulfuric acid cannot be titrated directly with potassium permanganate due to its low protolysis. However, it can titrated after reduction with iron(II) sulfate to sulfuric acid indirectly:

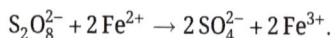

$$S_2O_8^{2-} + 2\,Fe^{2+} \rightarrow 2\,SO_4^{2-} + 2\,Fe^{3+}.$$

The peroxodisulfate ions are allowed to react with excess iron(II) sulfate solution, the content of which is precisely known, in the absence of air and the excess iron(II) ions are titrated back with potassium permanganate solution after the reaction.

3.3.2.9 Determination of nitrite

The oxidation of nitrite by $KMnO_4$ according to

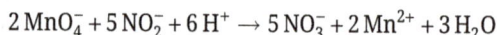

$$2\,MnO_4^- + 5\,NO_2^- + 6\,H^+ \rightarrow 5\,NO_3^- + 2\,Mn^{2+} + 3\,H_2O$$

only proceeds slowly at room temperature. The nitrous acid is decomposed in the heat and can also escape undecomposed from the sulfuric acid solution. Titration is therefore carried out according to Lunge (1891, 1904, see also [89]), and the nitrite solution is therefore titrated in reverse by allowing it to flow into a warm permanganate solution acidified with dilute sulfuric acid until it has become colorless.

Procedure: To test the purity of potassium nitrite, dissolve 2 g of the sample in water and make up the solution to 250 ml in the volumetric flask. This solution is poured into the burette. Then add 20 ml H_2SO_4 (c = 2 mol/l) to 25 ml of a $KMnO_4$ solution (c = 0.02 mol/l), dilute the solution to approximately 300 ml, heat to 40 °C (not higher!) and titrate very slowly and carefully with the nitrite solution until decolorization occurs. You must constantly swirl around and titrate slowly, especially near the end point. If titration is too fast, the error can be over 1% according to Kolthoff [89]. The method is used to determine the N_2O_3 content of the nitrose:

1 ml $KMnO_4$ solution, $c(\frac{1}{5}\,KMnO_4)$ = 0.1 mol/l, corresponds to 4.255 mg KNO_2 or 1.900 mg N_2O_3.

3.3.2.10 Determination of hydroxylamine

Hydroxylamine is heated in an acidic solution with an excess of iron(III) salt solution to boiling point, after

$$2\,NH_2OH + 4\,Fe^{3+} \rightarrow N_2O\uparrow\ + 4\,Fe^{2+} + H_2O + 4\,H^+.$$

The hydroxylamine is oxidized to nitrous oxide and water, while an equivalent amount of Fe(III) is reduced to Fe(II). The Fe(II) ions can be titrated with permanganate solution.

However, a sufficiently large excess of iron(III) sulfate must be used; otherwise the side reaction

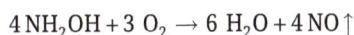

$$4\,NH_2OH + 3\,O_2 \rightarrow 6\,H_2O + 4\,NO\uparrow$$

makes itself noticeable.

3.3.2.11 Determination of manganese(IV)

The determination of manganese(IV) oxide is based on the fact that manganese(IV) is reduced to manganese(II) when heated with excess oxalic acid in a sulfuric acid solution:

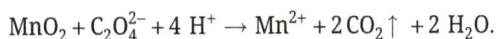

$$MnO_2 + C_2O_4^{2-} + 4\,H^+ \rightarrow Mn^{2+} + 2\,CO_2\uparrow\ + 2\,H_2O.$$

The excess oxalic acid is titrated back with permanganate solution. The reduction of the MnO_2 can also be carried out with an adjusted sulfuric acid iron(II) sulfate solution.

Procedure: The manganese dioxide is powdered as fine as dust (very important!). Weigh out about 0.5 g and place it in an Erlenmeyer flask using a funnel. Add 100 ml oxalic acid solution (0.05 mol/l) of known titer and 10 ml sulfuric acid (2 mol/l) from a pipette. Heat on the water bath until the manganese dioxide has decomposed and only the gangue remains. Then dilute with hot water and retitrate the excess oxalic acid with permanganate solution.

The consumption $V(H_2C_2O_4)$ of oxalic acid solution is calculated from the volume added (100 ml), the consumption $V(KMnO_4)$ of potassium permanganate solution and the titres of the two solutions to give

$$V(H_2C_2O_4) = [100 \cdot t(H_2C_2O_4) - V(KMnO_4) \cdot t(KMnO_4)]\ ml.$$

One milliliter of oxalic acid solution, $c(\tfrac{1}{2}\,H_2C_2O_4) = 0.1$ mol/l, corresponds to 4.347 mg MnO_2. The MnO_2 content of the brownstone is given as a percentage by mass.

The determination of manganese in pig iron, steel and manganese-containing iron ores is based on the possibility of oxidizing manganese(II) to manganese(IV) oxide using potassium chlorate (Hampe, 1883/85) or potassium peroxodisulfate (Knorre, 1901):

$$Mn^{2+} + S_2O_8^{2-} + 2H_2O \rightarrow MnO_2 \downarrow + 2SO_4^{2-} + 4H^+.$$

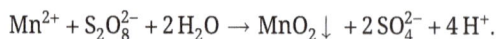

The precipitated MnO_2 is filtered off, washed and titrated as described. The method is suitable for determining the manganese in pig iron, ferromanganese, manganese alloys and manganese iron ores.

3.3.2.12 Determination of manganese(II)

If potassium permanganate is added to a very weakly acidic, almost neutral manganese(II) salt solution at about 80–90 °C, the permanganate oxidizes the manganese(II) to manganese(IV) oxide hydrate and is itself reduced to the same compound, which precipitates as a dark brown precipitate (Guyard, 1863):

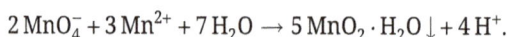

$$2MnO_4^- + 3Mn^{2+} + 7H_2O \rightarrow 5MnO_2 \cdot H_2O \downarrow + 4H^+.$$

However, since the manganese(IV) oxide hydrate is a jelly by nature and a weak acid by chemical character, it easily absorbs components from the solution, in particular, divalent cations. Too little $KMnO_4$ consumption is observed, which is due to the fact that as yet unoxidized manganese(II) ions form poorly soluble manganese(II) manganate(IV) of the formula $Mn(HMnO_3)_2$ with the manganese(IV) oxide hydrate so that a certain proportion of the Mn^{2+} is not permanganometrically detected. However, according to Volhard (1879), this problem can be remedied by adding plenty of foreign cations, e.g. Zn^{2+}, from the outset. The zinc(II) manganates(IV) then precipitate instead of the manganese(II) manganates. If the analysis involves the dissolution of iron-containing manganese ores or manganese alloys (in sulfuric acid), the iron previously oxidized to Fe(III) must be removed from the solution, which is almost completely neutralized with sodium carbonate at room temperature. This is best done by adding slurried zinc oxide:

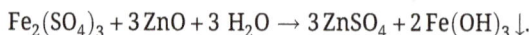

$$Fe_2(SO_4)_3 + 3ZnO + 3H_2O \rightarrow 3ZnSO_4 + 2Fe(OH)_3 \downarrow.$$

Procedure (Fischer, 1909): The neutralized manganese(II) salt solution is mixed with 1–1.5 g of slurried zinc oxide and a solution of 5–10 g $ZnSO_4$ in a 500 ml volumetric flask and the flask is filled with water. After mixing well, either filter through a dry filter and a dry funnel into a dry beaker, discard the first 10–20 ml of the filtrate and use 100–200 ml of the filtrate that runs off later for the titration, depending on the manganese content, or allow to settle and carefully pipette 100–200 ml of the clear liquid above the bottom of the flask into an Erlenmeyer flask. Dilute the solution slightly, heat to boiling point and titrate quickly with permanganate solution, swirling constantly, until the supernatant solution remains faintly pink after the precipitate has settled. Experience has shown that – probably due to a low adsorption of Mn^{2+} on the excess zinc oxide – you are still close to the actual end point. It is therefore acidified with 1 ml glacial acetic acid, which dissolves the zinc oxide and releases the Mn^{2+} ions. The titration is now completed. Zinc oxide, zinc sulfate and glacial acetic acid must of course not consume permanganate. This is confirmed by a blank sample.

According to Reinitzer and Conrath (1926), the addition of ZnO and $ZnSO_4$ can be avoided by working in a weak acetic acid solution buffered with plenty of sodium acetate. Since – as the reaction equation shows – free mineral acid is formed during the titration, which inhibits the reaction process, the sodium acetate converts the mineral acid solution into an acetic acid solution with a low hydrogen ion concentration. This creates the most favorable conditions for the formation of a manganese(IV) oxide hydrate free of lower oxides. In addition, the high sodium concentration favors the formation of sodium manganates(IV); the large excess of electrolyte coagulates the initially colloidally dissolved oxide hydrate, making it easier to recognize the end point. Finally, any iron(III) ions present are precipitated as basic acetates in the boiling heat. In the presence of iron, care must be taken to ensure the absence of chloride ions; otherwise too much permanganate solution will be consumed. Recognizing the end point can be made much easier by adding potassium fluoride, especially if a lot of iron is present. Most of the iron then precipitates as heavy, white $K_3[FeF_6]$. In addition, the manganese(IV) oxide hydrate precipitates in the presence of potassium fluoride as a dense, dark-colored precipitate that separates relatively quickly.

Procedure: The analysis of a white iron with approximately 3% manganese serves as an example. Four to five grams of the substance is dissolved in sulfuric acid (ρ = 1.12 g/ml) and then heated to boiling point. To oxidize the iron(II) ions, 5 ml of concentrated nitric acid is added to the hot solution and heated to boiling until the nitrogen oxides are removed. After cooling, the solution is diluted to 1 l in a volumetric flask. Take an aliquot, e.g. 50 ml, neutralize it approximately with soda solution and allow it to run quickly into a very weak acetic acid solution of 5 g of the purest, carbonate-free sodium acetate and 5 g of potassium fluoride in about 400 ml of water, heated to boiling point. The resulting precipitate of $K_3[FeF_6]$ and basic iron(III) acetate settles quickly after a few swirls. Then titrate with potassium permanganate solution as described above. Toward the end of the titration, it is advisable to heat to boiling again in order to determine the stability of the decolorization. Here too, the reagents used must not consume any permanganate.

It is recommended to adjust the titer of the potassium permanganate solution in the same way with a manganese(II) salt solution of known content, depending on the application of the working method:

1 ml $KMnO_4$ solution, $c(\frac{1}{3} KMnO_4)$ = 0.1 mol/l, corresponds to 2.747 mg Mn or 3.547 mg MnO.

3.3.2.13 Determination of chromium in steel

To determine chromium in steel samples, the sample substance is first dissolved in a mixture of sulfuric acid and phosphoric acid. The resulting chromium(III) is then oxidized to dichromate with peroxodisulfate in the presence of silver ions as a catalyst:

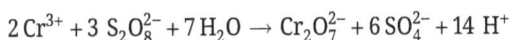

$$2\,Cr^{3+} + 3\,S_2O_8^{2-} + 7\,H_2O \rightarrow Cr_2O_7^{2-} + 6\,SO_4^{2-} + 14\,H^+$$

This is then boiled with NaCl so that permanganate formed from manganese(II) is destroyed and silver ions are precipitated. Both substances could otherwise react later as oxidizing agents and thus be captured. Excess peroxo oxygen decomposes under these conditions. An excess of ammonium iron(II) sulfate solution is then added in a strongly acidic environment so that the dichromate is reacted with part of the iron(II):

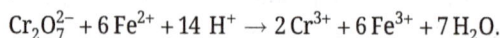

$$Cr_2O_7^{2-} + 6\,Fe^{2+} + 14\,H^+ \rightarrow 2\,Cr^{3+} + 6\,Fe^{3+} + 7\,H_2O.$$

The iron(II) unused in this reaction is titrated back to the first pink coloration with $KMnO_4$ solution.

> **Procedure:** A portion of steel containing approximately 5–15 mg chromium is weighed precisely and dissolved in 60 ml mixed acid (H_2SO_4 3 mol/l, H_3PO_4 6 mol/l) with careful heating. Dilute to about 200 ml with water in the titration flask. After adding 2 ml silver nitrate solution (25 g/l), add 20 ml fresh $(NH_4)_2S_2O_8$ solution (150 g/l) and boil for about 10 min until the solution has turned yellow-orange or violet (formation of dichromate or permanganate). After boiling again for 10 min, allow to cool and add a further 10 ml of mixed acid. After adding V_1 = 20 ml $(NH_4)_2Fe(SO_4)_2$ solution (0.1 mol/l) titrate with $KMnO_4$ solution ($c(\frac{1}{5}\,KMnO_4)$ = 0.1 mol/l). The volume of standard solution used here is V_2. $V_1 - V_2$ correspond to the volume of the $(NH_4)_2Fe(SO_4)_2$ solution reacted with dichromate before the titration (0.1 mol/l):
> 1 ml of this volume corresponds to 1.7332 mg chromium.

The procedure can also be carried out with a direct titration so that instead of adding an excess of iron(II), titration is carried out directly with a standard solution containing iron(II). This is then a ferrometric titration (see Section 3.3.5). The equivalence point is indicated, for example, with ferroin as an indicator or with a potentiometric measuring arrangement (Chapter 4).

3.3.2.14 Permanganate index (PMI)

Determining the oxidizability of dissolved organic substances in water is one of the classic methods of water analysis and is an important criterion for assessing water and wastewater. This sum parameter is referred to as chemical oxygen demand (COD) and is specified as the volume-related mass of oxygen to which the reacted oxidizing agent corresponds stoichiometrically [121a]. Potassium dichromate and potassium permanganate are used as oxidizing agents. The COD determination according to the DIN method is based on potassium dichromate as the oxidizing agent (see Section 3.3.3).

If such a determination is carried out permanganometrically, this parameter is referred to as the permanganate index (PMI) and is given in mg $KMnO_4$ per liter [113b].

The method is based on a double back titration. A defined quantity of potassium permanganate is added in excess to a water sample to be analyzed. The solution is acidified and heated for at least 10 min. The oxidizable substances present in the water sample are oxidized by permanganate:

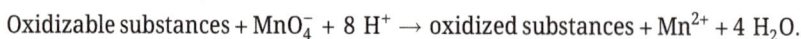

$$\text{Oxidizable substances} + MnO_4^- + 8\,H^+ \rightarrow \text{oxidized substances} + Mn^{2+} + 4\,H_2O.$$

The excess potassium permanganate is then mixed with a defined quantity of oxalate, also in excess. The remaining quantity of oxalate is determined by titration with a potassium permanganate solution.

The determination of the PMI is comparable to that of the COD, but has systematic errors when measuring special water samples such as wastewater. Therefore, the PMI is only used to evaluate drinking water and less-polluted natural waters. Otherwise, the dichromatometric method is preferred.

Procedure: About 50.0 ml of a water sample is placed in an Erlenmeyer flask using a volumetric pipette and acidified with 2.5 ml of 20 wt.% sulfuric acid. Add 5.0 ml of a $KMnO_4$ solution, $c(KMnO_4)$ = 0.005 mol/l, to the solution and heat for 10 min until boiling is just visible. For uniform heating, four to five boiling stones are added to the solution. Cover the flask with a watch glass. If the pink/violet color does not persist, the same amount of $KMnO_4$ must be added and heated again for 10 min.

Add 10.0 ml oxalate solution, $c(H_2C_2O_4)$ = 0.01 mol/l, to the still hot solution and titrate with $KMnO_4$ solution, $c(KMnO_4)$ = 0.005 mol/l, to a weak pink color.

Calculation: The amount of $KMnO_4$ used for oxidation can be calculated as follows:

$$n(KMnO_4)_{consumed} = n(KMnO_4)_{addition} - \tfrac{2}{5}\, n(H_2C_2O_4)_{addition} + n(KMnO_4)_{titration}.$$

$$PMI = n(KMnO_4)_{consumed} - M(KMnO_4)/V(sample) \text{ in mg } KMnO_4 \text{ per liter.}$$

Units of PMI values and COD values can be converted into each other: 1 mg $KMnO_4$ per liter (PMI unit) corresponds to 3.95 mg O_2 per liter (COD unit).

3.3.3 Dichromatometric determinations

Like permanganate, chromium of oxidation number VI is also able to oxidize a large number of reducing agents in acidic solution. This property has long been used in the cleaning of glass vessels with dichromate sulfuric acid (see p. 35 f.). The dichromate, which is stable in acidic solution, is reduced with the participation of hydrogen ions by taking up six electrons (three for each Cr(VI)), which are supplied by the respective reducing agent, according to

$$Cr_2O_7^{2-} + 14\,H^+ + 6\,e^- \rightarrow 2\,Cr^{3+} + 7\,H_2O.$$

Dichromatometry has a number of advantages over permanganometry. For example, by weighing potassium dichromate and dissolving it in a certain volume, a standard solution of the desired concentration is yielded without further titer adjustment (primary standard). The potassium dichromate solutions are titer-stable. Titrations with them in hydrochloric acid solutions do not cause any difficulties. Nevertheless, the importance of the method remained low for a long time and it was mainly used only for the determination of iron(II) salts.

The reason for this was the difficulty in recognizing the end point to be found. Green chromium(III) ions are formed from the orange dichromate ions during titration. For a long time, people had to rely on the spot reaction to determine the end point due to a lack of suitable indicators. It was necessary to remove a drop from the solution to be titrated (e.g. an iron(II) salt solution) from time to time toward the end of the determination and to remove it with the aid of a suitable standard solution, the spot indicator (here a $K_3[Fe(CN)_6]$ solution), to check whether it still contained the type of ion to be titrated (here Fe^{2+}).

Knop found a suitable indicator in 1915 in a colorless solution of 0.2 g diphenylamine in 100 ml nitrogen oxide-free concentrated sulfuric acid. If four drops of this solution are added to the sample solution, it turns deep violet with the slightest excess of dichromate over an intermediate green color. Just 0.1 ml of a potassium dichromate solution with a concentration of $\frac{1}{60}$ mol/l is sufficient to color 100 ml of water violet with four drops of indicator solution.

Diphenylamine is a typical redox indicator with reversible change, which does not depend on the character of the oxidizing or reducing agent, but on the position of the redox potentials of the system to be titrated and the indicator. Redox indicators in general are easily reversible, oxidizable or reducible organic dyes whose reduced form is usually colorless. Such compounds are suitable as indicators for determining the end point of a redox titration if the redox potential at the color change corresponds to the potential at the equivalence point of the titration or is close to it. If H^+ ions are also involved in the reaction underlying the color change, the changeover potential is pH-dependent. Oxidizing acids, such as nitric acid or nitrous acid, must not be present during the titration, as they would affect the indicator. In acidic solutions, the diphenylamine according to Kolthoff and Sarver (1930), diphenylamine is first irreversibly oxidized to diphenylbenzidine by strong oxidizing agents:

The diphenylbenzidine is then further reversibly oxidized to diphenylbenzidine violet:

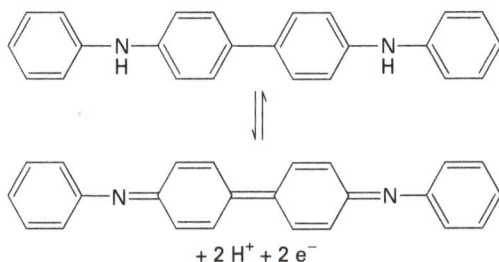

$$+ 2\,H^+ + 2\,e^-$$

The **reversal potential** of the indicator, where the oxidized and reduced forms are present in the same concentration, is +0.76 V.

The potentiometric indication of the end point, which has been used to an increasing extent in recent decades (see Section 4.4) for reactions that can be analyzed by measurement has given dichromatometry, like bromatometric methods, increased importance and wider distribution.

3.3.3.1 Preparation of the dichromate solution

The reaction equation on which dichromatometry is based shows that a potassium dichromate solution of the concentration $c(\frac{1}{6}\,K_2Cr_2O_7) = 0.1$ mol/l must contain $\frac{1}{60}$ mol/l, i.e. 4.9031 g $K_2Cr_2O_7$.

If the available potassium dichromate is not pure for analysis but contaminated by potassium sulfate, it can easily be made titer-pure by recrystallizing it three to four times from hot water. A glass frit nutsche is used for filtration. Drying takes place in a drying oven at 130 °C until constant weight is reached.

About 4.9031 g $K_2Cr_2O_7$ are weighed precisely and the standard solution is prepared by dissolving in a volumetric flask and filling to 1 l. It has a practically unlimited shelf life.

3.3.3.2 Determination of iron by spot reaction

Mineral acid iron(II) salt solutions are immediately oxidized quantitatively at room temperature after

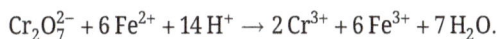

$$Cr_2O_7^{2-} + 6\,Fe^{2+} + 14\,H^+ \rightarrow 2\,Cr^{3+} + 6\,Fe^{3+} + 7\,H_2O.$$

The titration is completed when a drop taken from the solution no longer responds to a detection reaction specific for iron(II) ions. The reaction with $K_3[Fe(CN)_6]$, which indicates the presence of Fe^{2+} ions extremely sensitively by forming the intensely colored Turnbull blue, serves as such a reaction:

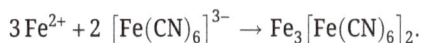

$$3\,Fe^{2+} + 2\,\left[Fe(CN)_6\right]^{3-} \rightarrow Fe_3\left[Fe(CN)_6\right]_2.$$

Fe^{3+} ions, on the other hand, form a brownish soluble compound with $K_3[Fe(CN)_6]$ of the formula $Fe[Fe(CN)_6]$.

The $K_3[Fe(CN)_6]$ used must be absolutely free of $K_4[Fe(CN)_6]$ because this is converted with the Fe^{3+} ions formed during the titration approximately according to

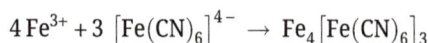

$$4\,Fe^{3+} + 3\,\left[Fe(CN)_6\right]^{4-} \rightarrow Fe_4\left[Fe(CN)_6\right]_3$$

the Berlin blue would form. This would always result in a blue coloration. Use only the purest $K_3[Fe(CN)_6]$, rinse it thoroughly with water before dissolving and check with a sample of Fe(III) salt.

Procedure: The acidic iron(II) salt solution, which should contain about 0.1 g iron in 100 ml, is titrated with $K_2Cr_2O_7$ solution (⅙₀ mol/l). A drop is taken from the sample solution several times with a glass rod shortly before the end point. The drop is placed on a white porcelain plate next to a drop of an approximately 2% $K_3[Fe(CN)_6]$ solution. If no more blue or green coloration occurs when the two drops are combined with a clean glass rod, the titration is complete. The first determination should only be regarded as a preliminary sample. In the second titration, add almost all of the $K_2Cr_2O_7$ solution required to reach the end point and try to reach the end point by spotting only two or three times:

1 ml $K_2Cr_2O_7$ solution, $c(⅙\ K_2Cr_2O_7)$ = 0.1 mol/l, corresponds to 5.585 mg Fe.

It seems obvious that spotting increases the inaccuracy of the determination. However, a rough calculation shows how large the error is that is made by taking several drops toward the end of the titration in relation to the overall result: 10 drops of 0.05 ml each, i.e. 0.5 ml in total, may have been taken. With a liquid volume of 250–300 ml, this would only result in a proportion of about 0.2% of the total iron present that is not covered by the titration. However, the error does not relate to the total iron, but only to the iron still present as Fe(II) toward the end of the titration, e.g. the last tenth. The proportion would then only be around 0.02%, i.e. hardly noticeable. In addition, it is constantly reduced as the oxidation of the Fe(II) progresses. The main titration does not require 10 drops, but only three to four, which means that the proportion caused by taking the spot samples can be kept below 0.005%.

3.3.3.3 Determination of iron with redox indicators

Indication with diphenylamine: A green to blue-green coloration already occurs during the titration, which is due to the fact that the changeover potential of the indicator of +0.76 V has approximately the same value as the standard potential of the redox system Fe^{2+}/Fe^{3+}, i.e. the color change already occurs before the start of the potential jump at the equivalence point. In order to bring the indicator change and the equivalence point of the titration into agreement, the redox potential of the Fe^{2+}/Fe^{3+} system must be lowered. This is achieved by adding phosphoric acid, which reduces the Fe(III) concentration sufficiently through complex formation (see p. 166 f.).

Procedure: To 25 ml of the iron(II) salt solution, add 10 ml of sulfuric acid (2 mol/l) or hydrochloric acid (4 mol/l) and four drops of indicator solution (see p. 176 f.) (in hydrochloric acid solution only toward the end of the titration). Then titrate with the $K_2Cr_2O_7$ solution. The green color of the solution deepens near the end point. When it has turned dark green – approximately at $\tau = 0.985$ – 5 ml of 25% phosphoric acid is added. The solution is carefully titrated further until its color suddenly turns blue-violet.

Indication with diphenylamine-4-sulfonic acid sodium salt: In the presence of tungstic acid, titration of iron with diphenylamine as a redox indicator is not possible. Knop and Kubelková-Knopová (1941) introduced N-methyldiphenylamine-p-sulfonic acid or its sodium salt as a redox indicator instead of diphenylamine. In addition to their insensitivity to the presence of tungstic acid, these substances are also characterized by better solubility in water and a sharper color change. Instead of these compounds, which are no longer commercially available, it is now possible to work with diphenylamine-4-sulfonic acid sodium salt, or alternatively with the barium salt. The color changes from colorless to red-violet. The reaction mechanism is the same as for diphenylamine. The indicator is used in the form of the sodium salt in 0.1% aqueous solution.

Procedure: Add 10–15 ml of an acid mixture (150 ml H_3PO_4, $\rho = 1.7$ g/ml, and 150 ml H_2SO_4, $\rho = 1.84$ g/ml, in 700 ml water) to 100–200 ml of the iron(II) salt solution, which should contain approximately 0.05–0.2 g iron, and 10–20 ml sulfuric acid (1:4) or hydrochloric acid (1:1) and 0.1–0.2 ml of 0.1% indicator solution and titrated with the $K_2Cr_2O_7$ solution. The end of the titration is indicated by a gray-violet discoloration. At the equivalence point, the color is a rich gray-red-violet. Due to the influence of H^+ ions on the color change, the H^+ concentration of the solution should not be less than 0.5 mol/l and not greater than 2 mol/l:

If iron(III) salts are present, reduce with $SnCl_2$ (see p. 166 f.) and titrate as indicated.

3.3.4 Cerimetric determinations

Cerium(IV) sulfate is a strong oxidizing agent. The redox potential in sulfuric acid, 0.5–4 mol/l, is 1.44 V. It can only be used in an acidic solution. When the solution is neutralized, cerium(IV) hydroxide or basic salts precipitate. The solution is intensely yellow. In hot, not too dilute solutions, the end point can be recognized without an indicator, but a blank titration is then necessary.

Cerium(IV) sulfate as an oxidizing agent in volumetric analysis has the following advantages: The solutions are stable over a long period of time. They do not need to be protected from light and do not change in concentration when heated to boiling for a short time. Stable solutions are obtained by adding 10–40 ml of concentrated sulfuric acid per liter. They are therefore superior to permanganate solutions in terms of stability. Cerium(IV) sulfate can be used to determine reducing agents in the presence of high HCl concentrations. Ce^{4+} solutions with a concentration of 0.1 mol/l are not so strongly colored that the reading of the meniscus in burettes is disturbed.

Cerimetry is based on the redox equilibrium

$$Ce^{4+} + e^- \rightleftharpoons Ce^{3+}.$$

Unlike permanganate, different reaction products are not formed depending on the reaction conditions. The cerium(III) ion is colorless. Most of the titrations for which permanganate is used can be carried out with cerium(IV) sulfate as well as a few others. As redox indicators for cerimetric titrations, N-phenylanthranilic acid, ferroin and 5,6-dimethylferroin can be used (see [114, 115]).

Ferroin indicator: The indicator ferroin is a chelate complex of iron(II) ions and phenanthroline. It is produced by dissolving iron(II) sulfate and phenanthroline HCl in water.

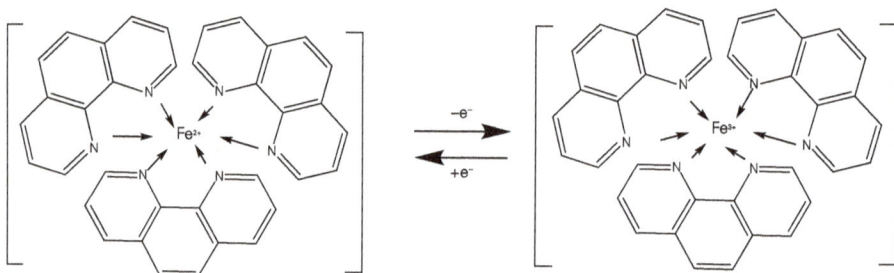

The excess of the cerium(IV) solution oxidizes the iron(II) ion (in ferroin) to the iron(III) ion (in ferriin) and the color changes from red to blue. The normal potential of the complex is given as 1.14 V, the transition range as 1.08–1.2 V.

An alternative to ferroin is diphenylamine, which changes from colorless to blue. The diphenylamine is presumably first oxidized in acidic solution to diphenylbenzidine, which is then reversibly oxidized further to the blue dye (see Section 3.3.3).

Ph. Eur. uses this redox indicator in the cerimetric determination of tocopherol (vitamin E) [8a].

3.3.4.1 Preparation of the cerium(IV) sulfate solution

Forty-two grams of cerium(IV) sulfate tetrahydrate, $Ce(SO_4)_2 \cdot 4\,H_2O$, or 65 g ammonium cerium(IV) sulfate dihydrate, $2\,(NH_4)_2SO_4 \cdot Ce(SO_4)_2 \cdot 2\,H_2O$, is dissolved in 500 ml sulfuric acid (1 mol/l) with moderate heating on a water bath and made up to 1,000 ml. Suitable reagents for adjustment are, for example, arsenic(III) oxide, oxalic acid, metallic iron or potassium iodide. Of course, titration can also be carried out with suitable standard solutions, such as iron(II) sulfate solution.

Adjustment with potassium iodide: In a sulfuric acid solution containing acetone, iodide ions are oxidized by cerium(IV) ions to form iodoacetone:

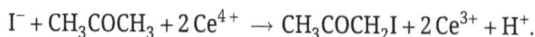

$$I^- + CH_3COCH_3 + 2\,Ce^{4+} \rightarrow CH_3COCH_2I + 2\,Ce^{3+} + H^+.$$

Since some acetone and iodoacetone are also oxidized by cerium(IV) ions near the equivalence point, the titer found when setting a standard solution (0.1 mol/l) is slightly too low, by approximately 0.05%. With a concentration of the cerium(IV) sulfate solution of 0.01 mol/l, on the other hand, the error is 0.8–1%.

Procedure: Approximately 0.35 g of potassium iodide dried at 250 °C is dissolved in 100 ml sulfuric acid (1.5 mol/l). After adding three drops of ferroin solution, 0.025 mol/l (commercially available solution), and 25 ml acetone, shake well and titrate from red to colorless with cerium(IV) sulfate solution. The transition point is reached when the red coloration disappears for at least 30 s after adding a drop of the standard solution.

3.3.4.2 Determination of iron

Iron(II) is smoothly oxidized to iron(III) in a hydrochloric or sulfuric acid solution:

$$Fe^{2+} + Ce^{4+} \rightarrow Fe^{3+} + Ce^{3+}.$$

If iron(III) is present in whole or in part, it must be reduced before the determination. The reduction can be carried out, for example, with silver in the reducer burette described on p. 163.

Procedure: Acidify 100 ml of the sample solution, which should contain approximately 0.5–1 g/l Fe, with 10 ml conc. hydrochloric acid. After adding one drop of ferroin solution (0.025 mol/l) titrate with cerium(IV) sulfate solution (0.1 mol/l), until the color changes sharply from orange-red to yellow-green. If 5 ml conc. sulfuric acid is used instead of hydrochloric acid, the color changes from red to pale blue:
  1 ml cerium(IV) sulfate solution, $c(Ce(SO_4)_2) = 0.1$ mol/l, corresponds to 5.5845 mg Fe.

3.3.4.3 Determination of nitrite

In acidic solution, nitrite is oxidized to nitrate by cerium(IV):

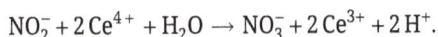

$$NO_2^- + 2\,Ce^{4+} + H_2O \rightarrow NO_3^- + 2\,Ce^{3+} + 2\,H^+.$$

The reaction is quantitative and initially fast, but slower near the equivalence point. During titration, a measured volume of the cerium(IV) sulfate solution is added and the sample solution from the burette is added.

Procedure: Twenty-five milliliters of cerium(IV) sulfate solution (0.1 mol/l) is mixed with 5 ml conc. nitric acid in a 250-ml beaker (wide form), diluted to 100–150 ml and heated to 50 °C. The sample solution should contain 0.05–0.1 mol/l. Titrate with the sample solution, which should contain 0.05–0.1 mol/l NO_2^-, until the yellow coloration of the solution provided has almost disappeared. The tip of the burette must be just immersed in the solution. Then add a drop of ferroin solution (0.025 mol/l) and titrate slowly until the color changes from blue-gray to light pink:

1 ml cerium(IV) sulfate solution, $c(Ce(SO_4)_2) = 0.1$ mol/l, corresponds to 2.3003 mg NO_2^-.

3.3.4.4 Determination of hexacyanoferrate(II)

Cerium(IV) oxidizes hexacyanoferrate(II) to hexacyanoferrate(III) in acidic solution:

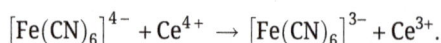

$$[Fe(CN)_6]^{4-} + Ce^{4+} \rightarrow [Fe(CN)_6]^{3-} + Ce^{3+}.$$

Procedure: Add 10 ml conc. hydrochloric acid (7 mol/l) and 1 drop of ferroin solution (0.025 mol/l) to 100 ml sample solution, which should contain approximately 2 g/l hexacyanoferrate(II). Titrate at room temperature with cerium(IV) sulfate solution (0.1 mol/l) until the color changes sharply from orange-brown to yellow-green:

1 ml cerium (IV) sulfate solution, $c(Ce(SO_4)_2) = 0.1$ mol/l, corresponds to 21.195 mg $Fe[(CN)_6]^{4-}$ or 36.835 mg $K_4[Fe(CN)_6]$.

3.3.4.5 Determination of nifedipine

Nifedipine contains a dihydropyridine ring system that can be oxidized to a pyridine ring system using cerium(IV):

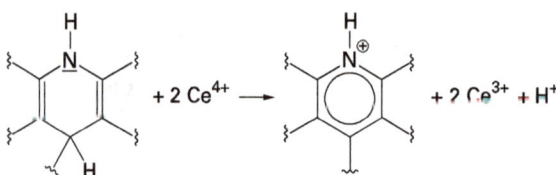

As nifedipine is practically insoluble in water, it is dissolved in *tert*-butanol as a non-reducing alcohol. Cerium(IV) sulfate is used as the titer substance and ferroin as the indicator. As the titration is carried out in acidic conditions, no cerium salts can precipitate. Because the reaction does not proceed very quickly, titration must be slow enough, especially near the equivalence point. A blank titration is recommended.

Procedure: Accurately weigh about 0.13 g of the sample substance and dissolve in a mixture of 25 ml *tert*-butanol and 25 ml perchloric acid (0.6 mol/l). After adding a little ferroin solution (¼₀ mol/l) titrate with $Ce(SO_4)_2$ solution (0.1 mol/l) until the pink coloration disappears. Titrate slowly enough near the equivalence point. A blank titration should be carried out:
   1 ml $Ce(SO_4)_2$ solution, $c = 0.1$ mol/l, corresponds to 17.32 mg $C_{17}H_{18}N_2O_6$ [8a].

Procedure: The purity test of nifedipine with regard to basic impurities is carried out by anhydrous acid-base titration. Approximately 4 g of the sample substance is accurately weighed and dissolved in 160 ml anhydrous acetic acid on the ultrasonic bath. The titration is carried out with perchloric acid

(0.1 mol/l) as the standard solution and some naphthol solution as an indicator. The equivalence point is indicated by a change from brownish yellow to green. A maximum of 0.48 ml perchloric acid (0.1 mol/l) may be used [8a].

3.3.4.6 Determination of paracetamol

The pharmaceutical content of paracetamol is determined cerimetrically after acid-catalyzed saponification of the compound at the amide bridge. This produces *p*-aminophenol (*p*-hydroxyaniline) and acetic acid:

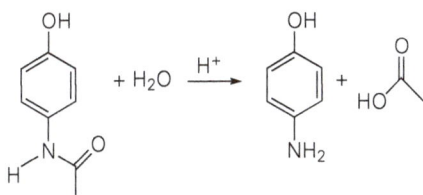

p-Aminophenol is oxidized by cerium(IV) ions to *p*-quinoneimine. The stoichiometry applicable here is characterized by a ratio of two cerium(IV) ions to one *p*-aminophenol molecule, which in turn corresponds to one molecule of paracetamol [28]:

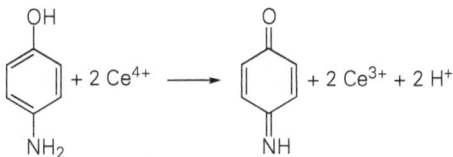

Procedure: About 0.3 g of the sample substance is accurately weighed and dissolved in a mixture of 10 ml water and 30 ml dilute sulfuric acid (1.0 mol/l). The saponification now takes place under reflux and should last 1 h. Allow to cool, transfer without loss to a 100-ml volumetric flask and make up to the mark with water. Place 20 ml of this solution in the titration flask as an aliquot. Now add 40 ml water, 40 g ice and 15 ml diluted hydrochloric acid (2.0 mol/l). After adding 0.1 ml ferroin solution (1/40 mol/l), titrate with $Ce(SO_4)_2$ solution (0.1 mol/l) until a greenish-yellow color is obtained (disappearance of the pink color). A blank titration should be carried out:

   1 ml $Ce(SO_4)_2$ solution, $c = 0.1$ mol/l, corresponds to 7.558 mg $C_8H_9NO_2$ [8a].

3.3.4.7 Determination of vitamin K_3 (menadione) according to Ph. Eur

Quinones can be titrated cerimetrically after prior reduction with zinc and hydrochloric acid to hydroquinone. The determination of menadione (vitamin K_3) according to Ph. Eur. is listed here as an example [8a]:

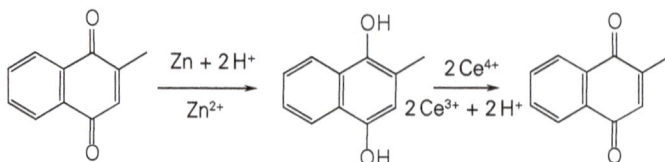

Procedure: About 0.150 g substance is dissolved in 15 ml acetic acid 99% in a flask fitted with a Bunsen valve. After adding 15 ml of dilute hydrochloric acid and 1 g of zinc dust, the flask is sealed and left to stand for 60 min under light protection, shaking occasionally. The solution is filtered through absorbent cotton and the absorbent cotton is washed thrice with 10 ml CO_2-free water each time. After adding 0.1 ml ferroin solution, the total filtrate is immediately titrated with ammonium cerium (IV) nitrate (0.1 mol/l):

  1 ml ammonium cerium (IV) nitrate, $c = 0.1$ mol/l, corresponds to 8.61 mg $C_{11}H_8O_2$.

3.3.5 Ferrometric determinations

Just as iron(II) can be titrated with potassium dichromate, chromium(VI) can also be reduced quantitatively to chromium(III) by means of iron (II). A series of further reductometric titrations can be carried out with an Fe(II) solution. However, the low titer stability, which is due to the easy oxidizability of the iron(II) ions by atmospheric oxygen, is a disadvantage. Schäfer (1949) achieved the constancy of the titer of an $FeSO_4$ solution by a simple measure. He worked with a reductor burette consisting of a burette filled with the standard solution and an upstream Jones reductor (see p. 163 f.). The iron(II) sulfate solution, oxidized to a small extent by the atmospheric oxygen, runs over the reductor metal (in this case silver) before dripping into the solution to be titrated, whereby the Fe^{3+} ions present are reduced back to Fe^{2+} ions.

The basis of ferrometry is the equilibrium:

$$Fe^{2+} \rightleftharpoons Fe^{3+} + e^-.$$

Redox indicators are also suitable for ferrometric titrations.

Because the transition from an oxidizing environment to a reducing environment takes place at the equivalence point in ferrometric titrations, it can be considered advantageous if the redox indicator shows a color depression. This is not the case with diphenylamine and its derivatives, with ferroin (tris(1,10-phenanthroline)iron(II) sulfate), however, as the color change is from a rather weak blue to deep red.

3.3.5.1 Preparation of the iron(II) sulfate solution

A iron(II) sulfate solution with a concentration of 0.1 mol/l contains 27.801 g $FeSO_4 \cdot 7\,H_2O$ or 39.214 g $(NH_4)_2Fe(SO_4)_2 \cdot 6\,H_2O$ (Mohr's salt). Only Mohr's salt is suitable for preparing. However, $FeSO_4$ has the advantage that it is oxidized less quickly in hydrochloric acid or sulfuric acid solutions.

Approximately 28 g $FeSO_4 \cdot 7\,H_2O$ is dissolved in a mixture of 100 ml water, 25 ml concentrated sulfuric acid and 50 ml hydrochloric acid, and the solution is made up to 1 l with water in the volumetric flask. The addition of hydrochloric acid is necessary for use in the reducer burette with silver filling.

As reductor silver is used, which has been prepared by precipitation with metallic copper from a silver nitrate solution and washed several times with sulfuric acid and water. A few milliliters of standard solution are poured into the reducer chamber of the burette, which is sealed against the tap with a plug of quartz or glass wool, and 5–6 g of silver are added. After rinsing the approximately 8 cm high bubble-free layer with the standard solution, the burette is ready for use. The reducer layer does not interfere with the outflow. The iron(II) salt solution must be acidic. During the reduction of the Fe^{3+} ions, silver chloride, which is difficult to dissolve, is formed from the metallic silver in the presence of Cl^- ions.

For regeneration, the used reducer silver is shaken with zinc rods in diluted sulfuric acid. Any precipitated basic zinc salts are removed by treatment with hydrochloric acid (6 mol/l) under heat.

Various types of reducer burettes are described by Flaschka (1950), Weber and Hahn (1952) as well as by Karsten, Kies and Bergshoeff (1952).

The iron(II) sulfate solution is adjusted with potassium dichromate using a drop of ferroin solution (0.025 mol/l) as a redox indicator, with a color change from green to red.

3.3.5.2 Determination of chromate(VI) and chromium(III)

The reaction between the $Cr_2O_7^{2-}$ ions present in acidic solution and the Fe^{2+} ions proceeds according to the equation given on p. 177. Chromium(III) salts are oxidized to chromium(VI) compounds in the presence of Ag^+ before titration with ammonium peroxodisulfate. Ferroin has also proven itself as a redox indicator here.

Procedure (based on Blasius and Wittwer [116]): The determination is carried out as described for the iron(II) titration (see p. 178), except that a drop of ferroin solution (0.025 mol/l) is added as an indicator and titrated to the first red tint.

Add 20 ml of a silver nitrate solution (0.02 mol/l) and 15 ml of 15% ammonium peroxodisulfate solution to sulfuric acid chromium(III) salt solutions with a content of about 50–55 mg and a maximum volume of 100 ml at room temperature and heat until the dichromate color appears (do not heat to boiling!). The excess $S_2O_8^{2-}$ is then destroyed by boiling for about 15 min. After cooling, 5 ml of 5% NaCl solution is added. The solution is made up to 150 ml and titrated as described after adding phosphoric acid mixture:

1 ml iron(II) sulfate solution, $c(FeSO_4) = 0.1$ mol/l, corresponds to 1.733 mg Cr, 2.533 mg Cr_2O_3 or 4.903 mg $K_2Cr_2O_7$.

3.3.5.3 Determination of chromium in steel

Chromium can be determined in steel samples by dissolving the sample substance in mixed acid and oxidizing the resulting chromium(III) with peroxodisulfate to the dichromate and titrating it with an iron(II)-containing standard solution after boiling off the peroxo oxygen. Further information on sample preparation can be found in Section 3.3.2.

> **Procedure:** Add about 1 g of steel sample to a 150-ml beaker (corresponding to about 10 mg of chromium) and dissolve in about 60 ml of mixed acid, stirring and boiling if necessary. After cooling to room temperature, transfer to a 400-ml beaker with 20 ml ultrapure water! Then add 1 ml silver nitrate solution (2.5%) and 20 ml ammonium peroxodisulfate solution (15%). Boil for 10 min until the solution turns yellow-orange or – in the presence of manganese – violet! After adding 5 ml of sodium chloride solution (5%), boil again for 10 min – a precipitate of silver chloride precipitates – and then allow to cool. After adding 20 ml of mixed acid (see below) and three drops of the ferroin indicator (see below), wait for 10 min. Titrate directly – slowly enough – with an iron(II) sulfate solution ($c(FeSO_4) = 0.1$ mol/l). The color changes from orange to red via a gradual green color and a brief blue color at the equivalence point:
>   1 ml iron(II) sulfate solution, $c(FeSO_4) = 0.1$ mol/l, corresponds to 1.733 mg Cr, 2.533 mg Cr_2O_3 or 4.903 mg $K_2Cr_2O_7$.
>   Mixed acid: 150 ml 18 mol/l H_2SO_4 and 150 ml 15 mol/l H_3PO_4 are carefully added to 700 ml H_2O and mixed.
>   Ferroin indicator: 0.7 g iron(II) sulfate heptahydrate and 1.76 g 1,10-phenanthroline hydrochloride are dissolved in 70 ml demineralized water. The solution is diluted with water to 100 ml.

3.3.5.4 Determination of vanadium

In acidic solution, vanadium(V) is reduced from iron(II) to vanadium(IV):

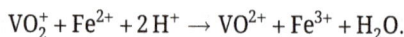

$$VO_2^+ + Fe^{2+} + 2\,H^+ \rightarrow VO^{2+} + Fe^{3+} + H_2O.$$

Instead of vanadium(V) oxide ions, the solution may also contain polyvanadate(V) ions, as the equilibrium between the various condensation forms of vanadate(V) ions depends on the hydrogen ion concentration. However, this is irrelevant for the reduction process. When using ferroin, vanadates(V) can be titrated with iron(II) sulfate in the same way as dichromates.

> **Procedure:** To the solution containing 0.05–0.1 g vanadium, add 10 ml phosphoric acid mixture (see p. 173) and enough diluted sulfuric acid so that the acid concentration is between 0.25 and 1 mol/l after dilution to 100 ml. After adding a drop of indicator solution, titrate with iron(II) sulfate solution (0.1 mol/l) until the first red tint appears:
>   1 ml iron(II) sulfate solution, $c(FeSO_4) = 0.1$ mol/l, corresponds to 8.294 mgVO_2^+ or 9.094 mg V_2O_5.

3.3.6 Bromatometric determinations

Potassium bromate is an oxidizing agent and can be used for the titrimetric determination of arsenic(III), antimony(III), tin(II), copper(I), thallium(I) and hydrazine. The bromate is reduced to bromide in an acidic solution:

$$BrO_3^- + 6\,H^+ + 6\,e^- \rightarrow Br^- + 3\,H_2O.$$

The standard potential of the system BrO_3^-/Br^- is $E° = 1.44$ V. With $z^* = 6$, the molar mass of the potassium bromate in terms of equivalents is M (⅙ $KBrO_3$) = 27.833 g/mol.

To recognize the titration end point is based on the fact that the bromide formed during the titration (or intentionally added beforehand) comproportions with excess bromate to form elemental bromine:

$$BrO_3^- + 5\,Br^- + 6\,H^+ \rightarrow 3\,H_2O + 3\,Br_2.$$

Once the end point has been exceeded, the previously colorless solution turns slightly yellow due to bromine precipitation. However, as the coloration is difficult to detect due to the low bromine concentration, the oxidation effect of bromine on organic dyes such as methyl orange or methyl red is utilized. The dyes are irreversibly oxidized (decolorized) by bromine. The reaction requires a certain amount of time. Therefore, the bromate solution must be added slowly and drop by drop toward the end of the titration. Shortly before reaching the end point, another drop of indicator solution is added to the solution because a local excess of bromate at the drop-in point sometimes causes the indicator to fade prematurely. The temperature should be 40–60 °C.

Quinoline yellow is also well suited for detecting the end point. Its decolorization can even be reversed by adding a small excess of bromate and a few drops of arsenic(III) solution.

3.3.6.1 Preparation of the potassium bromate solution

Potassium bromate is available in a very pure form (primary standard). It can also be produced in sufficient purity by repeated recrystallization and drying at 180 °C. To prepare the standard solution, c(⅙ $KBrO_3$) = 0.1 mol/l, dissolve ⅟₆₀ of the molar mass, 2.7833 g $KBrO_3$, in water and make up to 1 l in a volumetric flask. The solution has a good shelf life and is titer-stable. However, it must not contain any bromide ions. To test for purity, acidify 5 ml of the bromate solution with 2 ml of sulfuric acid, $c(H_2SO_4) = 2$ mol/l, and add one drop of 0.1% methyl orange solution. According to Kolthoff [89], the pink color must not have disappeared after 2 min.

3.3.6.2 Determination of arsenic and antimony

The titration described as early as 1893 by St. Györy is based on the oxidation of the ions from oxidation number III–V:

$$BrO_3^- + 3\,As^{3+} + 6\,H^+ \rightarrow Br^- + 3\,As^{5+} + 3\,H_2O,$$
$$BrO_3^- + 3\,Sb^{3+} + 6\,H^+ \rightarrow Br^- + 3\,Sb^{5+} + 3\,H_2O.$$

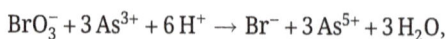

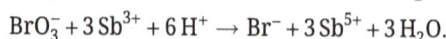

Procedure: The solution of arsenate(III) or antimonate(III) is strongly acidified with hydrochloric acid so that the acid concentration is 1–2 mol/l.

After adding one to two drops of methyl red or methyl orange solution (1%), heat to approximately 50 °C and titrate with $KBrO_3$ solution ($\frac{1}{60}$ mol/l) until decoloration occurs. The titration beaker must be constantly inverted. The bromate solution must be allowed to drip in slowly toward the end of the titration. Shortly before the end point, add one drop of indicator solution:

1 ml $KBrO_3$ solution, $c(\frac{1}{6}\,KBrO_3) = 0.1$ mol/l, corresponds to 3.7461 mg As or 6.0880 mg Sb or 4.9460 mg As_2O_3 or 7.288 mg Sb_2O_3.

3.3.6.3 Determination of bismuth

The process described by Reißaus in 1927 is based on the reduction of bismuth(III) in hydrochloric acid solution by metallic copper producing an equivalent amount of copper(I), which is titrated bromatometrically:

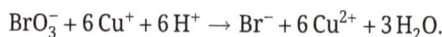

$$Bi^{3+} + 3\,Cu \rightarrow Bi + 3\,Cu^+,$$

$$BrO_3^- + 6\,Cu^+ + 6\,H^+ \rightarrow Br^- + 6\,Cu^{2+} + 3\,H_2O.$$

The difficulty lies in the extraordinarily high sensitivity of copper(I) ions to atmospheric oxygen.

Procedure: The bismuth is first separated as bismuth oxychloride in order to separate it from accompanying metals such as copper, arsenic and antimony. The salt is filtered off and dissolved in 30 ml of concentrated hydrochloric acid in a 500-ml Erlenmeyer flask. Dilute to 200 ml with water, heat to boiling while passing CO_2 through the flask and add bare electrolytic copper filings after the air has been completely displaced. If, after boiling for 15 min, a freshly added copper chip remains bright, the solution is filtered through a glass filter crucible into a liter flask filled with carbon dioxide and rinsed with hot water containing hydrochloric acid in the absence of air in a CO_2 atmosphere. The solution (400–500 ml) is mixed with two to three drops of 1% methyl orange solution and titrated hot with $KBrO_3$ solution. The color changes from pink to blue at the end point:

1 ml $KBrO_3$ solution, $c(\frac{1}{6}\,KBrO_3) = 0.1$ mol/l, corresponds to 6.9660 mg Bi or 6.3546 mg Cu.

3.3.6.4 Determination of hydroxylamine

Hydroxylamine is oxidized to nitrate by bromate in the presence of hydrochloric acid:

$$NH_2OH + BrO_3^- \rightarrow NO_3^- + Br^- + H^+ + H_2O.$$

For the determination, add a measured volume of $KBrO_3$ solution ($\frac{1}{60}$ mol/l) to the hydroxylamine solution so that there is an excess of 10–30 ml and add 40 ml HCl (5 mol/l). After 15 min, determine the excess bromate by adding KI solution and titrating the resulting iodine with thiosulfate solution (see Section 3.3.7):

1 ml KBrO$_3$ solution, c(⅙ KBrO$_3$) = 0.1 mol/l, corresponds to 0.5505 mg NH$_2$OH.

The method just described and the following methods combine the bromatometric conversion of the analyte with one of the variants of iodometry (see Section 3.3.7).

3.3.6.5 Determination of metal ions as oxinato complexes

Numerous metal ions can be determined indirectly by bromatometry via their hydroxyquinoline complexes. The method is based on the precipitation of the metal complexes under suitable pH conditions, release of the 8-hydroxyquinoline (oxine) by dissolving in dilute hydrochloric acid and bromination of the oxine to 5,7-dibromo-8-hydroxyquinoline:

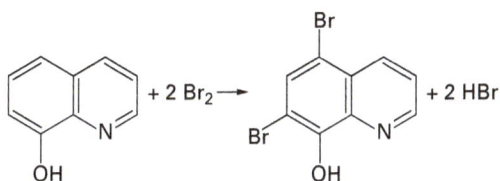

The bromine is produced by adding KBrO$_3$ solution to excess bromide. The substitution reaction proceeds slowly. Therefore, an excess of bromine is used. The unreacted residue is reduced with KI and the resulting iodine is titrated against starch as an indicator with a thiosulfate solution (see Section 3.3.7).

Divalent metal ions bind two, trivalent three oxine ligands.

Calculation: 1 mol oxine corresponds to 2 mol Br$_2$ or 4 mol Br. One ion of a divalent metal therefore corresponds to 4 mol Br$_2$ (8 mol Br) of a trivalent metal to 6 mol Br$_2$ (12 mol Br). The reaction of 1 mol KBrO$_3$ and 5 mol KBr produces 3 mol Br$_2$ (6 mol Br).

The determination of aluminum is described as an example.

3.3.6.6 Determination of aluminum

The standard solution is prepared by dissolving 8-hydroxyquinoline in acetic acid (c = 2 mol/l) to a 2% solution, adding ammonia solution until a small precipitate appears and heating until redissolution.

Procedure: A volume of 25 ml of sample solution containing approximately 0.02 g of aluminum is mixed with 125 ml of water in an Erlenmeyer flask and heated to 50–60 °C. A 20% excess of oxine solution is then added. Then add a 20% excess of oxine solution (1 ml precipitates 0.001 g Al). The precipitation is completed by adding a solution of 4 g ammonium acetate in as little water as possible. Allow the solution to cool while stirring, then filter through a glass or porcelain filter crucible (porosity 4) and rinse with warm water.

The complex is dissolved in warm concentrated hydrochloric acid and mixed with a few drops of methyl red or methyl orange solution (1%) and 0.5–1 g KBr. Then titrate slowly with $KBrO_3$ solution ($\frac{1}{60}$ mol/l) until the color turns pure yellow. The exact end point is not easy to recognize. It is therefore best to add an excess of $KBrO_3$ solution (approx. 2 ml) so that the solution contains free bromine. To prevent precipitation of 5,7-dibromoxine during the titration, dilute the solution strongly with hydrochloric acid (2 mol/l), add 10 ml of a 10% KI solution after 5 min and titrate the released iodine with sodium thiosulfate solution (0.1 mol/l) using starch as an indicator (see Section 3.3.7):

1 ml $KBrO_3$ solution, $c(\frac{1}{6} KBrO_3) = 0.1$ mol/l, corresponds to 0.22485 mg Al.

Remarks: Washing the precipitate with warm water serves to remove the excess oxine. In this way, complications due to adsorption of iodine can be avoided. The brown additional compound of iodine with the dibromo derivative of oxine can precipitate during the titration. However, it generally dissolves again during the subsequent titration with thiosulfate, producing a yellow color so that the end point can be recognized with starch. Adding 10 ml of carbon disulfide to the solution before adding the KI prevents the dark brown compound from dissolving quickly enough (carbon disulfide is very flammable and a strong lung and skin poison).

The method is suitable for determining the following metals: Al, Fe, Cu, Zn, Cd, Ni, Co, Mn and Mg.

3.3.6.7 Determination of 8-hydroxyquinoline (oxine)

The bromination of oxine described above can, of course, be used not only for metal determination but also – using a simpler method – for the determination of the oxine itself, namely for the analysis of products containing oxine as the main component, e.g. quinosol.

Procedure: Just enough sample substance is weighed out precisely so that – based on an estimated value – approximately 0.16 g oxine can be calculated. Then dissolve in 50 ml hydrochloric acid (3.0 mol/l) and 25 ml water. After adding 1.0 g potassium bromide and one to two drops of methyl red solution 0.2%, potassium bromate solution [$c(\frac{1}{6} KBrO_3) = 0.1$ mol/l] is added until the solution is light yellow, and then a further 1–2 ml is added. The volume used is V_1.

The irreversible destruction of the methyl red stained red in the acidic environment up to the time of the light yellow coloration indicates the occurrence of excess elemental bromine. The systematic error (overfinding) of the final result caused by the indicator is estimated to be 0.1% of the analyte mass.

To produce elemental iodine, 1 g of potassium iodide is added to 5 ml of water. Titration is carried out with sodium thiosulfate 0.1 mol/l as the standard solution and – after brightening the mixture – with starch solution as the indicator. The equivalence point is indicated by the disappearance of the blue coloration. The volume used is V_2:

$V_1 - V_2$ correspond to the volume of the $KBrO_3$ solution consumed during bromination [$c(\frac{1}{6} KBrO_3) = 0.1$ mol/l];

1 ml of this volume corresponds to 3.629 mg oxine [28].

3.3.6.8 Bromatometric determination of phenols and anilines

Anilines and phenols can be determined by bromatometric titration. This method goes back to Koppeschaar, W. F. and has been used in pharmacopoeias (e.g. DAB 7) for over 50 years. Elemental bromine can be used to quantitatively brominate unsubstituted phenols and anilines in the *o*- and *p*-positions (positions 2, 4 or 6):

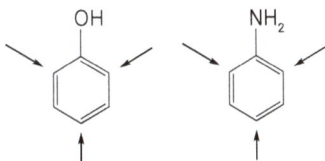

This can be used for the quantification of both structural elements. In some cases, phenols and anilines can be titrated directly with a potassium bromate solution in the presence of bromide. Ethoxychrysoidine or methyl red can be used as indicators. More common is the evaluation via an iodometric back titration, as in some cases further bromination occurs with elemental bromine, which then reacts back with excess potassium iodide.

3.3.6.9 Determination of salicylic acid

Salicylic acid can also be determined in the same way as oxine, as this is also a phenolic compound. However, waiting times are required due to the slower reaction rate. In addition, the stoichiometry is different, as not only two hydrogen atoms but also the carboxylic acid group are substituted by bromine atoms [28]:

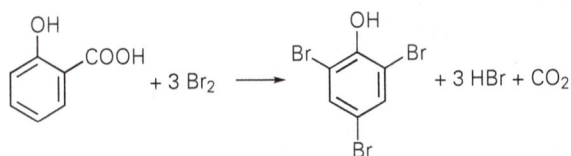

Acetylsalicylic acid can only be determined in a similar way after saponification [28]. Finally, the mesomeric effect of the RCOO group on the aromatic ring is considerably reduced by the fact that a mesomeric effect on the carbonyl group is to be expected. Thus, direct bromination of acetylsalicylic acid would not direct it to the ortho positions and to the para position in a sufficiently selective manner, as the phenolic OH group of salicylic acid does.

Procedure: Just enough of the sample substance is weighed out precisely so that – based on an estimate – around 35 mg of salicylic acid can be expected. Then dissolve in 25 ml of water in an iodine flask or dilute to such a volume. After adding V_1 = 25 ml potassium bromate solution [$c(\frac{1}{6} KBrO_3)$ = 0.1 mol/l] and 1.0 g potassium bromide, wait until the latter has dissolved. Then add 5 ml hydrochloric acid (3 mol/l) and leave the mixture to stand in the sealed flask for 25 min without swirling.

To produce elemental iodine, 1 g of potassium iodide is added, stirred briefly and left to stand for 5 min. Make up to about 200 ml with water. Titration is carried out with sodium thiosulfate 0.1 mol/l as the standard solution and – after brightening the mixture – with starch solution as the indicator. The equivalence point is indicated by the disappearance of the blue coloration. The volume used is V_2:

$V_1 - V_2$ correspond to the volume of the KBrO$_3$ solution consumed during bromination [c(⅙ KBrO$_3$) = 0.1 mol/l];

1 ml of this volume corresponds to 2.302 mg C$_7$H$_6$O$_3$ [28].

3.3.6.10 Determination of phenol

The determination of phenol content in pharmaceuticals is described in a similar way [8a]. The stoichiometry here is analogous to the determination of salicylic acid. For one molecule of phenol there are also three molecules of elemental bromine so that three hydrogen atoms of the phenol are substituted by bromine atoms. Here too, waiting times are necessary due to the slow reaction.

Procedure: Approximately 2 g of the substance to be tested is accurately weighed and dissolved in a volume of 1 l of water. Place exactly 25 ml of this solution in an iodine number flask and add V_1 = 50 ml potassium bromate standard solution [c(⅙ KBrO$_3$) = 0.1 mol/l]. According to the European Pharmacopoeia, this standard solution is prepared in such a way that it also contains 13 g of potassium bromide per liter of standard solution. This corresponds to a mass of potassium bromide of 0.65 g in 50 ml standard solution. Then add 5 ml of hydrochloric acid 10 mmol/l and leave the mixture to stand for 30 min in a closed flask while swirling occasionally. Then leave to stand in the sealed flask for a further 15 min without swirling.

To produce elemental iodine, 1 g of potassium iodide is added. According to the pharmacopoeia, a 200 g/l potassium iodide solution is prepared for this purpose, from which 5 ml is pipetted into the mixture. After shaking, titration is carried out with sodium thiosulfate (0.1 mol/l) as the standard solution and – after brightening the mixture – with starch solution as the indicator. The equivalence point is indicated by the disappearance of the blue coloration. The volume used is V_2:

$V_1 - V_2$ corresponds to the volume of the KBrO$_3$ solution consumed during bromination [c(⅙ KBrO$_3$) = 0.1 mol/l];

1 ml of this volume corresponds to 1.569 mg C$_6$H$_6$O [8a].

3.3.6.11 Determination of sulfonamides according to DAB 7

The German Pharmacopoeia 7 determined numerous sulfonamide antibiotics bromatometrically according to Koppeschaar with subsequent iodometric back titration. Today, this substance class is mostly titrated nitritometrically (see Section 4.5.4).

3.3.6.12 Determination of sulfanilamide

Procedure: About 0.100–0.110 g of substance, weighed accurately, are dissolved hot in 10.0 ml acetic acid and 10.0 ml hydrochloric acid in an iodine flask of about 200 ml capacity. After adding a solution

of 1.0 g potassium bromide in 2.0 ml water and 20.0 ml acetic acid, 30.0 ml potassium bromate solution (0.0166 mol/l) is added to the cooled solution while swirling. After 2 min, 0.5 g KI is added. After a
further minute, titrate back against starch with sodium thiosulfate solution (0.1 mol/l):
   1.0 ml potassium bromate solution, c = 0.0166 mol/l, corresponds to 4.305 mg $C_6H_8O_2N_2S$:

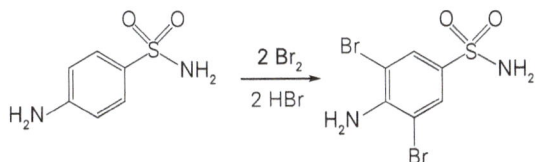

3.3.6.13 Determination of cystine according to Ph. Eur

Cystine is the oxidation product of the amino acid cysteine. The substance is determined bromatometrically in the Ph. Eur. [8a]. Cystine is used as a food additive (flour
treatment), is a component of various dietary supplements against hair loss and is
used as a starting material for various syntheses.

   The disulfide in cystine is oxidatively cleaved by an excess of bromine to form
two sulfonic acids. The excess bromine is then converted into iodine with iodide and
this is titrated with sodium thiosulfate:

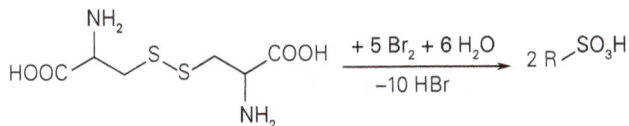

Procedure: In an Erlenmeyer flask with ground-glass stopper, dissolve 0.100 g substance in a mixture
of 2 ml dilute sodium hydroxide solution (85 g/l) and 10 ml water. After adding 10 ml of a solution of
potassium bromide (200 g/l), 50.0 ml of potassium bromate solution (0.0167 mol/l) and 15 ml of dilute
hydrochloric acid, the flask is sealed, cooled in a mixture of ice and water and allowed to stand for 10
min under light protection. After adding 1.5 g potassium iodide, titrate the solution after 1 min with
sodium thiosulfate solution (0.1 mol/l), adding 2 ml starch solution toward the end of the titration. A
blank titration is carried out:
   1 ml potassium bromate solution, c = 0.0167 mol/l, corresponds to 2.403 mg $C_6H_{12}N_2O_4S_2$.

3.3.7 Iodometric determinations

Iodometry is one of the most versatile methods of redox titration. This versatility is
based on the oxidizing effect of iodine on the one hand and the reducing effect of iodide ions on the other. The underlying process

$$I_2 + 2\,e^- \rightleftharpoons 2\,I^-$$

is fully reversible.

Basically, this results in two possibilities for the use of iodometry:

1. Reducing agents can be titrated directly with iodine solution. They are oxidized to iodide by reducing the iodine, e.g.

$$S^{2-} + I_2 \rightleftharpoons 2\,I^- + S.$$

2. Oxidizing agents are reduced with acidified potassium iodide solution in excess, whereby the iodide is oxidized to elemental iodine, e.g.

$$2\,Fe^{3+} + 2\,I^- \rightleftharpoons I_2 + 2\,Fe^{2+}.$$

The resulting iodine is then titrated with the standard solution of a suitable reducing agent: sodium sulfite, arsenious acid or sodium thiosulfate.

Du Pasquier carried out the first iodometric determination in 1840. He titrated the iodine with sulfurous acid. The water requirement of this reaction is utilized in the later-developed Karl Fischer method for water determination (see Section 4.5.4). Bunsen did the same in 1853 who drew the attention of chemists to the iodometric methods he had systematically worked on. In the same year, Schwarz introduced sodium thiosulfate instead of sulfite into analytical practice, which is now used almost exclusively for titrating iodine. Arsenious acid is only used in more alkaline solutions.

When titrating the iodine with sodium thiosulfate in a neutral to slightly acidic solution, the thiosulfate is oxidized to tetrathionate:

$$2\,S_2O_3^{2-} + I_2 \rightarrow S_4O_6^{2-} + 2\,I^-.$$

The reaction only occurs quantitatively according to the equation given in neutral or weakly acidic solutions. Alkaline iodine solutions partially oxidize the thiosulfate further to sulfate, which is shown by the equation:

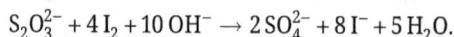

$$S_2O_3^{2-} + 4\,I_2 + 10\,OH^- \rightarrow 2\,SO_4^{2-} + 8\,I^- + 5\,H_2O.$$

In an alkaline solution, the reduction of the same amount of iodine requires much less thiosulfate than in a neutral solution. A lower consumption is therefore observed. This is due to the fact that alkaline iodine solutions have a higher redox potential than neutral iodine solutions due to the presence of hypoiodous acid, HIO. When titrating iodine with thiosulfate, the hydrogen ion concentration of the sample solution must never fall below a minimum value. According to Kolthoff [89], this lower limit for iodine solutions with a concentration of 0.05 mol/l is located at approximately $2.5 \cdot 10^{-8}$ mol/l (pH = 7.6), in the case of a concentration of 0.005 mol/l at approximately $3 \cdot 10^{-7}$ mol/l (pH = 6.5) and assuming a concentration of 0.0005 mol/l at approximately 10^{-5} mol/l (pH = 5).

The required H^+ concentration therefore increases sharply with dilution. Furthermore, salts whose solution reacts alkaline, such as sodium and ammonium carbonate, sodium hydrogen phosphate and borax, must not be present. Conversely, if the iodine solution is allowed to flow into the sodium thiosulfate solution, these restrictions do not apply. The disturbing side effect of the hydroxide ions is almost completely eliminated because the iodine is immediately reduced by the excess thiosulfate before hypoiodous acid is formed. When titrating thiosulfate with iodine solution, low hydroxide concentrations therefore hardly interfere.

Whether the reaction

$$I_2 + 2\, e^- \rightleftharpoons 2\, I^-$$

takes place to the left or right side of the equation depends on the magnitude of the redox potential of the substance to be determined. The iodine solution has an oxidizing effect if the redox potential of the reactant is lower than that of the iodine. Iodide acts as a reducing agent if, conversely, the redox potential of the partner is higher than that of the iodide. As the magnitude of the redox potential is strongly dependent on the hydrogen ion concentration, the temperature and other factors, it is possible for one and the same reaction to take place quantitatively in the direction of the oxidation process on the one hand and in the direction of the reduction process on the other by selecting suitable experimental conditions. For example, arsenic acid in a strongly acidic solution can be quantitatively reduced to arsenic acid by iodide ions, while arsenic acid in a neutral or weakly alkaline solution is quantitatively oxidized to arsenic acid by iodine. The processes can be described by the equation

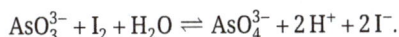

$$AsO_3^{3-} + I_2 + H_2O \rightleftharpoons AsO_4^{3-} + 2\,H^+ + 2\,I^-.$$

The use of arsenious acid to titrate iodine in a weak alkaline solution is based on the same reaction.

3.3.7.1 End point detection

The end point of iodometric titrations is characterized by the appearance or disappearance of iodine. As elemental iodine is poorly soluble in water, the iodine solutions used in iodometry always contain potassium iodide in addition to iodine. This forms a deep brown triiodide ion that is easily soluble in water:

$$I_2 + I^- \rightleftharpoons I_3^-.$$

The solutions are therefore still yellow in strong dilution (up to about 10^{-5} mol/l iodine) so that the inherent color could be sufficient as an indicator. For better recognition of the iodine, however, some starch solution is added as an indicator. Starch forms a deep blue compound with iodine, which can still be used to recognize iodine concentrations of 10^{-5} mol/l. The color strength of the blue iodine-starch compound considerably exceeds that of free iodine alone. Important for analytical practice is the fact that the high sensitivity of the iodine-starch reaction is linked to the presence of iodide ions.

You can see this for yourself by carrying out the following experiment: If you add saturated iodine water to approximately 200 ml of water to which contains only a little starch solution from a burette (the saturated solution of iodine in pure water contains about (1/1,400) mol/l iodine), a weak blue coloration of the water is observed only after the addition of several milliliters. However, if a little potassium iodide solution is added to the water in addition to the starch solution, an intense blue coloration is observed immediately after the addition of a few drops of iodine water.

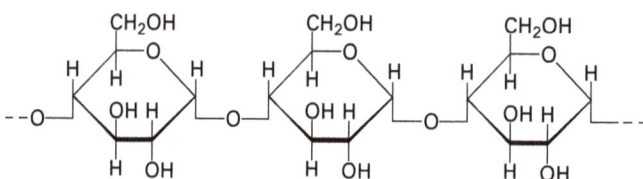

Fig. 3.15: Part of the chain structure of amylose.

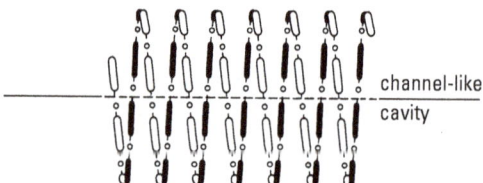

Fig. 3.16: Helical structure of amylose.

According to studies by Cramer, the constitution of the blue iodine-starch compound can be described as follows [117]: the soluble component of the starch consists of amylose unbranched chains of glucose molecules in an α-$(1 \rightarrow 4)$-glycosidic bond (Fig. 3.15).

Iodine forms a blue inclusion compound with it. This refers to compounds whose structure and composition are largely determined by spatial relationships and not by bonding relationships. The glucose chains of amylose are coiled up like a helix (Fig. 3.16), creating channel-like cavities inside.

The iodine is embedded in these channel-like cavities in the form of linear polyiodide chains e.g. with I_5^- units, in which the iodine atoms have an average I–I distance of 310 pm. The units are linked to each other by bonds, which facilitates electron delocalization along the chain and explains the appearance of the deep blue color (absorption maximum 620 nm) [118].

3.3.7.2 Preparation of the starch solution

Three grams of soluble starch is rubbed with a little cold water in a grating bowl until a paste of uniform consistency is obtained. Add this paste to 600 ml of boiling water and boil for a few minutes. Then leave to cool in a small container and remove any undissolved particles. The clear solution above the sediment is poured or siphoned off into a clean bottle and a few milliliters of mercury(II) iodide solution are added to eliminate fungal or bacterial infestation and increase the shelf life of the starch solution. If no soluble starch is available, use potato starch (not wheat starch) to prepare the solution. For determination, add 1–3 ml of starch solution. The color with iodine should be pure blue.

Occasionally, the property of iodine to dissolve with red color in organic solvents that are not miscible with water and whose molecule does not contain oxygen, such as carbon tetrachloride (very toxic!) or chloroform, is used to detect the end point. This method is very sensitive and can be advantageous in certain cases. Titration is then carried out in iodine number flasks which are wide-necked flasks that can be closed with a ground-glass stopper. Add about 5–10 ml of the organic solvent and shake the closed flask vigorously after each addition of reagent.

For the iodometric determinations the following standard solutions are required:
- iodine solution, $c(\frac{1}{2} I_2) = 0.1$ mol/l,
- sodium thiosulfate solution, $c(Na_2S_2O_3) = 0.1$ mol/l,
- potassium iodide solution, $c(KI) = 0.2$ mol/l (3.3%) and
- starch solution.

3.3.7.3 Preparation of the sodium thiosulfate solution

From the reaction equation

$$2\,S_2O_3^{2-} + I_2 \rightarrow S_4O_6^{2-} + 2\,I^-$$

follows that a solution with a concentration of 0.1 mol/l based on equivalents must contain $\frac{1}{10}$ of the molar mass of $Na_2S_2O_3 \cdot 5\,H_2O$ per liter. The salt must be free of impurities such as carbonate, chloride, sulfate, sulfite, sulfide and elemental sulfur. If necessary, it can be purified by repeated recrystallization and drying over calcium chloride. It is commercially available in high purity (p.a.). Nevertheless, a standard solution containing only about 0.1 mol/l is prepared because experience has shown that thiosulfate solutions are not titer-stable in the first 1–2 weeks. There are several possible reasons for this. The assumption that carbon dioxide dissolved in the water causes the decomposition could not be confirmed by Kolthoff [89]. Oxidation by atmospheric oxygen may be responsible for the reduction in the titer (no increases were observed). It can be catalytically accelerated or triggered by traces of heavy metals (e.g. Cu^{2+}) or by metabolic processes of certain microorganisms, especially *Bacillus thiooxydans*. Oxidation products are tetrathionate and sulfate. The presence of hydrogen ions can also cause a reduction in titer, possibly after

$$5\,S_2O_3^{2-} + 6\,H^+ \rightarrow 2\,S_5O_6^{2-} + 3\,H_2O.$$

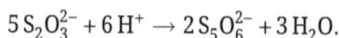

The resulting pentathionate ions decompose into tetrathionate and sulfur. The contamination of the sodium thiosulfate with polythionates could affect its titer stability.

For the preservation of thiosulfate solutions, the following were recommended: Addition of 1 g amyl alcohol/l and 0.1 g mercury(II) cyanide/l; introduction of steam for thorough sterilization.

Weigh out approximately $\frac{1}{10}$ of the molar mass, $M(Na_2S_2O_3 \cdot 5\,H_2O) = 248.19$ g/mol, i.e. approximately 25 g, of the purest thiosulfate and dissolve it in 1 l of boiled, double-distilled water. After standing for about a week, the titer of the solution is determined. It is stored away from light.

3.3.7.4 Adjusting the sodium thiosulfate solution

Primary standards such as iodine, potassium iodate or potassium dichromate can be used. An adjusted potassium permanganate solution can also be used:

1. Adjustment with iodine: Commercially available iodine can be contaminated with chlorine, bromine and water. It is purified by double sublimation and dried. Precisely weighed portions of the purified iodine are dissolved in potassium iodide solution and titrated with the sodium thiosulfate solution.

Procedure: 10 g of pure iodine is mixed with 1 g KI and approximately 2 g CaO and triturated, then sublimated from a dry beaker under the dry bottom of a round-bottomed flask filled with cold water, which closes the beaker at the top. The sublimation is repeated without the addition of KI and CaO. The purified iodine is stored in a desiccator with a fat-free lid over calcium chloride. Special precautions are required for weighing the iodine samples due to the high vapor pressure of iodine. Dissolve about 2 g of pure iodate-free potassium iodide in 2 ml of water in a weighing jar, weigh the jar after temperature equalization about 15 min after dissolution, quickly add about 0.3 g of iodine, close the weighing jar and weigh it again to determine the exact amount of iodine added. The iodine dissolves immediately in the concentrated KI solution. The weighing jar is then placed in an Erlenmeyer flask charged with about 300 ml of water and about 1 g of KI by opening it while sliding it into the inclined flask and throwing the lid on. Then titrate immediately with the sodium thiosulfate solution from a burette. When the iodine solution is only very faintly yellow, add 2 ml of starch solution and titrate the dark blue solution until it is completely decolored. The titer is calculated on the basis of three to four titrations:

1 ml sodium thiosulfate solution, $c(Na_2S_2O_3) = 0.1$ mol/l, corresponds to 12.690 mg iodine.

2. Adjustment with potassium iodate: Potassium iodate is reduced in acidic solution by excess potassium iodide to iodine, which is titrated with thiosulfate:

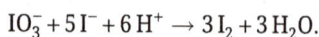

$$IO_3^- + 5\,I^- + 6\,H^+ \rightarrow 3\,I_2 + 3\,H_2O.$$

If necessary, the commercially available potassium iodate can be further purified by repeated recrystallization from water and drying at 180 °C.

Procedure: Weigh out 0.08–0.1 g of KIO_3 and dissolve it in about 200 ml of water in an Erlenmeyer flask. After adding about 1 g of KI, acidify with dilute hydrochloric acid and titrate with $Na_2S_2O_3$ solution as described:
1 ml sodium thiosulfate solution, $c(Na_2S_2O_3) = 0.1$ mol/l, corresponds to 3.567 mg KIO_3.

3. Adjustment with potassium dichromate:

Potassium dichromate is reduced to chromium(III) by concentrated hydrochloric acid, whereby an equivalent amount of chloride is oxidized to chlorine:

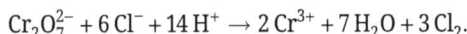

$$Cr_2O_7^{2-} + 6\,Cl^- + 14\,H^+ \rightarrow 2\,Cr^{3+} + 7\,H_2O + 3\,Cl_2.$$

Bunsen who developed this method in 1853, carried out the reaction in a small closed distillation apparatus in which the developed chlorine could be overdriven and collected in a receiver with excess KI solution. The resulting I_2 is titrated with $Na_2S_2O_3$:

$$Cl_2 + 2\,I^- \rightleftharpoons I_2 + 2\,Cl^-.$$

The setting is best made using the apparatus illustrated on p. 211 (Fig. 3.18).

Procedure: Accurately weigh about 0.1 g of $K_2Cr_2O_7$ and transfer it to the decomposition flask. Instead, you can also pipette in 5–10 ml of a $K_2Cr_2O_7$ solution with a concentration of 1.5–3.0 mol/l and a precisely known titer. The dropping funnel is filled with 40 ml of concentrated hydrochloric acid. Add a total of 40 ml of KI solution ($c \approx 0.2$ mol/l) to the Erlenmeyer flask and the Péligot tube, both of which are cooled from the outside with ice water. Then connect the dropping funnel to a CO_2 developer, open the stopcock and press the hydrochloric acid through the CO_2 into the cuvette. During the determination, a very slow stream of CO_2 should constantly pass through the apparatus. Finally, heat slowly until boiling begins and distil for 30–40 min. The contents of the Péligot tube are then transferred to the Erlenmeyer flask and the released iodine is titrated with the $Na_2S_2O_3$ solution:
1 ml sodium thiosulfate solution, $c(Na_2S_2O_3) = 0.1$ mol/l, corresponds to 4.903 mg $K_2Cr_2O_7$.

According to Zulkowski (1868), distillation of the chlorine is not necessary. He adds the solutions of $K_2Cr_2O_7$, HCl and KI together and titrates the released iodine according to

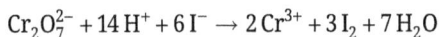

$$Cr_2O_7^{2-} + 14\,H^+ + 6\,I^- \rightarrow 2\,Cr^{3+} + 3\,I_2 + 7\,H_2O$$

in the solution mixture directly with the $Na_2S_2O_3$ solution.

When working according to this method, care must be taken to eliminate a number of possible errors. The reaction between Cr(VI) and I^- only proceeds quantitatively in concentrated and sufficiently acidified solutions. Above all, it requires some time to complete. The titration must therefore not be carried out immediately after adding the dichromate solution to the acidified KI solution. Otherwise the chromic acid still present will cause an additional consumption of thiosulfate. It is assumed that if the acid concentration is too high, the oxidation of iodide by atmospheric oxygen is responsible for the additional consumption. The titration is therefore often carried out in a CO_2 atmosphere, which is created by adding $NaHCO_3$ to the solution.

Procedure: Add 40 ml of concentrated hydrochloric acid and 40 ml of KI solution, $c \approx 0.2$ mol/l, to an Erlenmeyer flask cooled from the outside with ice water. Accurately weigh out about 100 mg $K_2Cr_2O_7$, dissolve in a little water and add the solution to the hydrochloric acid KI solution. After 15–20 min, the released iodine is titrated with the $Na_2S_2O_3$ solution. The end point is reached when the solution is no longer blue due to iodine strength, but bluish-green due to Cr^{3+} ions, which can be recognized quite well after some practice.

When carried out correctly, Bunsen's distillation method and Zulkowski's method produce consistent results.

According to Kolthoff (1920), titration is simpler – without achieving significantly worse results (1920) immediately after adding the KI solution and the hydrochloric acid, if the hydrochloric acid concentration is at least 0.6 mol/l.

The titer of the $Na_2S_2O_3$ solution should be found to be no more than 0.5‰ too high if a precisely weighed $K_2Cr_2O_7$ portion is dissolved in enough water so that the concentration is approximately $c(⅙ \ K_2Cr_2O_7) = 0.1$ mol/l. Add 12–13 ml of 25% hydrochloric acid and 10 ml of KI solution (1 mol/l) to 50 ml of this solution and titrate with $Na_2S_2O_3$ solution after mixing.

4. Adjustment with potassium permanganate: According to Volhard (1879), permanganate is used after

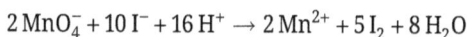

$$2 \, MnO_4^- + 10 \, I^- + 16 \, H^+ \rightarrow 2 \, Mn^{2+} + 5 \, I_2 + 8 \, H_2O$$

is quantitatively reduced to manganese(II) by an acidified KI solution with separation of an equivalent amount of iodine. The iodine can then be titrated with sodium thiosulfate solution.

5. Adjustment with potassium bromate according to Ph. Eur.
According to Ph. Eur., potassium bromate is used after

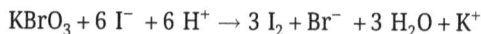

$$KBrO_3 + 6 \, I^- + 6 \, H^+ \rightarrow 3 \, I_2 + Br^- + 3 \, H_2O + K^+$$

is quantitatively reduced to bromide by an acidified KI solution with a release of an equivalent amount of iodine. The iodine can then be titrated with sodium thiosulfate solution.

Procedure: About 10.0 ml potassium bromate solution (0.033 mol/l) is mixed with 40 ml water, 10 ml potassium iodide solution (166 g/l) and 5 ml hydrochloric acid (0.37 g/l) and titrated with sodium thiosulfate solution. Toward the end of the titration, 1 ml starch solution (10 g/l) is added [8a].

3.3.7.5 Preparation of the iodine solution

Dissolve 20–25 g of pure iodate-free potassium iodide in about 40 ml of water in a liter flask and add 12.7–12.8 g of iodine. The sealed flask is shaken without adding any more water until all the iodine has dissolved. Then make up to the mark with water. If water is added too early, the undissolved iodine will dissolve extremely slowly. The iodine solution prepared in this way has an approximate concentration of $c(\frac{1}{2}\,I_2) = 0.1$ mol/l. It is adjusted with sodium thiosulfate solution of known concentration or with arsenic acid.

Adjustment of the iodine solution with sodium thiosulfate according to Ph. Eur.

In order to avoid the toxic arsenic(III) oxide, the factor of a standard iodine solution in Ph. Eur. is determined against an adjusted standard sodium thiosulfate solution under strength indication [8a].

Procedure: 2 ml of the iodine solution (0.05 mol/l) is titrated with sodium thiosulfate solution (0.1 mol/l) after adding 1 ml of diluted acetic acid and 50 ml of water using starch solution:
   1 ml iodine solution, c = 0.05 mol/l corresponds to 1 ml sodium thiosulfate solution, c = 0.1 mol/l.

3.3.7.6 Determination of sulfides

Sulfide ions and iodine react with each other according to

$$S^{2-} + I_2 \rightarrow 2\,I^- + S.$$

The direct titration of H_2S water with I_2 solution leads to fluctuating values that are always too low because hydrogen sulfide partly volatilizes during the titration and because interfering side reactions take place. Titration is therefore carried out in reverse and a certain volume of H_2S water from a pipette is allowed to run into excess iodine solution. The excess iodine solution is titrated back with $Na_2S_2O_3$ solution.

Procedure: About 10–20 ml of H_2S water of medium concentration is pipetted into 50 ml iodine solution, $c(\frac{1}{2}\,I_2)$ = 0.1 mol/l. The excess is titrated back to 200 ml with sodium thiosulfate solution (0.1 mol/l) after diluting the solution. If the released sulfur is brown in color, it contains iodine that has escaped the titration. The sulfur, which floats on the surface as a continuous skin, is then removed and shaken in a glass stoppered vial containing 5 ml carbon disulfide (caution: very toxic and flammable!). This turns purple when the iodine dissolves. By adding the $Na_2S_2O_3$ solution drop by drop from a burette until the CS_2 decolors, the iodine trapped in the sulfur can also be detected and taken into account in the calculation:
   1 ml iodine solution, $c(\frac{1}{2}\,I_2)$ = 0.1 mol/l, corresponds to 1.704 mg H_2S.

Solutions of alkali metal sulfides are titrated in the same way. However, as the sulfides are often contaminated by alkali metal hydroxides, which would cause additional consumption of iodine solution (see p. 194), a little acetic acid is added to the excess iodine solution.

Poorly soluble sulfides can be decomposed with hydrochloric acid in the heat. The escaping hydrogen sulfide is quantitatively converted into excess iodine solution, $c(\frac{1}{2} I_2) = 0.1$ mol/l, by an indifferent gas stream, e.g. nitrogen or carbon dioxide, which is then retitrated with sodium thiosulfate solution, $c(Na_2S_2O_3) = 0.1$ mol/l. A suitable apparatus is shown in Fig. 3.18 (see p. 211).

3.3.7.7 Determination of sulfites

Direct titration with iodine solution also results in incorrect values here. On the one hand, this is due to the volatility of the sulfurous acid on the other hand due to interfering side reactions, such as the oxidation of the sulfite to sulfate caused by atmospheric oxygen, which is noticeably induced and accelerated by the titration reaction (oxidation of the sulfite by iodine). However, correct values are obtained if a measured volume of the sulfite solution, which is not too concentrated, is allowed to flow into excess iodine solution and the unused iodine is titrated back with $Na_2S_2O_3$-standard solution. The reaction proceeds as follows:

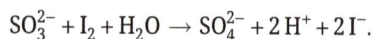

$$SO_3^{2-} + I_2 + H_2O \longrightarrow SO_4^{2-} + 2H^+ + 2I^-.$$

Procedure: Dilute 10 ml of the sulfurous acid to be determined to 1 l in the volumetric flask. Pipette 50 ml of this solution into 50 ml iodine solution, $c(\frac{1}{2} I_2) = 0.1$ mol/l. The solution is diluted to 200 ml. The excess iodine is titrated back with sodium thiosulfate solution, $c(Na_2S_2O_3) = 0.1$ mol/l. If sulfites are to be determined, the iodine solution is weakly acidified with hydrochloric acid before the sulfite solution is added:
1 ml iodine solution, $c(\frac{1}{2} I_2) = 0.1$ mol/l, corresponds to 3.203 mg SO_2.

3.3.7.8 Determination of hydrazine

Hydrazine hydrate and its salts are oxidized to nitrogen in a solution containing hydrogen carbonate according to Stollé (1902), and they are quantitatively oxidized to nitrogen by iodine solution:

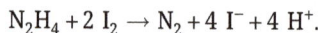

$$N_2H_4 + 2 I_2 \longrightarrow N_2 + 4 I^- + 4 H^+.$$

The titration should be carried out immediately after adding the hydrogen carbonate:
1 ml iodine solution, $c(\frac{1}{2} I_2) = 0.1$ mol/l, corresponds to 0.8011 mg N_2H_4.

3.3.7.9 Determination of arsenic and antimony

The determination of arsenic(III) oxide is carried out in the same way as already described for the titration of the iodine solution.

In all cases where arsenic is obtained as arsenic(III) chloride in the distillate by distillation with concentrated hydrochloric acid in the presence of suitable reducing agents, its determination can be carried out iodometrically. Hydrogen carbonate must be added to the solution provided:

$$AsCl_3 + 6\,NaHCO_3 \rightarrow Na_3AsO_3 + 3\,NaCl + 6\,CO_2 + 3\,H_2O.$$

This determination is important for the analysis of arsenic-containing ores or alloys.

The determination of antimony(III) oxide is also carried out in a similar way in a solution containing $NaHCO_3$:

$$SbO_2^- + I_2 + 4\,H_2O \rightarrow 2\,I^- + 2\,H^+ + [Sb(OH)_6]^-.$$

To prevent basic antimony salts from precipitating as a result of protolysis, it is necessary to add tartaric acid or potassium sodium tartrate (Seignette salt).. Soluble antimony tartrate complexes are formed.

Procedure

To determine the antimony content of crushed tartar emetic (potassium antimonyl tartrate sesquihydrate), $KSb[C_4H_2O_6] \cdot \frac{3}{2}\,H_2O$, dissolve 0.3–0.4 g of the preparation in about 100 ml water, add about 0.5 g $NaHCO_3$ and titrate with iodine solution, $c(\frac{1}{2}\,I_2) = 0.1$ mol/l, after adding starch solution, until the first permanent blue coloration appears:

1 ml iodine solution, $c(\frac{1}{2}\,I_2) = 0.1$ mol/l, corresponds to 6.089 mg Sb or 7.289 mg Sb_2O_3 or 16.697 mg $KSb[C_4H_2O_6] \cdot \frac{3}{2}\,H_2O$.

3.3.7.10 Determination of tin

The determination of tin(II) can be carried out not only in a solution containing hydrogen carbonate but also in an acidic solution:

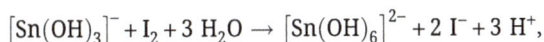

$$[Sn(OH)_3]^- + I_2 + 3\,H_2O \rightarrow [Sn(OH)_6]^{2-} + 2\,I^- + 3\,H^+,$$

respectively

$$Sn^{2+} + I_2 \rightarrow 2\,I^- + Sn^{4+}.$$

To avoid interference due to oxidation by air when titrating in an acidic solution, work with excess iodine solution and titrate back with $Na_2S_2O_3$ solution. At the same time, a carbon dioxide atmosphere is created above the solution by throwing in a piece of marble.

Procedure

1 In alkaline solution: 1 g of potassium sodium tartrate and excess sodium hydrogen carbonate are added to the tin(II) chloride solution. After adding starch solution, titrate with iodine solution until blue coloration.

2 In acidic solution: The hydrochloric acid tin(II) chloride solution, which should contain about 0.12–0.15 g tin, is diluted with 100 ml water. A piece of marble is thrown in, and after a few mi-

nutes 50 ml of iodine solution, $c(\frac{1}{2} I_2) = 0.1$ mol/l, is pipetted in. The unused iodine is then titrated back with $Na_2S_2O_3$ solution, $c(Na_2S_2O_3) = 0.1$ mol/l:

1 ml iodine solution, $c(\frac{1}{2} I_2) = 0.1$ mol/l, corresponds to 5.935 mg Sn or 6.735 mg SnO.

The titration of tin in acidic solution can also be carried out in the presence of iron(II) salts, antimony salts, iodides and bromides. It allows, for example, the determination of tin in tin-antimony alloys.

3.3.7.11 Determination of mercury

Mercury(I) salts are converted to $K_2[HgI_4]$ by iodine solution in the presence of excess potassium iodide:

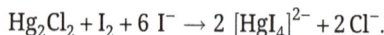

$$Hg_2Cl_2 + I_2 + 6\ I^- \rightarrow 2\ [HgI_4]^{2-} + 2\ Cl^-.$$

An excess of iodine solution is used and the unused iodine is titrated back with $Na_2S_2O_3$ solution.

Procedure: About 0.2–0.25 g calomel, Hg_2Cl_2, are shaken in an Erlenmeyer flask with 1 g potassium iodide and 50 ml iodine solution, $c(\frac{1}{2} I_2) = 0.1$ mol/l, until a clear yellow solution is obtained. Then titrate back the excess iodine with sodium thiosulfate solution, $c(Na_2S_2O_3) = 0.1$ mol/l, using starch as an indicator:

1 ml iodine solution, $c(\frac{1}{2} I_2) = 0.1$ mol/l, corresponds to 23.604 mg Hg_2Cl_2 or 20.059 mg Hg.

Mercury(II) salts are reduced to metallic mercury after conversion to the complex tetraiodomercurate(II), e.g. with formaldehyde:

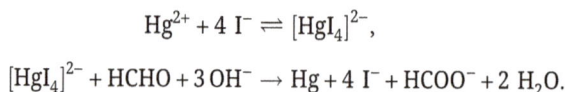

$$Hg^{2+} + 4\ I^- \rightleftharpoons [HgI_4]^{2-},$$

$$[HgI_4]^{2-} + HCHO + 3\ OH^- \rightarrow Hg + 4\ I^- + HCOO^- + 2\ H_2O.$$

The metallic mercury can then be oxidized again by excess iodine solution in the presence of potassium iodide:

$$Hg + I_2 + 2\ I^- \rightarrow [HgI_4]^{2-}.$$

The iodine consumed for this purpose is determined by back titration of the excess iodine with thiosulfate solution.

Procedure (according to Rupp, 1905/07): Add 1–2 g potassium iodide to a sublimate solution containing approximately 0.2 g $HgCl_2$ in a wide-necked bottle with a ground-in glass stopper and swirl until a clear yellow solution is obtained. This is diluted with 30 ml water. Add 20 ml sodium hydroxide solution (2 mol/l) and pour a mixture of 3 ml pure 40% formaldehyde solution and 10 ml water into the alkaline solution, swirling continuously. Close the bottle and shake vigorously for 2–3 min without interruption. Then acidify with 20 ml glacial acetic acid and add 30 ml iodine solution, $c(\frac{1}{2} I_2) = 0.1$ mol/l. Shake vigorously again until all the mercury has dissolved. The excess iodine is then titrated back with sodium thiosulfate solution, $c(Na_2S_2O_3) = 0.1$ mol/l. The determination procedure can be carried out as described because formaldehyde is not oxidized by iodine in acidic solution:

1 ml iodine solution, $c(\frac{1}{2} I_2) = 0.1$ mol/l, corresponds to 13.575 mg $HgCl_2$ or 10.030 mg Hg.

3.3.7.12 Determination of ascorbic acid

Ascorbic acid – vitamin C – can be determined both directly and indirectly by titrimetry. The oxidation of the two hydroxyl groups of the C(2) and C(3) atoms (endiol) leads to the formation of dehydroascorbic acid.

In the direct method, ascorbic acid is reacted with elemental iodine. The equivalence point is recognized by the blue color of the starch-iodine complex when starch is used as an indicator:

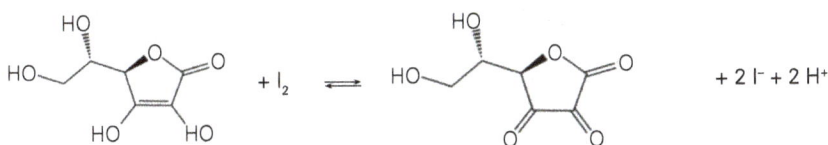

In the indirect titration developed by Stevens – so-called double back titration – a known excess of iodine standard solution and then a known excess of sodium thiosulfate standard solution are added to the sample solution. The unused amount of sodium thiosulfate is then back-titrated with iodine standard solution [118a].

C.R. Silva et al. showed in 1999 that ascorbic acid can also be used as a primary standard for an iodine solution using direct titration [118b]. Care should be taken to ensure that the titration is carried out quickly in order to minimize oxidation of the ascorbic acid by atmospheric oxygen.

The accuracy of these methods can be limited by possible ingredients of a real sample, which can also be oxidized by iodine. This can be shown by an increased consumption of iodine standard solution. However, an interference-free ascorbic acid determination is possible in some fruit juices [118a, 118c]. Juices should be prepared as fresh as possible. The solid components can be removed using large-pored filter paper, for example.

Direct determination of ascorbic acid in orange juice:

Procedure: About 20.0 ml of orange juice is placed in an Erlenmeyer flask using a volumetric pipette, diluted with 50 ml water and acidified with approximately 20 ml (1 mol/l) sulfuric acid. Add exactly 10 drops of starch solution to the sample solution and titrate with the iodine solution, $c(\frac{1}{2} I_2) = 0.005$ mol/l, until the solution has a greenish color.

Note: the green color at the equivalence point results from the mixture of the starch solution colored blue at the equivalence point and the yellow color of the orange juice:

1.0 ml iodine solution, $c(\frac{1}{2} I_2) = 0.005$ mol/l, corresponds to 1.761 mg $C_6H_8O_6$.

3.3.7.13 Determination of iodide

Iodides are determined according to the method of Duflos (1845). The iodide ions are oxidized to iodine by iron(III) sulfate:

$$2\ I^- + 2\,Fe^{3+} \rightarrow I_2 + 2\,Fe^{2+}.$$

The released iodine is overdistilled into excess potassium iodide solution and titrated in the template with sodium thiosulfate solution. Bromides do not interfere as, unlike iodides, they are not oxidized by iron(III) salts.

Procedure (analysis of potassium iodide): Using the distillation apparatus in Fig. 3.18, weigh about 0.3 g of KI into the small distillation flask and dissolve in a little water.

Add about 1 g solid ammonium iron(III) sulfate and 10 ml H_2SO_4, $c(\frac{1}{2} H_2SO_4) = 2$ mol/l, dilute with water to about 50 ml and close the flask. The Erlenmeyer flask and the Péligot tube contain a total of about 30 ml potassium iodide solution, $c(KI) = 0.2$ mol/l, which is further diluted with water. The Erlenmeyer flask is cooled from the outside with ice water. While carbon dioxide is now introduced very slowly through the gas supply tube, the contents of the distillation flask are carefully heated to boiling point and distilled until no more iodine vapors are released. The contents of the receiver are then titrated with $Na_2S_2O_3$ solution:

1 ml sodium thiosulfate solution, $c(Na_2S_2O_3) = 0.1$ mol/l, corresponds to 12.690 mg I_2 or 16.600 mg KI.

3.3.7.14 Determination of hypochlorite

Hypochlorite can be determined iodometrically because it reacts with iodide in an acidic solution to form chloride and iodine:

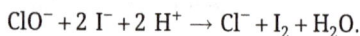

$$ClO^- + 2\ I^- + 2\ H^+ \rightarrow Cl^- + I_2 + H_2O.$$

After the addition of hydrochloric acid or sulfuric acid, chlorite is quantitatively captured in a strongly acidic solution:

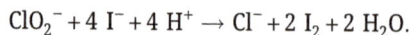

$$ClO_2^- + 4\ I^- + 4\ H^+ \rightarrow Cl^- + 2\ I_2 + 2\ H_2O.$$

As both ions have a strong oxidizing effect and this analysis is primarily interested in the oxidation effect of the mixture, it makes sense to record the ions as a common parameter. In addition, chlorite is formed during the aging of hypochlorite and is therefore occasionally present in the corresponding samples, e.g. chlorinated lime. If the acidity is too high, chlorate can partially coreact and thus falsify the result. The iodine released by the redox reaction is titrated with a sodium thiosulfate solution, whereby starch is added shortly before the equivalence point [28].

Procedure: Approximately 5 g of sample substance is precisely weighed and ground into a paste in a mortar with water, which is then transferred to a 500-ml volumetric flask without loss while rinsing with water. The mixture is filled up to the mark with water and shaken vigorously. Pipette off 50.00 ml from this mixture and then add 20.00 ml fresh KI solution, 100 g/l and 5.00 ml hydrochloric acid (6 mol/l). Titrate with $Na_2S_2O_3$ solution (0.1 mol/l) and add a little starch solution shortly before the equivalence point. The equivalence point is reached when the blue coloration disappears for the first time:
1 ml $Na_2S_2O_3$ solution, c = 0.1 mol/l, corresponds to 3.545 mg effective chlorine [28].

3.3.7.15 Determination of chlorate, bromate, iodate and periodate

Chlorates are reduced in the presence of potassium bromide in a strongly hydrochloric acid solution:

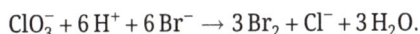

$$ClO_3^- + 6\,H^+ + 6\,Br^- \rightarrow 3\,Br_2 + Cl^- + 3\,H_2O.$$

After the addition of potassium iodide solution, the bromine precipitates an equivalent amount of iodine, which is titrated with sodium thiosulfate solution. The determination of chlorates using the distillation method by Bunsen is more precise (see p. 209 ff.).

Procedure: A volume of 10 ml of a potassium chlorate solution, c(⅙ $KClO_3$) ≈ 0.2 mol/l, is mixed with 1 g of pure potassium bromide and 20 ml of concentrated hydrochloric acid in a wide-necked bottle with a ground-in stopper. Leave the bottle closed for 10 min. Then add 30 ml potassium iodide solution, c(KI) = 0.2 mol/l, dilute and titrate with sodium thiosulfate solution:
1 ml sodium thiosulfate solution, c($Na_2S_2O_3$) = 0.1 mol/l, corresponds to 2.0425 mg $KClO_3$, 1.4074 mg $HClO_3$ or 1.391 mg ClO_3^-.

Bromates are better determined argentometrically according to Volhard [101] (see p. 142 ff.) after they have been reduced with nitrous acid. If they are to be determined iodometrically, then according to Kolthoff (1921), the concentration of hydrochloric acid must be quite high (at least 0.5 mol/l). It is also necessary to wait some time before titrating. The addition of three drops of ammonium molybdate solution (1 mol/l) to the strongly hydrochloric bromate-potassium iodide mixture considerably accelerates the adjustment of the equilibrium.

Iodates can be excellently determined iodometrically. The reaction

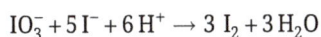

$$IO_3^- + 5\,I^- + 6\,H^+ \rightarrow 3\,I_2 + 3\,H_2O$$

is proceeding at great speed.

Procedure: Dissolve approximately 0.1 g KIO_3 together with 3 g KI in approximately 200 ml water and then add 20 ml hydrochloric acid (2 mol/l). After shaking well, titrate with $Na_2S_2O_3$ solution (0.1 mol/l):
  1 ml sodium thiosulfate solution, $c(Na_2S_2O_3)$ = 0.1 mol/l, corresponds to 3.567 mg KIO_3 or 2.932 mg HIO_3 or 2.782 mg I_2O_5.

Periodates react in acidic solution with iodide according to the equation

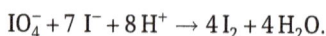

$$IO_4^- + 7\ I^- + 8\,H^+ \rightarrow 4\,I_2 + 4\,H_2O.$$

The titration is carried out as described for iodates:
  1 ml sodium thiosulfate solution, $c(Na_2S_2O_3)$ = 0.1 mol/l, corresponds to 2.399 mg HIO_4.

3.3.7.16 Determination of hydrogen peroxide
Hydrogen peroxide reacts slowly with KI in acidic solution according to the equation:

$$H_2O_2 + 2\ I^- + 2\ H^+ \rightarrow I_2 + 2\ H_2O.$$

The reaction is accelerated by tungstate or molybdate ions (Brode, 1901).

The iodometric determination of hydrogen peroxide has the advantage over the manganometric method that certain organic preservatives, such as glycerol or salicylic acid, which may be contained in technical hydrogen peroxide, have no interfering influence.

Procedure: A volume of 10 ml of an approximately 3% solution of hydrogen peroxide is made up to 250 ml in a volumetric flask. Use 25 ml of this solution for analysis. Add 30 ml of KI solution, 0.2 mol/l (3.3%), to a wide-necked bottle with a glass stopper, acidify with sulfuric acid (1 mol/l) and then allow 25 ml of diluted hydrogen peroxide to flow slowly and drop by drop from a pipette into the acidic iodide solution while constantly swirling the bottle. Close the bottle and leave the mixture to stand for about a quarter to half an hour to complete the reaction. Then titrate slowly and carefully with sodium thiosulfate solution (0.1 mol/l), until the solution is only slightly yellow. After adding 1–3 ml of starch solution, titrate to the end. If you want to titrate immediately, add three drops of ammonium molybdate solution (0.1 mol/l) (Kolthoff, 1921):
  1 ml sodium thiosulfate solution (0.1 mol/l) corresponds to 1.7007 mg H_2O_2.

Alkali and alkaline earth peroxides, percarbonates and perborates are titrated in the appropriate manner.

3.3.7.17 Determination of the peroxide value
The peroxide content of fatty foods and pharmaceuticals can be used to assess their level of decomposition. Peroxides are formed by the interaction of atmospheric oxygen with fat so that their content increases as decomposition progresses. The peroxide value – based on one kilogram of sample substance – is defined as the amount of substance in mmol ¼ O_2 that iodide is able to oxidize to elemental iodine ("milliequiva-

lents of active oxygen" [8a]). The resulting iodine is finally titrated with sodium thiosulfate solution, and starch is added as indicator shortly before the equivalence point.

Procedure [8a]: Approximately 5.00 g of sample substance are accurately weighed, placed in a 250 ml
iodine value flask and dissolved in a mixture of two parts by volume chloroform and three parts by
volume glacial acetic acid. After adding 0.5 ml saturated KI solution, shake for exactly 1 min and then
immediately add 30 ml water and titrate with $Na_2S_2O_3$ solution (0.01 mol/l). Shortly before reaching
the equivalence point, add some starch solution. The equivalence point is reached when the blue coloration disappears. The consumption of standard solution is V_1. A blank test is carried out under the
same conditions and V_2 is obtained. The peroxide value, related to the sample weight m_s, is obtained
from the following equation:

$$\text{Peroxide value} = \frac{10(V_1 - V_2)}{m_s}$$

with V_1 and V_2 in ml and m_s in g.

3.3.7.18 Determination of higher oxides

To determine a series of substances that can occur in two different, well-defined oxidation states, Bunsen (1853) specified a distillation process. It is based on the fact that
when concentrated solutions of hydrogen halides are exposed to the substances in the
higher oxidation state – e.g. higher oxides such as lead dioxide – halogen is set free,
distilled off in a suitable apparatus and then collected in a cooled, excess potassium
iodide solution:

$$PbO_2 + 4\,HCl \rightarrow PbCl_2 + 2\,H_2O + Cl_2 \uparrow.$$

That according to the equation

$$Cl_2 + 2\,I^- \rightarrow I_2 + 2\,Cl^-$$

released iodine is then titrated with sodium thiosulfate solution.

As an example of the historical development of an apparatus, the modification
that the original apparatus used by Bunsen has undergone over the course of time is
described in more detail here.

Bunsen used a simple apparatus for these determinations, consisting of a round
decomposition flask of about 50–80 ml capacity with a not too narrow and short neck.
The neck is connected by a piece of rubber tubing to a longer transfer tube, bent
twice, which is inserted into the belly of an inverted retort, the neck of which has one
or more spherical extensions. The belly of the retort is completely filled with potassium iodide solution, the neck with the spherical extensions only to a small extent.
During distillation, it should be able to completely absorb the potassium iodide solution displaced from the retort belly by the passing air. Fig. 3.17 illustrates the apparatus described. When working with this apparatus, the following possible errors must
be taken into account: On the one hand, the receiver liquid can very easily rise into

the decomposition bulb during distillation, but especially toward the end of the determination. On the other hand, iodine losses can occur, both as a result of evaporation from the retort neck, which is difficult to cool, and after the distillation has been completed by pouring the receiver liquid from the retort into a vessel suitable for the titration. Finally, a small loss of chlorine can already be expected during the charging of the apparatus if higher oxides are to be determined, which are already quickly attacked by cold, concentrated hydrochloric acid. Many attempts have therefore been made to redesign the apparatus and create a more suitable one (Ullmann, 1894; Marc, 1902; Farsoe, 1907).

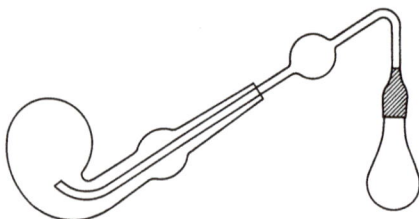

Fig. 3.17: Apparatus for the determination of higher oxides according to Bunsen.

But Rupp (1918, 1928) recognized that, in addition to the shortcomings of the apparatus, the method could also have a fundamental flaw: the chlorine distilling into the sample at the same time as the water vapors can be noticeably reduced according to the equation:

$$2\,H_2O + 2\,Cl_2 \rightleftharpoons 4\,HCl + O_2$$

can be converted back into hydrogen chloride and thus partially loses its iodometric effect, as the oxygen that is produced at the same time only has a very slow and inert effect on an acidified potassium iodide solution with iodine deposition. For the most part, it passes through the receiving liquid without any effect.

Based on the consideration that the proportion of chlorine that is lost for the determination due to the aforementioned reaction with the water vapor is all the smaller: (1) the higher the concentration of hydrogen chloride in the decomposition flask and thus also in the vapors that pass over (law of mass action!) and (2) the shorter the time that water vapor and chlorine are present next to each other or the smaller the space between the liquid surface in the decomposition flask and the end of the transfer tube at the point of contact with the feed liquid, the improved distillation apparatus illustrated in Fig. 3.18 was later created and a method of operation was specified which allows the chlorine reduction to be avoided and also prevents the troublesome rising back of the potassium iodide solution introduced (Jander and Beste, 1924).

The analysis substance is decomposed in a small, pear-shaped distillation flask with a capacity of around 60–80 ml, to which a short, only 40 cm long transfer tube, bent downward at a right angle, is attached at the side. A gas-tight dropping funnel with a glass stopcock and a capacity of about 20 ml is ground into the neck of the bulb, the outlet tube of which extends to a point almost at the bottom of the bulb. At the top, the dropping funnel is connected to a carbon dioxide developer by a short glass tube bent at right angles and passing through a rubber stopper. The transfer tube extends from the decomposition bulb to the bottom of a 200–300 ml Erlenmeyer flask into the neck of which it is inserted through a double-bored rubber stopper. A short, rectangularly bent glass tube leads through the other opening of the rubber stopper and is connected to a Péligot tube – glass to glass – by a short piece of tubing.

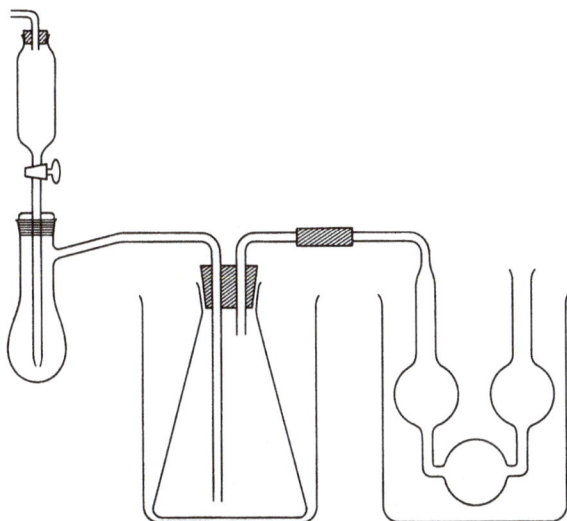

Fig. 3.18: Apparatus for the determination of higher oxides according to Jander and Beste.

The Erlenmeyer flask and Péligot tube are used to hold the excess potassium iodide solution and are kept in ice water during distillation. The substance to be determined is weighed into the decomposition flask or pipetted in as concentrated a solution as possible. The dropping funnel takes up the concentrated hydrochloric acid required for decomposition, i.e. about 40 ml. At the start of the analysis, the hydrochloric acid is carefully introduced into the decomposition flask under the pressure of carbon dioxide, and the liquid in the flask is heated to a very low boil and distilled for about 30 min while slowly passing carbon dioxide through it. After the distillation is complete, the contents of the Péligot tube, which should show at most a very slight iodine coloration, are rinsed into the Erlenmeyer flask and its contents titrated with sodium thiosulfate solution (0.1 mol/l). The method gives excellent results. It has already been described in connection with the titration of the sodium thiosulfate solution with potassium dichromate (see p. 199 ff.).

Hahn (1930) overdrives the chlorine formed with carbon tetrachloride, also using a modified Bunsen apparatus.

According to the Bunsen method, the following analytes can be determined (see Tab. 3.9).

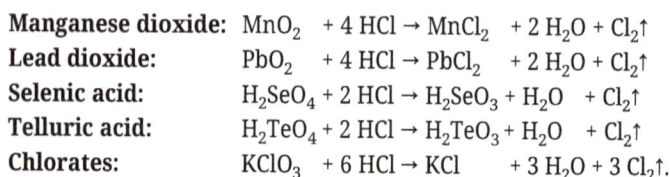

Manganese dioxide:	MnO_2	$+ 4\,HCl \rightarrow MnCl_2 + 2\,H_2O + Cl_2\uparrow$
Lead dioxide:	PbO_2	$+ 4\,HCl \rightarrow PbCl_2 + 2\,H_2O + Cl_2\uparrow$
Selenic acid:	$H_2SeO_4 + 2\,HCl \rightarrow H_2SeO_3 + H_2O + Cl_2\uparrow$	
Telluric acid:	$H_2TeO_4 + 2\,HCl \rightarrow H_2TeO_3 + H_2O + Cl_2\uparrow$	
Chlorates:	$KClO_3 + 6\,HCl \rightarrow KCl + 3\,H_2O + 3\,Cl_2\uparrow.$	

Tab. 3.9: Determination of higher oxides using the Bunsen method.

Substance	Appropriate weight about	1 ml sodium thiosulfate solution, $c(Na_2S_2O_3) = 0.1$ mol/l, corresponds to		
MnO_2	0.2 g	4.348 mg MnO_2	Or	2.747 mg Mn
PbO_2	0.5 g	11.960 mg PbO_2	Or	10.360 mg Pb
K_2SeO_4	0.3 g	11.058 mg K_2SeO_4	Or	3.948 mg Se
K_2TeO_4	0.4 g	13.490 mg K_2TeO_4	Or	6.380 mg Te
$KClO_3$	0.05 g	2.043 mg $KClO_3$	Or	1.391 mg ClO_3^-
V_2O_5	0.2 g	9.094 mg V_2O_5	Or	5.094 mg V
CeO_2	0.4 g	17.212 mg CeO_2	Or	14.012 mg Ce

Vanadium(V) oxide can, according to Holverscheidt (1890), only be uniformly reduced to blue vanadium(IV) oxide bromide with hydrogen bromide solution instead of hydrochloric acid:

$$V_2O_5 + 6\,HBr \rightarrow 2\,VOBr_2 + 3\,H_2O + Br_2.$$

Bromine is therefore overdistilled here. Add 2 g of potassium bromide to the vanadate weighed into the decomposition flask and proceed as for the other determinations.

Cerium(IV) compounds: Pure cerium(IV) oxide can only be reduced by hydrogen iodide:

$$2\,CeO_2 + 2\,KI + 8\,HCl \rightarrow 2\,KCl + 2\,CeCl_3 + 4\,H_2O + I_2.$$

Here, iodine is distilled off. Add 2 g of potassium iodide to the cerium(IV) salt weighed into the decomposition flask and distil in the usual way after adding 40 ml of hydrochloric acid.

3.3.7.19 Determination of copper

The iodometric determination of copper(II) salts according to de Haën (1854) and Low (1905) is based on the equilibrium reaction:

$$2\,Cu^{2+} + 4\,I^- \rightleftharpoons 2\,CuI\downarrow + I_2.$$

CuI is poorly soluble and precipitates as a yellowish-white precipitate. Due to this and the fact that the released iodine is continuously removed during titration with sodium thiosulfate, it is possible to place the reaction predominantly on the right-hand side. However, it is only practically complete if the concentration of iodide ions is also very high due to the addition of a large excess of KI.

Although the excess KI dissolves the precipitated CuI again, the resulting iodocuprate(I) complex ions have no influence on the equilibrium position. The titration must be carried out in a weak sulfuric acid solution. At higher concentrations, mineral acids, especially hydrochloric acid, partially dissolve the precipitated CuI: Cu(I) ions are then present in the solution, which induce the oxidation of the hydrogen iodide by atmospheric oxygen and can lead to an additional consumption of sodium thiosulfate solution.

> **Procedure:** Approximately 0.6 g of the copper(II) salt to be determined is accurately weighed in a sealable bottle and dissolved in 50 ml of water. After acidification with 2 ml concentrated sulfuric acid, 2 g iodate-free potassium iodide is added to the solution, and the bottle is closed and shaken for a short time. Then titrate with sodium thiosulfate solution (0.1 mol/l), until the solution is only slightly yellow. After adding 2 ml of starch solution (see p. 197), titrate slowly to the end with continuous swirling. The end point is reached when the bluish hue has just disappeared and the cloudy liquid only appears yellowish to brownish white. Iron and arsenic must not be present:
> 1 ml sodium thiosulfate solution, $c(Na_2S_2O_3) = 0.1$ mol/l, corresponds to 6.3546 mg Cu.

The method described by de Haën in 1854 [119] for the determination of copper requires a lot of potassium iodide, the price of which is quite high. Bruhns [120] has therefore proposed a variant that allows iodide to be saved by adding potassium thiocyanate to the solution in addition to less potassium iodide. As copper(I) thiocyanate is more difficult to dissolve than copper(I) iodide, the solutions prepared according to

$$2\,Cu^{2+} + 2\,I^- \rightleftharpoons 2\,Cu^+ + I_2$$

Copper(I) ions are not precipitated as CuI, but as CuSCN:

$$Cu^+ + SCN^- \rightarrow CuSCN\downarrow.$$

If the released iodine is titrated with thiosulfate solution, the iodide ions are released after

$$I_2 + 2\,S_2O_3^{2-} \rightarrow 2\,I^- + S_4O_6^{2-}$$

and can reduce Cu again^{2+}. Much less potassium iodide is therefore required, as it only plays an intermediary role in the overall reaction:

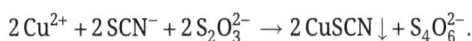

$$2\,Cu^{2+} + 2\,SCN^- + 2\,S_2O_3^{2-} \rightarrow 2\,CuSCN\downarrow + S_4O_6^{2-}.$$

However, this seemingly elegant method has the disadvantage that underdetection often occurs as a result of side reactions, which has led to numerous investigations. Accordingly, iodine can oxidize some thiocyanate to form H_2SO_4, and HCN and HCN can further react with iodine to form ICN. The interference can only be kept to a minimum if the acidification is sufficiently strong and the titration is carried out immediately after adding the iodide-thiocyanate mixture. Thus, according to the instructions given below, there is a constant underconsumption of thiosulfate solution of 0.5%, which can, however, be eliminated by adjusting the thiosulfate solution against a copper(II) salt solution of known content. According to work by Bastius [121], the acid concentration is of decisive importance: the titration is successful in sulfuric, hydrochloric or nitric acid solutions (but not in acetic acid) if the acid concentration is greater than 1 mol/l. If you are working in nitric acid solution, add urea and keep c $(HNO_3) < 2$ mol/l.

To analyze a copper-containing alloy, the sample is dissolved in a mixture of sulfuric acid and nitric acid. The resulting nitrous acid is removed by adding urea:

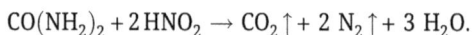

$$CO(NH_2)_2 + 2\,HNO_2 \rightarrow CO_2 \uparrow + 2\,N_2 \uparrow + 3\,H_2O.$$

If the alloy contains more than 0.2% iron, this must be complexed by adding sodium diphosphate. Mercury and silver must not be present. Lead, on the other hand, makes it easier to detect the end point, as in its presence the dirty gray-violet copper(I) thiocyanate takes on a faint yellowish hue, from which the faint blue color of the iodine strength just before the end point is reached stands out better.

Procedure: The initial weight of the alloy, which should contain about 0.2 g copper, is heated to boiling with 10 ml of a mixed acid (500 ml H_2SO_4, diluted 1: 1, 200 ml HNO_3, $\rho = 1.40$ g/ml and 300 ml H_2O) until everything is dissolved and no more brown vapors appear. Then add 10 ml of a solution containing 100 g urea, 1.5 g lead nitrate and a little nitric acid per liter. The mixture is shaken vigorously and cooled to room temperature. Now add 10 ml of a solution containing 100 g KSCN and 10 g KI per liter, shake thoroughly and titrate the dirty green solution immediately with thiosulfate standard solution. The precipitate takes on a violet-gray color. Toward the end of the titration, 5 ml of starch solution is added. The now dark blue solution is slowly titrated to the end drop by drop with thiosulfate solution. Recognition of the end point is facilitated by the fact that the yellowish gray precipitate clumps together and begins to settle.

Procedure (variant): About 80–100 mg of chips are weighed out (containing about 50–70 mg of copper) and dissolved in about 10–15 ml of semiconcentrated nitric acid. The resulting nitrogen oxides are then boiled away. Five milliliters of 20% urea solution is added to destroy the nitrous acid formed, followed by several milliliters of diluted NH_3 solution until a deep blue color appears (pH approx. 8–9). The pH value is now brought to 3–4 with semiconcentrated acetic acid and finally cooled. Add 1 g of sodium fluoride to the buffered solution to mask the iron and 2.5 g of potassium iodide to produce a quantity of elemental iodine equivalent to the oxidizing effect of copper(II). To ensure that the reaction is complete, leave the sample in a beaker with a watch glass cover in the dark and sealed for 10 min. Now titrate with sodium thiosulfate solution $c(Na_2S_2O_3) = 0.1$ mol/l. Toward the end of the titration (indicated by the brown color becoming lighter), 1.5 g of ammonium thiocyanate and 2.0 ml of freshly prepared starch solution are added. The color changes from a brownish flocculent suspension via dark blue to a pure white (possibly beige).

3.3.7.20 Determination of oxygen according to Winkler

The method is used to determine the dissolved oxygen in water samples. It is based on the fact that the oxidation capacity of the oxygen is transferred to the manganese through a redox reaction with manganese(II) and thus fixed. The reaction takes place in an alkaline solution and produces a sediment of manganese(IV) oxide hydrate:

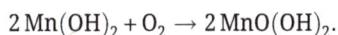

$$2\,Mn(OH)_2 + O_2 \rightarrow 2\,MnO(OH)_2.$$

When acid is added, manganese(III) ions are released from this precipitate by the action of excess manganese(II) through comproportionation, which react with iodide:

$$2\,Mn^{3+} + 2\,I^- \rightarrow 2\,Mn^{2+} + I_2.$$

The iodine released in this way is now titrated with sodium thiosulfate solution, whereby starch is added shortly before the equivalence point [121a].

Procedure [121a, 121b]: The samples are taken using a special "oxygen bottle" and stored free of air bubbles with the stopper closed. Very soon after sampling, 1 ml of $MnCl_2$ solution ($MnCl_2$ tetrahydrate 667 g/l) and 1 ml of precipitating reagent (NaOH 410 g/l, KI 350 g/l; if necessary 12 g/l NaN_3 to destroy NO_2^-) are added to the bottle with the pipette tip immersed so that only excess water can run over the edge of the bottle when the stopper is subsequently closed. The sample can now be stored for a maximum of 2 days in the dark at a constant temperature or can be treated immediately. One option is to titrate the sample later in the sampling bottle. To do this, carefully pipette off about a third of the solution above the settled precipitate. Make sure that the iodide present in the solution is sufficient for the reaction with the manganese(III). Add 5 ml of phosphoric acid 50% with the pipette tip immersed, close the bottle again, mix by shaking and keep the mixture in the dark for 10 min. Titrate with $Na_2S_2O_3$ solution (0.01 mol/l); add a little starch solution shortly before the equivalence point. The equivalence point is reached when the blue coloration disappears for the first time. The consumption of standard solution is V_M. The filling volume of the oxygen bottle is V_P.

The mass concentration of dissolved oxygen in the water sample is calculated according to the following equation:

$$\beta(O_2) = \frac{V_M \cdot 80}{V_P - 2} \quad \text{in mg/l}$$

with V_M and V_P in ml.

3.3.7.21 Determination of the iodine value

The storage behavior and nutritional properties of fats are strongly influenced by the content of unsaturated bonds. One measure of this is the iodine value, which is the mass of iodine in grams that can be bound by 100 g of sample substance. The iodine value increases with the number of unsaturated bonds in the molecule, albeit not linearly, but disproportionately.

Several variants based on the addition of halogen atoms to the unsaturated bonds are proposed for titrimetric determination. The addition reagents used are, for example, elemental bromine (Kaufmann method), iodine monochloride (Wijs method) or iodine monobromide (Hanus method). A sufficient quantity of one of these reagents is added to the sample substance; after the reaction has taken place, an excess of potassium iodide is added and then titrated back with sodium thiosulfate solution, whereby starch is added shortly before the equivalence point.

Procedure [8a]: First, a suitable initial weight is determined, e.g. 15 times the reciprocal of the expected iodine number as a mass in grams. Approximately this mass of sample substance is weighed out precisely and dissolved in a 250-ml iodine value flask rinsed with glacial acetic acid dissolved in 15 ml chloroform. Gradually add 25.0 ml IBr solution (20 g/l in glacial acetic acid), close the flask and shake for 30 min in the dark. Finally, add 10 ml KI solution 100 g/l and 100 ml water and titrate with $Na_2S_2O_3$ solution (0.1 mol/l). Shortly before reaching the equivalence point, add a little starch solution. The equivalence point is reached when the blue coloration disappears. The consumption of standard solution is V_1. A blank test is carried out under the same conditions and V_2 is obtained. The iodine value, related to the sample weight m_s, is obtained from the following equation:

$$\text{Iodine value} = \frac{1.269 \ (V_2 - V_1)}{m_s},$$

with V_1 and V_2 in ml and m_s in g.

The iodometric determination methods are characterized by their elegance of execution and high reliability, as the titration end point can be easily identified by the iodine starch reaction. The relatively high price of iodine and potassium iodide has always stood in the way of their applicability, especially for serial determinations. For this reason, efforts have often been made to replace iodometry with **bromometry**. However, difficulties arose due to the considerably higher vapor pressure of bromine solutions compared to iodine solutions and the resulting lower titer stability as well as the fact that the titration end point is not so easy to detect.

Some bromatometric procedures lead to the formation of elemental bromine, which oxidizes iodide to iodine, which is titrated with a thiosulfate solution (see Section 3.3.6).

3.3.7.22 Determination of acetylcysteine according to Ph. Eur

Mercaptans such as in the drugs acetylcysteine (see formula), captopril or in the amino acid cysteine can be quantitatively oxidized to disulfides by iodine solution.

The secretolytic agent acetylcysteine is quantified by direct iodometric titration in Ph. Eur [8a].

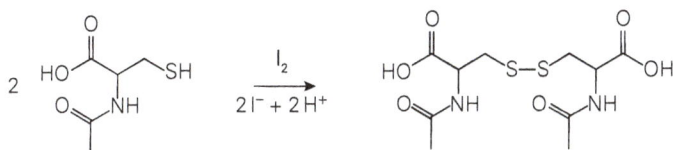

Procedure: Dissolve 0.140 g substance in 60 ml water and add 10 ml diluted hydrochloric acid (20 g hydrochloric acid 36% in 100 ml water). After cooling, 10 ml of potassium iodide solution (166 g/l) is added to the solution in an ice-water mixture. This solution is titrated with iodine solution (0.05 mol/l) with the addition of 1 ml starch solution (10 g/l)·

The addition of mercury(II) iodide for preservation as prescribed in pharmacopoeias should be omitted for environmental toxicological reasons:

1 ml iodine solution, $c = 0.05$ mol/l, corresponds to 16.32 mg $C_5H_9NO_3S$.

3.3.7.23 Determination of cysteine HCl according to Ph. Eur

The salt cysteine hydrochloride is determined in Ph. Eur. by iodometric back titration [8a].

Procedure: In an Erlenmeyer flask with ground-glass stopper, dissolve 0.300 g substance and 4 g potassium iodide in 20 ml water. After cooling in an ice-water mixture, 3 ml hydrochloric acid (70 g hydrochloric acid 36% in 100 ml water) and 25.0 ml iodine solution (0.05 mol/l) are added to the solution. The flask is closed and left to stand in the dark for 20 min. Titrate the solution with sodium thiosulfate solution (0.1 mol/l), adding 3 ml starch solution (10 g/l) toward the end of the titration. A blank titration is carried out.

The addition of mercury(II) iodide for preservation as prescribed in pharmacopoeias should be omitted for environmental toxicological reasons:

1 ml iodine solution, $c = 0.05$ mol/l, corresponds to 15.76 mg $C_3H_8ClNO_2S$.

3.3.7.24 Determination of phenazone

The analgesic phenazone can be quantified by iodometric back titration. Phenazone is substituted to iodophenazone by an excess of iodine standard solution. The excess is then back-titrated with sodium thiosulfate against starch.

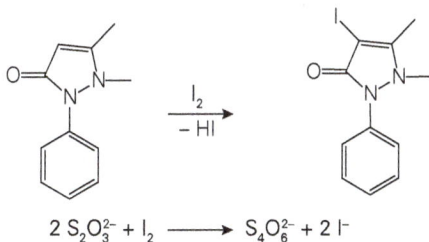

$$2 S_2O_3^{2-} + I_2 \longrightarrow S_4O_6^{2-} + 2 I^-$$

Procedure: About 0.150 g substance is dissolved in 20 ml water and, after adding 2 g sodium acetate, 1 ml dilute acetic acid and 25.0 ml iodine solution (0.05 mol/l) are added and left to stand for 30 min under light protection. Add 25 ml dichloromethane to the solution, shake until the precipitate dissolves and titrate with sodium thiosulfate (0.1 mol/l), adding 1 ml starch solution (10 g/l) toward the end of the titration. A blank titration is carried out:

    1 ml iodine solution, c = 0.05 mol/l, corresponds to 9.41 mg $C_{11}H_{12}N_2O$.

3.3.8 Determination of manganese in steel with arsenite standard solution

To determine the manganese in steel samples, the sample substance is first dissolved in semiconcentrated nitric acid. After boiling off the nitrogen oxides, the resulting manganese(II) is oxidized with peroxodisulfate in the presence of silver ions as a catalyst to form permanganate:

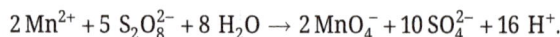

$$2\,Mn^{2+} + 5\,S_2O_8^{2-} + 8\,H_2O \rightarrow 2\,MnO_4^- + 10\,SO_4^{2-} + 16\,H^+.$$

Excess peroxo oxygen decomposes under these conditions. The silver ions are then precipitated with NaCl at room temperature so that they cannot be detected as oxidizing agents during the titration. Now titrate with sodium arsenite solution until the pink coloration disappears:

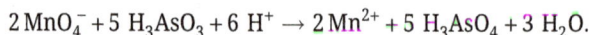

$$2\,MnO_4^- + 5\,H_3AsO_3 + 6\,H^+ \rightarrow 2\,Mn^{2+} + 5\,H_3AsO_4 + 3\,H_2O.$$

It is unlikely that chromium contained in the steel could be captured in this process, as the permanganate is first reduced due to its higher standard potential and then the pink coloration disappears before chromium can be converted.

3.3.8.1 Preparation of the sodium arsenite solution

Procedure: About 0.666 g As_2O_3 and 2 g $NaHCO_3$ are dissolved in hot water. The solution is cooled to room temperature and then made up to 1 l. In this way, a sodium arsenite solution with $c(\frac{1}{2}\,As_2O_3)$ = 0.00666 mol/l is obtained. The solution is adjusted using a standard steel.

3.3.8.2 Determination of manganese in steel

Procedure (Procter Smith): A portion of steel containing about 0.5–3.0 mg manganese is weighed accurately and dissolved in 15 ml nitric acid (7 mol/l) and 20 ml water. The nitrogen oxides are removed by boiling for 5–10 min. The solution is cooled, and 20 ml boiled nitric acid (7 mol/l), 50 ml silver nitrate solution (1.7 g/l) and 4 ml fresh $(NH_4)_2S_2O_8$ solution (500 g/l) are added. The mixture is left at 60–80 °C for about 5 min – violet coloration due to permanganate – and then cooled to room temperature. After adding 50 ml water and 3 ml sodium chloride solution (12 g/l), titrate with sodium arsenite solution until the pink coloration has disappeared:

One milliliter sodium arsenite solution, $c(½ As_2O_3) = 0.00666$ mol/l, corresponds to 0.07325 mg manganese [121c, 121d].

3.3.9 Determinations with formate standard solution

As already mentioned in the treatment of permanganometric determinations (Section 3.3.2), various analytes (e.g. phosphite, hypophosphite, methanol and formaldehyde) cannot be titrated permanganometrically in acidic solution, but undergo a defined redox reaction with permanganate in alkaline solution, whereby the latter is reduced to manganate:

$$MnO_4^- + e^- \rightarrow MnO_4^{2-}.$$

Further reduction of the manganate to manganese dioxide takes place much more slowly:

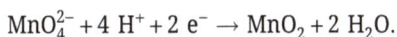

$$MnO_4^{2-} + 4\ H^+ + 2\ e^- \rightarrow MnO_2 + 2\ H_2O.$$

In order to obtain a suitable determination method based on the reduction of permanganate to manganate, the second reaction must be sufficiently inhibited. Stamm proposes several procedures for this [121e, 121f, 121g]:

- Because oxonium ions are required for the further reaction of the manganate, a high pH value > 14 is set, which both kinetically inhibits this reaction and shifts the equilibrium toward the manganate.
- A constant excess of permanganate during the entire reaction leads to a strongly oxidizing environment in the solution and thus hinders the further reaction of the manganate. To achieve this, the sample solution is added to such an excess of potassium permanganate solution that no more than 50% of the permanganate is consumed. However, at least 10% should be used so that the reading error does not have too great an effect on the precision of the analysis result.
- During the subsequent back titration of the unused permanganate, manganate is precipitated as sparingly soluble barium manganate with the aid of added barium chloride, thus hindering its further reaction.

This method therefore requires that the excess amount of permanganate is determined by back titration. This can be done using sodium formate solution in an alkaline environment or alternatively with oxalic acid in an acidic environment. The titration of permanganate with formate is based on a defined reaction to form manganate and carbonate:

$$2 \text{ MnO}_4^- + \text{HCOO}^- + 3 \text{ OH}^- \rightarrow 2 \text{ MnO}_4^{2-} + \text{CO}_3^{2-} + 2 \text{ H}_2\text{O}.$$

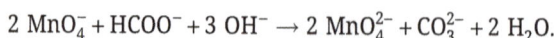

The equivalence point is recognized by the fact that the violet color of the permanganate disappears. Dark green barium manganate precipitates during the titration and settles quickly so that the precipitate does not interfere with the detection of the equivalence point. Sulfate interferes with the reaction by forming sparingly soluble barium sulfate, which absorbs permanganate in the form of solid solutions during the titration and thus prevents its reduction. Therefore, sulfate must be precipitated as barium sulfate before the analysis, which then no longer interferes with the further reactions. The speed of the titration reaction naturally slows down shortly before the equivalence point, so some nickel nitrate is added as a catalyst at this point to prevent overtitration.

3.3.9.1 Preparation of the sodium formate solution

Procedure [28]: About 3.400 g dried HCOONa and 5 g NaOH are dissolved in boiled and cooled water. When making up to 1 l, a standard solution of concentration $c = 0.05$ mol/l is obtained. The equivalent concentration $c(\text{eq})$ is 0.1 mol/l.

The solution is prepared as follows: $V_1 = 20.0$ ml KMnO$_4$ solution (0.1 mol/l) is mixed with 10 ml sodium hydroxide solution 300 g/l and 15 ml of a solution of barium chloride dihydrate 300 g/l and then diluted with water to approximately 100 ml in the titration flask. Now titrate with the standard solution to be adjusted until the supernatant solution is decolorized. After approximately 90% of the volume of standard solution used in the preliminary test has been consumed, some nickel nitrate is added repeatedly if necessary. The consumption of standard solution determined in this way is V_2. The titer results from $t = V_1/V_2$.

3.3.9.2 Determination of hypophosphite (phosphinate)
Hypophosphite reacts with permanganate as follows:

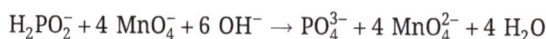

$$\text{H}_2\text{PO}_2^- + 4 \text{ MnO}_4^- + 6 \text{ OH}^- \rightarrow \text{PO}_4^{3-} + 4 \text{ MnO}_4^{2-} + 4 \text{ H}_2\text{O}$$

Procedure: To $V_1 = 20.0$ ml KMnO$_4$ solution (0.1 mol/l) add approximately 10 ml sodium hydroxide solution (300 g/l). Add the sample solution, which should contain 0.05–0.25 mmol hypophosphite in order of magnitude and in which any sulfate present has previously been precipitated with barium chloride. The mixture is allowed to stand for 10 min at 15–25 °C. Add 15 ml of a 300 g/l solution of barium chloride dihydrate. After dilution with water in the titration flask to about 100 ml, titrate with sodium formate solution 0.05 mol/l until the supernatant solution is decolorized. If less than 10 ml of standard solution is used, repeat with a smaller amount of analyte. After approximately 90% of the volume of standard solution used in the preliminary test has been consumed, some nickel nitrate is added repeatedly if necessary. The volume of standard solution consumed in this way is V_2. The difference $V_1 - V_2$ gives the volume of $KMnO_4$ solution consumed during the conversion of the analyte 0.1 mol/l:

1 ml of this corresponds to 1.650 mg H_3PO_2 [28].

3.3.9.3 Determination of phosphite (phosphonate)

Phosphite reacts with permanganate as follows:

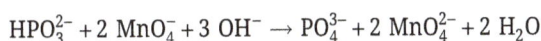

$$HPO_3^{2-} + 2\ MnO_4^- + 3\ OH^- \rightarrow PO_4^{3-} + 2\ MnO_4^{2-} + 2\ H_2O$$

Procedure: The determination is carried out in the same way as for hypophosphite. To $V_1 = 20.0$ ml $KMnO_4$ solution (0.1 mol/l), add approximately 10 ml sodium hydroxide solution (300 g/l). Add the sample solution, which should contain 0.1–0.5 mmol phosphite in order of magnitude and in which any sulfate present has previously been precipitated with barium chloride. The mixture is allowed to stand for 10 min at 15–25 °C. A volume of 15 ml of a solution of barium chloride dihydrate (300 g/l) is added. After dilution with water in the titration flask to about 100 ml, titrate with sodium formate solution (0.05 mol/l) until the supernatant solution is decolorized. If less than 10 ml of standard solution is used, repeat with a smaller amount of analyte. After approximately 90% of the volume of standard solution used in the preliminary test has been consumed, some nickel nitrate is added repeatedly if necessary. The volume of standard solution consumed in this way is V_2. The difference $V_1 - V_2$ gives the volume of $KMnO_4$ solution consumed during the conversion of the analyte:

1 ml $KMnO_4$ solution, $c = 0.1$ mol/l, corresponds to 4.100 mg H_3PO_3 [28].

3.3.10 Other possibilities for redox titration

In addition to the oxidizing and reducing agents discussed here, which are used in the standard solutions for redox titrations, numerous other reagents have been described in the literature that are used for volumetric analysis determinations. Potassium iodate and potassium periodate, for example, can be used as oxidizing agents. and potassium periodate [122], lead(IV) acetate and vanadate(V), iron(III) salts and cobalt(III) compounds [123]. Reducing agents that can be used include tin(II) chloride, chromium(II) sulfate, titanium(III) chloride and vanadium(II) sulfate [125]. These examples show the wide range of possibilities that lie in the applications of analytical redox methods for the determination of inorganic and organic compounds. For further information, reference must be made to more comprehensive works and monographs [27, 123].

In many cases, the methods only became more important after it was possible to eliminate the difficulties of end point detection with visual indicators by using the further developed physical indication methods, in particular electrical methods, to determine the titration end point. Electrochemical reagent generation in the sample solution using coulometry also made it possible to use reagents for analytical purposes whose standard solutions are unstable and therefore not titer-stable (see Chapter 4).

3.4 Complex formation titrations

In volumetric analysis, the use of chemical reactions involving the formation of complexes was for a long time essentially limited to the titration of cyanide with silver nitrate solution [100] introduced by Liebig in 1851 (see p. 146 f.).

The basis for further development 85 years later was the discovery that certain aminopolycarboxylic acids [126] – compounds with the grouping

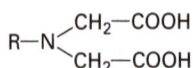

$$R-N\diagdown^{CH_2-COOH}_{CH_2-COOH}$$

– form stable water-soluble complex compounds with metal ions (also with alkaline earth metal ions). Initially, attempts were made to use the substances technically for water softening [127]. With regard to their analytical usability, they remained unnoticed for almost 10 years. Based on physicochemical investigations of these compounds, Schwarzenbach was the first who developed titration methods for metal ions with the complexing agents in 1945 [128, 129]. He introduced the name **complexones** for the aminopolycarboxylic acids and summarized the titration methods in which complexones are used in standard solutions under the term **complexometry** [130, 131]. Initially, the metal ions were determined by alkalimetric titration of the hydrogen ions released during complex formation. However, as early as 1946, the first indicator was introduced that responded directly to metal ions, the murexide [132]. The search for further **metal indicators** led to the azo dyes, first to the Eriochrome black T [132, 133]. Přibil and coworkers used the important indicator xylenol orange (XO) for the first time [134], with which titrations can be carried out at low pH values. Since the 1950s, this field of volumetric analysis has undergone rapid development – as can be seen from the number of published papers – which probably reached its peak at the beginning of the 1960s. Complexometry spread rapidly due to its versatility and is likely to have found its way into every analytical laboratory.

Today, most cations and some anions can be determined titrimetrically by direct or indirect methods with the help of complexones [135]. The appropriate choice of complexing agent and indicator, pH value and type of titration as well as the addition of auxiliary complexing agents numerous selective determinations can be carried out, whereby complicated separations can often be avoided and the analyses can be performed in a shorter time.

3.4.1 Basics of complex formation

3.4.1.1 Designations and definitions

A **complex** is a composite particle (ion or molecule) that is formed by the combination of simple molecules or ions that can exist independently of each other, e.g.

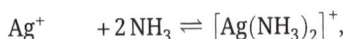

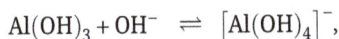

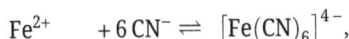

$$NH_3 \quad + HCl \quad \rightleftharpoons \quad [NH_4]^+ + Cl^-,$$

$$Ag^+ \quad + 2\,NH_3 \rightleftharpoons \left[Ag(NH_3)_2\right]^+,$$

$$Al(OH)_3 + OH^- \quad \rightleftharpoons \quad \left[Al(OH)_4\right]^-,$$

$$Fe^{2+} \quad + 6\,CN^- \rightleftharpoons \left[Fe(CN)_6\right]^{4-},$$

$$Ni \quad + 4\,CO \quad \rightleftharpoons \left[Ni(CO)_4\right].$$

The complex particles are usually placed in square brackets in formulas. As the examples show, complexes can occur as cations, anions or neutral particles.

According to a more comprehensive definition, the term "complex" is also extended to those composite particles whose formation can be imagined through the binding of atoms, ions or molecules by atoms or ions with free electron pairs, with electron gaps or with occupiable free d-orbitals. According to this, the anions NO_3^-, SO_4^{2-} and PO_4^{3-} are also complex ions. Further details can be found in textbooks on general or inorganic chemistry [32–35 and 136, 137].

Neutral complexes and compounds that contain ionic complexes are referred to as **complex compounds** or **coordination compounds**. Complex particles are also stable in solution and dissociate only slightly into their individual components. This distinguishes them from **lattice compounds**, which only exist as complex units in the crystalline state, but have largely disintegrated into their components in solution. Typical lattice compounds are alums, double sulfates with monovalent and trivalent cations, e.g. $KAl(SO_4)_2 \cdot 12\,H_2O$. There are transitions between the two types of compounds.

A characteristic feature of complex formation is that analytical reactions that are characteristic of one type of ion do not occur if the ions are complexed. For example, in a silver nitrate solution, no AgCl precipitate forms when chloride ions are added if the Ag^+ ions have previously been converted into the $[Ag(NH_3)_2]^+$ complex by adding ammonia solution. Such **masking** is often used when the effect of interfering ions is to be eliminated during separation, determination or detection operations.

3.4.1.2 Structure of the complexes

The center of a complex is the **central atom** or **central ion** which in the vast majority of cases is a cation and rarely an atom. As a rule, the central atom is referred to regardless of the charge carried by the particle. It is surrounded in a regular arrangement by the **ligands**. The ligands are generally anions, such as F^-, Cl^-, Br^-, I^-, OH^- and

CN^-, or neutral molecules such as NH_3, H_2O and CO. The cohesion between the ligands and the central ion is achieved by **coordinative bonds**. This is a **donor-acceptor bond** in which the ligand (donor) provides the binding electron pair and thus acts as a Lewis base, while the central atom (acceptor) accepts this electron pair and is therefore referred to as Lewis acid. The number of coordinative bonds in complex particles is called the **coordination number**. Coordination numbers 4 and 6 occur most frequently, less frequently 2 and 8. Complexes with odd coordination numbers are also known. Certain **molecular geometries** are assigned to the individual coordination numbers. For example, with a coordination number of 4, the four donor atoms in the ligands can be arranged in the form of a tetrahedron or a square around the central atom. Coordination number 6 corresponds to the geometric arrangement in the form of an octahedron, which often deviates from the regular structure and is distorted. In this case, the six ligand sites are no longer equivalent. A linear structure is found for the coordination number 2; the ideal arrangement of eight ligands is the corners of a cube. The coordination number is determined by steric factors, but above all by the number and type of atomic orbitals that are available for the formation of the coordinative bonds.

The **charge** of a complex is the sum of the charges of its components.

If there is only one coordinative bond between a ligand and the central atom, the ligand is called **monodentate** (unidentical). If two or more coordination sites are occupied by a ligand, it is referred to as **bidentate** or **multidentate** (bidental or multidental). If multidentate ligands occupy several coordination sites on the same central atom rings are formed. The complex is then referred to as a **chelate** or **chelate complex** and the ligand as a **chelating agent**. Not every ligand with two or more donor atoms can form a chelate ring. The donor atoms must be a minimum distance apart so that they can be bound to the same central atom. Hydrazine, for example, has, H_2N-NH_2, has two donor atoms but is only able to occupy one coordination site on the central atom, while ethylenediamine, $H_2N-CH_2-CH_2-NH_2$, acts as a bidentate ligand and occupies two coordination sites. The distance between the two donor atoms must allow stress-free ring formation. Therefore, certain ring sizes (five rings and six rings) are preferred.

Examples are

tris(ethylendiamine)platinum(IV)

tris(oxalato)ferrate(III)

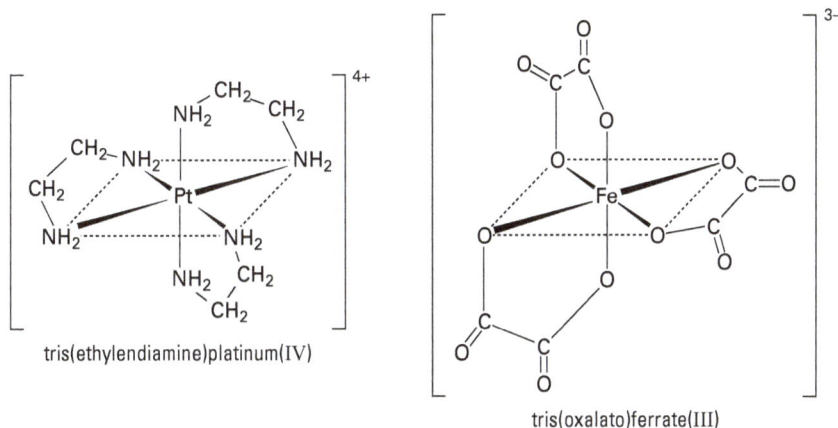

If a complex contains two or more central atoms, it is called **dinuclear** or **polynu-clear** (binuclear or polynuclear). The central atoms in the complex are connected via **bridging ligands.**

If the positive charges of the central ion in a chelate complex are compensated by the negative charges of the ligands, an **inner salt formation** occurs. Such neutral chelates are called **inner complexes** or **inner complex salts**. Bidentate ligands with one charged and one uncharged donor group form inner complexes if the coordination number is just twice as large as the positive charge of the central ion. If the positive charge is less than the maximum coordination number inner complex salts are formed in which the free coordination sites are occupied by neutral ligands.

3.4.1.3 Nomenclature rules
The following most important rules apply to the naming of complexes and complex compounds [8]:
1. In **formulas** and **names**, the cation is mentioned first, then the anion.
2. In **formulas** of complexes, the central atom is given first, followed by the ligands in the order anionic, neutral, cationic.
3. In **names** of complexes, the order (a) number of ligands, (b) type of ligands, (c) central atom and (d) oxidation number of the central atom must be observed:
   a) The number of identical ligands is indicated by the Greek numeral (di, tri, tetra, penta, hexa, etc.); for polydentate ligands and ligands whose names already begin with a numeral, the multiplicative numeral (bis, tris, tetrakis, etc.) is used.
   b) The ligands are listed in alphabetical order. Anionic ligands are given the suffix -o. The names of neutral ligands are not changed (exceptions: H_2O aqua, NH_3 ammine, CO carbonyl, NO nitrosyl).

c) In the case of a cationic complex, the English name of the element is used for the central atom; in the case of an anionic complex, the Latin root of the element is given the suffix -ate.

d) The oxidation number is expressed by a Roman numeral in brackets.

Example

$[Ag(NH_3)_2]Cl$	Diammine silver(I) chloride
$[CrCl_3(NH_3)_3]$	Triammine trichlorochrome(III)
$K_4[Fe(CN)_6]$	Potassium hexacyanoferrate(II)
$[CoCl_2(NH_3)_4]Cl$	Tetraamminedichlorocobalt(III) chloride
$Na_2[Fe(CN)_5NO]$	Sodium pentacyanonitrosylferrate(III)
$[Mg(C_9H_6NO)_2(H_2O)_2]$	Diaqua-bis(hydroxyquinolinato)-magnesium(II).

3.4.1.4 Stability constant

The formation and decomposition of complex compounds are equilibrium reactions to which the law of mass action can be applied, e.g.

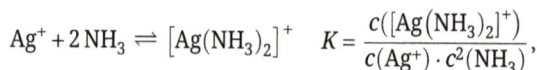

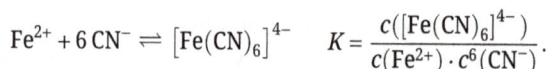

$$Ag^+ + 2\,NH_3 \rightleftharpoons \left[Ag(NH_3)_2\right]^+ \quad K = \frac{c([Ag(NH_3)_2]^+)}{c(Ag^+) \cdot c^2(NH_3)},$$

$$Fe^{2+} + 6\,CN^- \rightleftharpoons \left[Fe(CN)_6\right]^{4-} \quad K = \frac{c([Fe(CN)_6]^{4-})}{c(Fe^{2+}) \cdot c^6(CN^-)}.$$

The equilibrium constant K for the formation of complexes is known as the **complex formation** or **stability constant**, and the reciprocal value $K_D = 1/K$ is the **complex dissociation** or **instability constant**. The value K is a measure of the stability of a complex. If you compare complexes with the same coordination number the greater the stability constant of a complex, the more stable it is. Chelate complexes are particularly stable and therefore have very high K values. The stability constants are often given in the form of the **positive** decadic logarithms of their numerical values: $pK = \lg K$.

3.4.2 Basics of complexation titrations

Essential prerequisites for the suitability of a chemical reaction for volumetric analysis are the abrupt decrease in the concentration of the ion type to be determined near the equivalence point and the identification of suitable indication methods for this:

If, for example, a copper(II) salt solution is titrated with ammonia solution under conditions in which no precipitation of the hydroxide occurs, and the metal ion exponent $pCu = -\lg c(Cu^{2+})$ is applied against the mass ratio; then, due to the large stability constant of the copper tetraammine complex $K = 3.89 \cdot 10^{12}$, $pK = 12.59$, a clear jump in the equivalence point of the reaction should be expected:

$$Cu^{2+} + 4\,NH_3 \rightleftharpoons \left[Cu(NH_3)_4\right]^{2+}$$

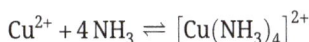

As shown in Fig. 3.19a, this is not the case. The reaction takes place in stages with the stability constants K_1–K_4 according to the following scheme:

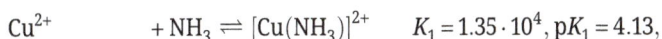

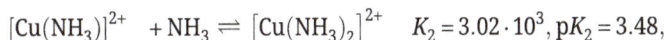

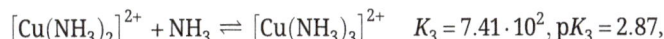

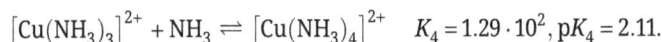

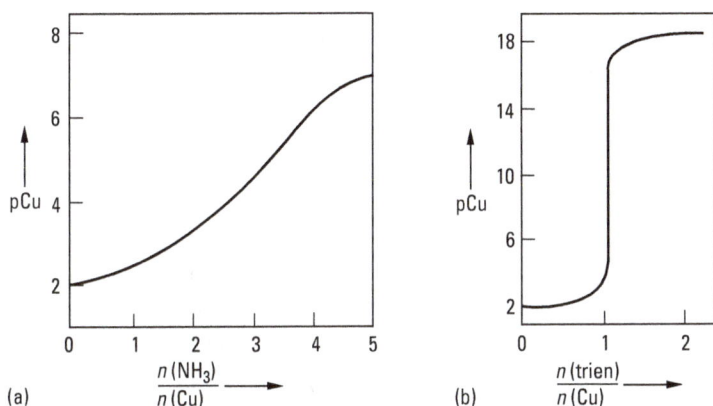

$$Cu^{2+} \qquad\quad + NH_3 \rightleftharpoons \left[Cu(NH_3)\right]^{2+} \qquad K_1 = 1.35 \cdot 10^4,\, pK_1 = 4.13,$$

$$\left[Cu(NH_3)\right]^{2+} \; + NH_3 \rightleftharpoons \left[Cu(NH_3)_2\right]^{2+} \qquad K_2 = 3.02 \cdot 10^3,\, pK_2 = 3.48,$$

$$\left[Cu(NH_3)_2\right]^{2+} + NH_3 \rightleftharpoons \left[Cu(NH_3)_3\right]^{2+} \qquad K_3 = 7.41 \cdot 10^2,\, pK_3 = 2.87,$$

$$\left[Cu(NH_3)_3\right]^{2+} + NH_3 \rightleftharpoons \left[Cu(NH_3)_4\right]^{2+} \qquad K_4 = 1.29 \cdot 10^2,\, pK_4 = 2.11.$$

Fig. 3.19: Titration of $CuSO_4$ solution (0.01 mol/l) with (a) ammonia solution and (b) triethylenetetramine solution (pCu = $-\lg c(Cu^{2+})$).

As can be seen from the constants, the values for K_1–K_3 cannot be neglected compared to K_4. The formation of the complex in one step at a ratio of $n(Cu) : n(NH_3)$ = 1:4 is therefore not to be expected. This would only be the case if the ratios K_1/K_2, K_2/K_3 and K_3/K_4 were very large, i.e. the first three stages were much more strongly dissociated in the solution compared to the fourth stage. Jumps in the mass ratios $n(Cu): n(NH_3)$ = 1:1, 1:2, 1:3 and 1:4 cannot occur either, because the values of the constants are not far enough apart. The ratios are similar to the titration of polybasic acids. The titration of dibasic sulfuric acid results in only one jump at the equivalence point, while that of tribasic phosphoric acid shows two jumps (see p. 123 f.).

As this observation shows, a complex formation reaction can only be analyzed if it is possible to eliminate the formation of intermediates. This is possible if the ligand is a compound that contains several oxygen or nitrogen atoms capable of coordination, which simultaneously bind the metal ion (formation of chelate complexes). The structure of such a compound is shown as an example in the formula for the reaction of Cu^{2+} with triethylenetetramine (triene):

$$H_2N - CH_2 - CH_2 - NH - CH_2 - CH_2 - NH - CH_2 - CH_2 - NH_2.$$

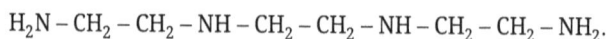

Resulting chelates:

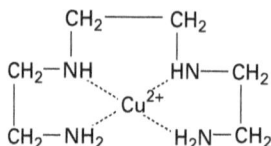

Copper ions react with triethylenetetramine in a ratio of 1:1, whereby the coordination number 4 of the copper(II) ion is fulfilled by the four nitrogen atoms of the amine. The chelate formation in the form of five-membered rings can be clearly seen in the formula image. The resulting large increase in the stability of the complex ($K = 3.16 \cdot 10^{20}$, pK = 20.5) is referred to as the **chelate effect**.

If a solution containing Cu^{2+} ions is titrated with a triethylenetetramine solution, the Cu^{2+} concentration actually decreases abruptly at a ratio of 1:1, as can be seen in Fig. 3.19b.

The reagents used for the complexation titrations are exclusively chelating agents; therefore the determination procedures carried out with them are referred to as **chelatometric titrations**. The aminopolycarboxylic acids already mentioned meet the requirements that must be met by a chelatometric titrant: good solubility, sufficiently high reaction rate with the ion to be determined, formation of an easily soluble stable chelate complex with the formation of several coordination bonds (multidentate ligand). The following compounds have proven to be suitable for universal use:

Nitrilotriacetic acid (NTE, H_3X)

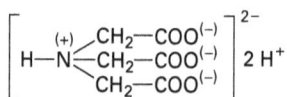

with four atoms capable of coordination (1N, 3O of the three monovalent O atoms of the carboxyl groups);

Ethylenediaminetetraacetic acid (EDTA, H_4Y)

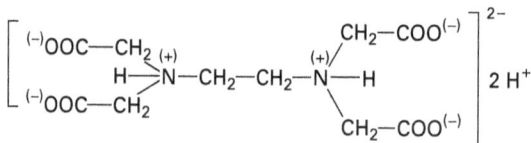

with six coordinating atoms (two N and four monovalent O atoms of the carboxyl groups). Due to its better solubility, the disodium salt of ethylenediaminetetraacetic acid is used in the form of the dihydrate.

Later, a whole series of other chelating compounds found their way into analytical practice. Mention should be made here of **diethylenetriaminepentaacetic acid** (DTPA), 1,2-cyclohexanediaminetetraacetic acid and *N*-(2-hydroxyethyl)-ethylenediaminetetraacetic acid. The compounds used in chelatometry are available under the trade names Titriplex® (Merck, Darmstadt) and Idranal® (Riedel-de Haën, Seelze).

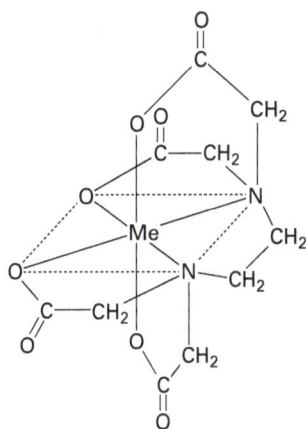

Fig. 3.20: Steric structure of a metal-EDTA complex.

Like most chelating compounds, EDTA is also a polyvalent acid. The neutral ethylenediaminetetraacetic acid is a four-proton acid (H_4Y) and, due to the two free electron pairs on the nitrogen atoms, can accept a further two protons at low pH values (H_6Y^{2+}). EDTA can therefore be present in seven different forms depending on the pH value: H_6Y^{2+}, H_5Y^+, H_4Y, H_3Y^-, HY_2^{2-}, HY^{3-} and Y^{4-}. The corresponding equilibrium constants are $pK_{a1} = 0.0$; $pK_{a2} = 1.5$; $pK_{a3} = 2.0$; $pK_{a4} = 2.69$; $pK_{a5} = 6.13$ and $pK_{a6} = 10.37$ [18], whereby the protons of the carboxyl groups are released first and then the protons are bound to the nitrogen atoms.

Although some protonated metal-EDTA complexes are known, such as $MgHY^-$ or $AlHY$, the formation constants of all metal ions with Y^{4-} are so dominant that the other complexes do not play a role.

Due to the six possible coordinative sites of the fully deprotonated species (Y^{4-}), EDTA forms a sixfold coordinated complex with almost all metal ions. The steric structure of a metal-EDTA complex is shown in Fig. 3.20. The metal ion as the central atom is enveloped octahedrally by the EDTA molecule. The formation of the five-membered chelate rings is clearly shown in the drawing.

Complex formation occurs with multivalent cations regardless of the ionic charge in a ratio of 1:1 and can be formulated as follows using nickel as an example:

$$Ni^{2+} + Y^{4-} \rightleftharpoons [NiY]^{2-}.$$

The corresponding complex formation constant is

$$K_F = \frac{c\left([NiY]^{2-}\right)}{c(Ni^{2+}) \cdot c(Y^{4-})}.$$

However, the concentration of the completely deprotonated Y^{4-} ligand depends on the prevailing environment. As the dynamic equilibrium between the EDTA species shows, the dissociation of EDTA is reduced with decreasing pH (increasing oxonium ion concentration) in accordance with the law
of mass action:

$$H_6Y^{2+} \overset{-H^+}{\rightleftharpoons} H_5Y^+ \overset{-H^+}{\rightleftharpoons} H_4Y \overset{-H^+}{\rightleftharpoons} H_3Y^- \overset{-H^+}{\rightleftharpoons} H_2Y^{2-} \overset{-H^+}{\rightleftharpoons} HY^{3-} \overset{-H^+}{\rightleftharpoons} Y^{4-}$$

In an acidic environment, the balance shifts in favor of H_3Y^- and in a very acidic solution also to H_4Y or HY_6^{2+}. The concentration of Y^{4-} decreases accordingly and may no longer be sufficient for the formation of a stable complex. The decisive factor here is the respective complex formation constant.

The pH-dependent proportion of Y^{4-} can be expressed via the degree of dissociation α:

$$\alpha_{Y4-} = \frac{c(Y^{4-})}{c_0(EDTA)},$$

where α_{Y4-} values of the EDTA equilibrium can be calculated for each pH value of a solution, taking into account the acid constants of the EDTA.

If the concentration of the ligand is replaced by the pH-dependent concentration, expressed by the respective α-value, an expression for the **effective complex formation constant (K')** is obtained, also known as the **conditional stability constant**:

$$K_K = \frac{c(NiY^{2-})}{c(Ni^{2+}) \cdot c(Y^{4-})} = \frac{c(NiY^{2-})}{c(Ni^{2+}) \cdot \alpha_{Y4-} \cdot c(EDTA)}$$

or with $K_K \cdot \alpha_{Y4-} = K'_K$

$$K'_K = \frac{c(NiY^{2-})}{c(Ni^{2+}) \cdot c(EDTA)}$$

In order for a metal cation to be quantified complexometrically with EDTA, i.e. for the complexation to be at least 99.9%, the minimum value of the conditional stability constant should be $10^6/10^x$, where 10^x corresponds to the approximate metal ion concentration in mol/l in the working solution. For example, if 10.0 ml of sample solution with $c(Me^{n+}) \approx 0.1$ mol/l is provided and diluted to 100 ml, the conditional stability constant (K') should be at least 10^8. This can be used to calculate the pH value from which titration of the respective metal ion is possible (Tab. 3.10).

Tab. 3.10: Complexation constant of some EDTA complexes as lg K [18] and minimum required pH value for 99.9% complexation ($K' > 10^8$).

Cation	lg K	pH value	Cation	lg K	pH value
Na^+	1.86	–	Al^{3+}	16.4	~3.7
Li^+	2.95	–	Zn^{2+}	16.5	~3.7
Ag^+	7.20	–	Ni^{2+}	18.4	~2.7
Ba^{2+}	7.88	–	Cu^{2+}	18.78	~2.6
Mg^{2+}	8.79	~9.5	Hg^{2+}	21.5	~1.5
Ca^{2+}	10.65	~7.5	Fe^{3+}	25.1	~3.2
Cd^{2+}	16.5	~3.7	Co^{3+}	41.4	< 0

The table can also be used as a guide for the selective titration of one ion in the presence of another (simultaneous determination). For example: If a Cu^{2+}/Mg^{2+} solution were titrated at pH $\approx$ 5, only copper would be detected, as magnesium forms a very unstable complex at this pH value. In practice, this means that chelates with small formation constants can only be determined at high pH values (alkaline aqueous solution), while those with large formation constants can also be determined in acidic solutions. In alkaline solutions, EDTA can be used to determine the sum contents of complex metal salt solutions (with the exception of alkali metal ions). In more acidic solutions, metal cations with smaller formation constants can only be determined incompletely or no longer at all.

It should also be noted that during complex formation, Y^{4-} is bound and the balance of all species shifts to Y^{4-}. This releases oxonium ions and lowers the pH, possibly reducing the effective formation constant of the metal-EDTA complex, which could lead to incomplete complex formation. Therefore, all complexometric titrations are performed in **buffered solution**. Although the resulting influence on the equilibrium is lowest in alkaline solution, buffering is carried out even at pH 10.5.

At a pH value greater than 10, all free metal ions that have a sufficiently high complex formation constant can generally be determined, unless they cannot be converted or can only be converted incompletely by the titration reaction due to precipitation or the formation of stable hydroxocomplexes. Such metal ions include Ti^{4+}, Fe^{3+}, Al^{3+}, Fe^{2+}, Zn^{2+}, Ni^{2+}, Co^{2+}, Mn^{2+}, Mg^{2+} and Pb^{2+}.

For example, at a magnesium ion concentration of 0.01 mol/l, magnesium precipitates as magnesium hydroxide from a pH value of 10.4:

$$Mg^{2+} + 2\,OH^- \rightarrow Mg(OH)_2 \downarrow ,$$

$$K_{sp}\left(Mg(OH)_2\right) = 6 \cdot 10^{-10} mol^3/l^3 = c\left(Mg^{2+}\right) \cdot c^2\left(OH^-\right),$$

$$c\left(OH^-\right) = \left(6 \cdot 10^{-10} mol^3/l^3 \,/\, 10^{-2} mol/l\right)^{0.5} = 2.45 \cdot 10^{-4} mol/l,$$

$$pH = 10.4.$$

At higher magnesium ion concentrations, magnesium precipitates at even lower pH values: at $c(Mg^{2+})$ = 0.05 mol/l from a pH value of approximately 10.0 and at $c(Mg^{2+})$ = 0.1 mol/l from a pH value of approximately 9.9.

In order to avoid this, so-called **auxiliary complexing agents** such as ammonia, are used to avoid this. An ammoniacal buffer (NH_3/NH_4^+) is used for this purpose in the magnesium determination (see p. 235 f.). Mg^{2+} ions remain in solution due to the formation of amine complexes:

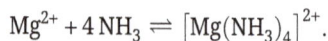

$$Mg^{2+} + 4\,NH_3 \rightleftharpoons \left[Mg(NH_3)_4\right]^{2+}.$$

It is important that the stability constant of the metal-EDTA complex is greater than that of the metal auxiliary complexing agent.

The formation of the sparingly soluble hydroxide can also be utilized analytically. Calcium determination alongside magnesium (löschen, see Section 3.4.4.4) is carried out at a pH value of 14 and without an ammoniacal buffer – i.e. without an auxiliary complexing agent. Under these conditions, magnesium forms a stable hydroxide and is therefore not detected during the titration.

The theoretically oriented works listed in the bibliography are recommended for in-depth study of this content.

The steric structure of a metal-EDTA complex is shown in Fig. 3.20. The metal ion as the central atom is enveloped octahedrally by the EDTA molecule. The formation of the five-membered chelate rings is clearly shown in the drawing.

3.4.3 Indication of the end point

The equivalence point of a chelatometric titration, e.g. with EDTA solution, is recognized by indicators which react to a change in the pMe value with a color change. Such metal indicators are, for example, Eriochrome black T, murexide, pyrocatechol violet, xylene orange, pyridylazoresorcinol and pyridylazonaphthol (PAN). These indicators also form chelate complexes with the metal ions, which have a different color than the free indicators. The color change at the equivalence point is caused by the decomposition of the metal-indicator complex and the appearance of the color of the free indicator. The stability of the metal-indicator complex must be lower than that of the metal-EDTA complex. However, it must be high enough to ensure a sharp color change.

Table 3.11 shows some typical metal indicators that are used in complexometry. Since most metal indicators are also acid-base indicators, the color of the free indicator is pH-dependent. Consequently, the indicator must be selected so that it harmonizes with the pH range that must be set due to the chelate complex formation.

Tab. 3.11: Some complexometric indicators.

Indicator	Color of the free indicator in the optimum pH working range	Color of the metal indicator complex	Application
Sulfosalicylic acid	Yellow (pH ≈ 2.5)	Red	Fe^{2+}
Tiron	Yellow (pH ≈ 2.5)	Blue-green	Fe^{2+}
Xylene orange (XO)	Yellow (pH < 5)	Red	Ba^{2+}, Sr^{2+}, Bi^{3+}, Pb^{2+}, Zn^{2+}, Al^{3+}, Th^{4+}, Cd^{2+}, lanthanides
Eriochrome black T	Blue (pH 6.5-11)	Wine red	Ba^{2+}, Cd^{2+}, In^{3+}, Mg^{2+}, Mn^{2+}, Pb^{2+}, Zn^{2+}, Zr^{2+}, lanthanides
Calcein (fluorexone)	Orange (pH ≈ 10), non-fluorescent	Yellow-green, fluorescent	Ca^{2+}, Mg^{2+}, Ba^{2+}, Sr^{2+}, Cr^{3+} and for the indirect determination of SO_4^{2-}
Calconcarboxylic acid	Violet (pH > 12)	Red	Ca^{2+} (interference by Sr^{2+}, Ba^{2+})
Murexide	Blue-violet (pH > 12)	Yellow red with Ca^{2+}	Cu^{2+}, Ni^{2+}, Co^{2+}, Ca^{2+}

Eriochrome black T (Erio T) belongs to the group of eriochrome black dyes (2,2'-dihydroxyazonaphthalenes) and has the following structure:

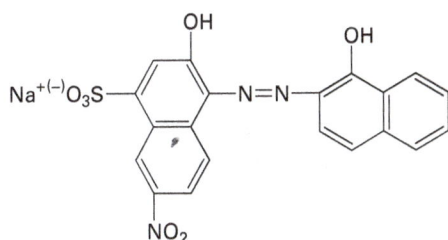

The H^+ ion of the strongly acidic sulfonic acid group is already split off in the pH range of interest (7–12) (abbreviated formula: H_2Ind^-). The color of the indicator changes according to

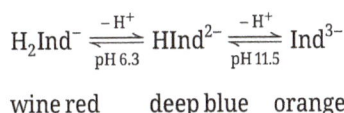

$$H_2Ind^- \underset{pH\,6.3}{\overset{-H^+}{\rightleftharpoons}} HInd^{2-} \underset{pH\,11.5}{\overset{-H^+}{\rightleftharpoons}} Ind^{3-}$$

$$\text{wine red} \qquad \text{deep blue} \quad \text{orange}$$

depending on the H^+ concentration. In the pH < 6 range, Erio T tends to polymerize with yellowish-brown coloration, which is accelerated by Na^+, K^+ or NH_4^+ in higher concentrations. The presence of acetone or alcohol and an increase in temperature counteracts polymerization. For this reason and because of the red color of the form H_2Ind^-, which is very similar to the color of the metal-indicator complex, the indicator is used above pH 6.5, for example, during the titration of Mg^{2+} to

$$[MgInd]^- + H^+ \rightleftharpoons Mg^{2+} + HInd^{2-}.$$
$$\text{red} \qquad\qquad\qquad \text{blue}$$

Solutions of Erio T in water or alcohol are unstable. In contrast, triethanolamine has a stabilizing effect through complex formation. Dissolve 0.5 g of the dye in 100 ml of triethanolamine. The solution can be kept for at least 1 month.

The indicator can also be added to the solution to be titrated in solid form mixed with common salt. If the indicator is rarely used, this avoids the decomposition of the solution, which cannot be stored indefinitely. The dye is triturated with pure sodium chloride in a ratio of 1:100 to form a powder as fine as dust. Each 100 ml of sample solution requires 3–7 mg of dye. If the trituration of Erio T with methyl red, known as Erio TM, is used, a color change from red to green is observed during titration.

Various metal ions, such as Co^{2+}, Ni^{2+}, Cu^{2+}, Al^{3+} and Ti^{4+}, form more stable complexes with Erio T than with EDTA. Contamination of the solution to be titrated with these metal cations prevents titration with Erio T as an indicator by blocking it.

Indicator buffer tablets contain Eriochrome black T as a "universal metal ion indicator". An optically sharp change occurs from red to green with slight gray color gradations. The tablets also buffer the solution. A common buffer solution in complexometry is the ammonia/ammonium chloride buffer (pH 10), which is used for determinations with Eriochromschwarz T.

The contrast of the transition can be improved by mixing Eriochrome black T with methyl red. In an alkaline environment, the metal ion in the sample changes from red to green.

The aminopolycarboxylic acids react with numerous metal ions to form complexes and are therefore not specific as reagents. If different metal ions are present in the solution at the same time, this must be taken into account.

3.4.4 Chelatometric determinations

3.4.4.1 Preparation of the EDTA solution
As the free acid is poorly soluble in water, the highly soluble dihydrate of the disodium salt is used, which is commercially available in very pure form. About 37.2239 g $C_{10}H_{14}N_2Na_2O_8 \cdot 2\ H_2O$ (molar mass 372.239 g/mol), which has been dried to constant mass at 80 °C, is weighed accurately and dissolved in water to make 1 l. The solution

has a concentration of 0.1 mol/l. The dihydrate can also be dehydrated between 120 and 140 °C. A solution with a concentration of 0.1 mol/l contains 33.6209 g of anhydrous salt per liter. However, the hygroscopicity of the anhydrous salt must be taken into account when weighing it out. Titration is not necessary, but can be carried out, for example, with a calcium salt solution of known content using the method described below. The calcium salt solution is prepared by dissolving pure calcium carbonate, annealed to constancy of mass, in hydrochloric acid. The water used to prepare the standard solution must be very pure, as the reagent reacts with calcium, magnesium and other cations that may be present in the water. These impurities can be easily removed using a cation exchanger. The solution can be stored for months in storage bottles made of polyethylene or borosilicate glass that have been well evaporated before use. Bottles made of ordinary glass are not suitable, as calcium is released into the solution to a noticeable extent over a longer period of time, gradually reducing the titer of the solution.

3.4.4.2 Determination of magnesium

Mg^{2+} ions can be determined by direct titration with EDTA solution and with Erio T as an indicator in alkaline solution. The stability constant of the Mg-Erio T complex ($K = 10^7$) is sufficient compared to that of the Mg-EDTA complex ($K = 10^{8.69}$) so that a sharp color change occurs at the equivalence point. The titration is carried out in a buffered solution at pH = 10 (ammonia-ammonium chloride buffer) to bind the hydrogen ions released during the titration and to ensure that the color of the pH-sensitive indicator changes from red (color of the Mg-Erio T complex) to blue (color of the free indicator). The reaction proceeds according to the equation

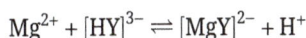

$$Mg^{2+} + [HY]^{3-} \rightleftharpoons [MgY]^{2-} + H^+$$

The alkaline earth ions are included in this determination and must be precipitated with ammonium carbonate solution. Co, Ni, Cu, Zn, Cd, Hg and the platinum metals can be masked with KCN. Mn, Cr, Al, Pb, Bi, Sb, Ti, Zr, Th, the lanthanides, Ta and Ga, on the other hand, must be removed, e.g. by precipitation as hydroxides, as must Fe^{3+}. However, the latter can also be reduced to Fe^{2+} and masked with KCN [138].

Procedure: Add 5 ml buffer solution (pH = 10, see below) to 100 ml of the solution to be titrated (acidic solutions are neutralized beforehand with diluted sodium hydroxide solution), which should not contain more than 0.01 mol/l Mg (0.24 g/l). After adding the indicator (two to four drops of solution or a spatula tip of saline trituration, see p. 235), titrate with EDTA solution (0.1 mol/l) until the color changes from red to pure blue. Due to the relatively slow rate of complex formation, titration must be slow near the equivalence point. The determination can also be carried out with EDTA solution with a concentration of 0.01 mol/l or even 0.001 mol/l.

Preparation of the buffer solution: Dissolve 5.40 g of ammonium chloride in 20 ml of water, add 35 ml of conc. ammonia solution and make up to 100 ml with water:

1 ml EDTA solution, c = 0.1 mol/l, corresponds to 2.4305 mg Mg.

3.4.4.3 Determination of calcium

Calcium can be dissolved at pH > 12 and can be titrated directly with EDTA solution with murexide or calconcarboxylic acid as an indicator. The determination is disturbed by the cations already listed in the magnesium determination, which must therefore be removed as described. Small amounts of barium and magnesium do not interfere, but strontium does. If larger quantities of magnesium are present, magnesium hydroxide precipitates at the pH value at which the titration is carried out, taking some calcium with it. In this case, calcium can be determined by back titration. To do this, a small excess of EDTA is added to the solution before it is made alkaline; pure, calcium-free magnesium hydroxide then precipitates out. The excess EDTA is back-titrated with a calcium solution of known content using murexide as an indicator. With Erio T as an indicator, calcium cannot be titrated directly, unlike magnesium, as the stability constant of the Ca-indicator complex is too small (K = $10^{5.4}$; in contrast, the Ca-EDTA complex: K = $10^{10.7}$), which is why the color change is slow. However, the determination is possible by substitution titration. A solution of the Mg-EDTA complex is added to the calcium solution (K = $10^{8.7}$). The magnesium is substituted by the calcium, which forms a more stable complex with EDTA, according to the equation:

$$[MgY]^{2-} + Ca^{2+} \rightleftharpoons [CaY]^{2-} + Mg^{2+}.$$

As with the magnesium determination, the end point is then indicated by the blue color of the free indicator formed at the equivalence point. The changeover is sharpest at a substance quantity ratio of $n(Ca) : n(Mg)$ = 10:1, but much smaller quantities of magnesium are also effective.

Procedure:
1. **Direct titration** with murexide as indicator: 100 ml of the sample solution, which should contain approximately 0.01 mol/l calcium, is neutralized with sodium hydroxide solution, if necessary, and 1 ml sodium hydroxide solution (2 mol/l) is added. After adding the indicator (freshly prepared saturated aqueous solution), titrate with EDTA solution (0.1 mol/l), until the color changes from red to violet. The titration must be carried out immediately after adding the sodium hydroxide solution, as otherwise $CaCO_3$ may precipitate due to absorption of CO_2 from the air.
2. **Substitution titration** with Erio T as indicator: 100 ml of the sample solution, which should contain approximately 0.01 mol/l calcium, is neutralized with diluted sodium hydroxide solution if necessary. Successively add 5 ml buffer solution (pH = 10, see p. 236), 20 ml Mg-EDTA solution

(0.1 mol/l) (see below) and the indicator (two to four drops or a spatula tip of saline trituration, see p. 234). The titration is carried out according to the magnesium determination:

$$n(\text{Ca}) = n(\text{Mg-EDTA})_{\text{consumed}} - n(\text{EDTA})_{\text{titration}}.$$

Preparation of the Mg-EDTA solution: Equivalent quantities of a magnesium sulfate solution and an EDTA solution are mixed. The pH value is adjusted between 8 and 9 with sodium hydroxide solution (change from phenolphthalein to red). Mg and EDTA are present in the correct ratio of 1:1 if, after adding a little buffer solution (pH = 10, see p. 236) and Erio T as an indicator, the solution has a dirty violet color, which changes to blue with a drop of EDTA solution (0.01 mol/l) and to red with a drop of $MgSO_4$ solution (0.01 mol/l). Finally, the concentration is adjusted to 0.1 mol/l by adding water. Titration is not necessary.

3.4.4.4 Determination of water hardness

The content of dissolved calcium and magnesium salts in water is referred to as its **hardness (calcium** and **magnesium hardness)**. It must be removed for many applications. If, for example, thermal power stations are operated with water that is too hard, some of the dissolved salts are deposited as **scale**. On the one hand, this narrows the cross-section of pipelines and, on the other hand, greatly reduces heat transfer. The result is increased fuel consumption and, if the deposits become too thick, overheating damage and even boiler explosions.

A distinction is made between temporary (transient) and permanent (lasting) hardness depending on how the water behaves when it boils.

The hydrogen carbonates $Ca(HCO_3)_2$ and $Mg(HCO_3)_2$ dissolved in the water together to form the **temporary hardness (carbonate hardness)**. They precipitate during boiling, e.g. according to the equation

$$Ca(HCO_3)_2 \rightarrow CaCO_3 \downarrow + H_2O + CO_2 \uparrow,$$

as carbonates (scale!), while the calcium and magnesium sulfates, chlorides, silicates, nitrates and humates remain in solution as **permanent hardness (noncarbonate hardness)**.

The determination of calcium and magnesium in water is carried out by atomic absorption spectrometry or chelatometrically. The metal ions that interfere with the titration and can be precipitated as hydroxides (see determination of magnesium) are normally only present in very low concentrations in water and are precipitated when the buffer solution is added. Iron(III) hydroxide remaining colloidally in solution interferes even in traces and must be removed with Na_2S. The other cations already mentioned in the magnesium determination can be masked with KCN. Phosphate ions are best removed using an anion exchanger.

The results of water hardness determinations are given in mol/m^3. In practice, the indication in degrees of hardness is still common. A German degree of hardness (1° DH) corresponds to 10 mg/l CaO or the equivalent mass concentration of 7.18 mg/l MgO.

Procedure:

1. **Determination of the sum of calcium and magnesium:** 50 ml of water, the hardness of which is to be determined, is boiled for about 1 min after adding 1 ml hydrochloric acid (2 mol/l) in order to decompose the hydrogen carbonate ions and expel the CO_2 formed from the solution. After cooling to approximately 50 °C, neutralize with diluted sodium hydroxide solution and add 5 ml buffer solution (pH = 10, see p. 235). If the water contains no Mg, add 1 ml Mg-EDTA solution (0.1 mol/l; see p. 237). After adding the indicator Erio T the titration is carried out in the same way as the magnesium determination (see p. 235 f.), but with EDTA solution of concentration c = 0.01 mol/l. If the consumption of standard solution is less than 4.5 ml, the titration is repeated with a larger volume of the sample solution; if the consumption is greater than 20 ml, the determination is repeated with a smaller volume of the sample solution and diluted to 50 ml with water:

   1 ml EDTA solution, c = 0.01 mol/l, corresponds to the amount of substance n(Ca + Mg) = 0.01 mmol. The result of the determination is given as the concentration of substance c(Ca + Mg) in mol/m^3.

2. **Determination of calcium:** 50 ml of the water sample is boiled with 1 ml hydrochloric acid (2 mol/l) for 1 min and neutralized with diluted sodium hydroxide solution after cooling. Add 2 ml sodium hydroxide solution (2 mol/l), and approximately 200 mg indicator (trituration of 0.2 g calconcarboxylic acid with 100 g NaCl) and titrate immediately with EDTA solution (0.01 mol/l) until the color changes from red to blue. The color should not change when a further drop of the standard solution is added. Titration must be slow near the changeover point. If more than 20 ml or less than 4.5 ml EDTA solution is used, proceed as described under 1:

   1 ml EDTA solution, c = 0.01 mol/l, corresponds to 0.01 mmol Ca. The result must be given in mol/m^3. The following applies to the magnesium concentration:

$$c(\text{Mg}) = c(\text{Mg} + \text{Ca}) - c(\text{Ca}).$$

3.4.4.5 Determination of zinc and cadmium

Similar to magnesium, zinc and cadmium can be titrated with Erio T as an indicator in an alkaline, buffered solution at pH = 10. The color change from red to blue is extremely sharp in the pH range between 7 and 10. The end point can still be determined well even in the presence of a lot of ammonia. Below pH = 7, Erio T is not suitable as an indicator due to the formation of the wine-red H_2Ind^-.

The determination is disturbed by numerous other metal ions. Fe^{3+}, Al^{3+}, Bi^{3+} and Pb^{2+} must be separated beforehand. However, small amounts of Al^{3+} can be removed with Tiron or can be masked with sodium fluoride. Metal ions that form stable cyano complexes can be masked with KCN. The cyanocomplexes of zinc and cadmium also formed in this process are decomposed in ammoniacal solution by reaction with formaldehyde to form cyanohydrin:

$$\left[\text{Zn(CN)}_4\right]^{2-} + 4\,\text{CH}_2\text{O} + 4\ \text{H}^+ \rightarrow \text{Zn}^{2+} + 4\,\text{HOCH}_2\text{CN}.$$

The cyano complexes of Fe, Hg, Cu, Ni and Co do not react or react only very slowly with formaldehyde. However, it should be noted that even traces of Cu^{2+}, Co^{2+} or Ni^{2+} block the indicator (for more details see [130]).

Due to the sharpness of the color change, micro- and ultra-microdeterminations are possible. Very pure zinc or zinc oxide are suitable as for the adjustment of EDTA solutions.

Procedure: A volume of 100 ml of the sample solution, which should not contain more than 250 mg/l Zn or 500 mg/l Cd, is neutralized with diluted sodium hydroxide solution if necessary. Add 5 ml buffer solution (pH = 10, see p. 236), add the indicator (two to four drops of solution or a spatula tip of saline trituration, see p. 234 and titrate with EDTA solution (0.1 mol/l) until the color changes from red-violet to blue. The reddish color disappears abruptly at the equivalence point after adding a drop of the standard solution:

1 ml EDTA solution, $c = 0.1$ mol/l, corresponds to 6.541 mg Zn or 11.241 mg Cd.

3.4.4.6 Determination of copper

The titration of copper with EDTA solution can be carried out in a weakly acidic, buffered solution with PAN (1-(2-pyridyl-azo)-2-naphthol) as an indicator.

Procedure: Add 5 ml acetate buffer solution (pH = 5, see below) and three to five drops of indicator solution (PAN in methanol, 1% by mass) to 100 ml of the sample solution, which has been almost neutralized with carbonate-free sodium hydroxide solution and should contain approximately 1 g/l Cu^{2+}. Titrate at boiling temperature with EDTA solution (0.1 mol/l) until a sharp change from violet to yellow occurs. If the solution contains 30–50% ethanol or acetone, titration can be carried out at room temperature. In solutions that do not contain alcohol, a fine skin of PAN is sometimes observed on the surface of the solution or on the wall of the titration beaker; as it can interfere with subsequent titrations, it must be removed with alcohol or acids.

Buffer solution: A solution of 27.3 g sodium acetate-3-hydrate is diluted to 1,000 ml with carbon dioxide-free water after addition of 30 ml hydrochloric acid (2 mol/l):

1 ml EDTA solution, $c = 0.1$ mol/l, corresponds to 6.355 mg Cu^{2+}.

3.4.4.7 Determination of aluminum

Aluminum can be determined chelatometrically by back titration. EDTA solution is added in excess and back-titrated with zinc sulfate or lead nitrate solution. When using Erio T as an indicator, the back titration must be carried out very quickly, as aluminum reacts slowly but irreversibly with Erio T to form the Al indicator complex. XO does not have this disadvantage and is therefore more suitable for this indication:

Procedure: Add exactly 20 ml EDTA solution (0.1 mol/l) to 100 ml of the sample solution, which should not contain more than 400 mg/l Al, and bring to the boil (1 min). After cooling, 4.0 g solid hexamethylenetetramine and three to five drops of indicator solution (50 mg xylenol orange (XO) in 100 ml water) are added. The excess EDTA is titrated with lead nitrate solution (0.1 mol/l) until the solution turns red. To prepare the lead nitrate solution 33.120 g of lead nitrate dried at 105 °C is made up to 1,000 ml with CO_2-free water.

Calculation: The consumption of EDTA solution is the difference between the added volume and the consumption of lead nitrate solution:

　1 ml EDTA solution, c = 0.1 mol/l, corresponds to 2.6982 mg Al.

3.4.4.8 Determination of bismuth

Due to the high stability (K = $10^{27.94}$) of the Bi-EDTA complex, bismuth can be titrated in a relatively acidic solution and therefore largely selectively. A suitable indicator is, for example, XO. Solutions that are too acidic are blunted with sodium hydrogen carbonate solution (1 mol/l) or with sodium acetate. If sodium hydroxide solution or ammonia is used, bismuth polycations are formed as a result of a locally too high base concentration, which react only slowly with EDTA and also decompose only slowly to Bi^{3+} ions when acidified. The direct titration of bismuth is disturbed by chloride ions, as bismuth oxychloride precipitates when the required pH value is adjusted. In this case, the determination can be carried out by back titration by first adding an excess of EDTA to the strongly acidic sample solution and only then increasing the pH to 2.5–3. The excess is back-titrated with bismuth nitrate solution titrated back. Bromide and iodide ions also interfere and are removed with nitric acid.

Procedure: A volume of 100 ml of the sample solution, which should contain approximately 4 g/l Bi, is mixed with dilute nitric acid or sodium hydrogen carbonate solution (1 mol/l) to a pH value of 2.5–3, if necessary. After adding XO (a few drops of 0.5% aqueous solution), titrate with EDTA solution (0.1 mol/l) until the color changes to yellow:

　1 ml EDTA solution, c = 0.1 mol/l, corresponds to 20.898 mg Bi.

3.4.4.9 Determination of iron

Iron(III) ions form a very stable complex with EDTA (K = $10^{25.1}$); they can therefore be determined in an acidic solution, whereby the pH value should be approximately 2.5. Sb and Bi interfere since the stability constants of their EDTA complexes are also very high. In contrast, divalent ions as well as Al^{3+} and Cr^{3+} are not detected at pH = 2.5. As Indicators, for example, Tiron or sulfosalicylic acid are suitable.

Procedure: The sample solution should contain around 200–700 mg/l Fe^{3+} and not too much free hydrochloric acid. If some of the iron is present with an oxidation number of +2, add a little ammonium peroxodisulfate and boil briefly. To adjust the pH value to around 2.5, add solid p-chloroaniline until it no longer dissolves. The excess acid can also be blunted with sodium or ammonium acetate. However, if the concentration of acetate becomes too high, the conversion will be slower due to the formation of iron acetate complexes. After adding the indicator, titrate with EDTA solution (0.1 mol/l):

1. **Indication with tiron 2% aqueous solution:** Addition of 2 ml indicator solution per 100 ml: titration at 40–50 °C until pure yellow coloration.
2. **Indication with sulfosalicylic acid 2% aqueous solution:** Addition of 2 ml indicator solution per 100 ml; change from violet-red (Fe(III) indicator complex) to yellow (Fe(III) EDTA complex):

1 ml EDTA solution, c = 0.1 mol/l, corresponds to 5.585 mg Fe.

3.4.4.10 Determination of phosphate

Phosphate is determined indirectly via an equivalent amount of magnesium. For this purpose, the phosphate is precipitated as $Mg(NH_4)PO_4 \cdot 6\ H_2O$. The precipitate is washed, filtered and dissolved in hydrochloric acid. EDTA solution is added in excess, a pH value of approximately 10 is set and titrated back with $MgCl_2$ solution. The interfering polyvalent cations can be masked before precipitation with EDTA solution (1 mol/l) or removed using cation exchangers.

Procedure: Dilute 20 ml of the sample solution, which should contain a maximum of 0.1 mol/l phosphate, to 50 ml in a 250-ml beaker and add 1 ml conc. hydrochloric acid and a few drops of methyl red (0.1% in ethanol). After adding 3 ml magnesium sulfate solution (1 mol/l), heat to boiling point and add conc. ammonia solution dropwise while stirring vigorously until the indicator turns yellow, and then add a further 2 ml. After standing for several hours (or overnight), the precipitate is filtered through a glass filter crucible G4 and washed with approximately 100 ml ammonia solution (1 mol/l). Rinse the precipitation vessel with 25 ml hydrochloric acid (1 mol/l) and add this liquid in portions to the filter crucible, removing the vacuum in the suction bottle before each addition. The process is repeated with a further 10 ml of the hydrochloric acid and then with 75 ml of water. Add 25 ml EDTA solution (0.1 mol/l) to the hydrochloric acid solution in the suction flask and neutralize with sodium hydroxide solution (1 mol/l). After adding 5 ml buffer solution (pH = 10, see p. 236) and Erio T (two to four drops of solution or a spatula tip of saline triturition, see p. 234), titrate the excess EDTA with $MgCl_2$ solution (0.1 mol/l) until the color changes from blue to wine red.

Preparation of the $MgCl_2$ solution, 0.1 mol/l: About 2.4305 g of pure magnesium shavings are dissolved in dilute hydrochloric acid. The solution is almost neutralized with sodium hydroxide solution (1 mol/l) and made up to 1,000 ml with deionized water.

Calculation: The consumption of EDTA solution is the difference between the added volume and the consumption of $MgCl_2$ solution:
   1 ml EDTA solution, c = 0.1 mol/l, corresponds to 9.4971 mg PO_4^{3-}.

3.4.4.11 Determination of sulfate

Sulfate can be determined using the back titration method with DTPA and Erio T as an indicator. The sulfate ions are precipitated with barium chloride solution and the excess of barium ions is back titrated with DTPA solution. The solubility of barium sulfate can be reduced by adding ethanol. Polyvalent cations interfere with the determination and must be removed, preferably with a cation exchanger.

Procedure: Add 1 ml hydrochloric acid (0.1 mol/l), 20 ml $BaCl_2$ solution (0.1 mol/l, see below), 30 ml ethanol, 10 ml buffer solution (pH = 10, see p. 236) and 5 ml Mg-DTPA solution (0.1 mol/l) to 100 ml of the sample solution, which should contain between 0.5 and 1.5 g/l sulfate. (The Mg-DTPA solution is prepared in the same way as the Mg-EDTA solution, see p. 237). After adding the indicator Erio T (1 spatula tip of saline trituration 1:100), titrate to blue with DTPA solution (0.1 mol/l) to the point of transition.

To prepare the DTPA solution, dissolve 39.336 g DTPA in 150 ml sodium hydroxide solution (1 mol/l) in a 1 l volumetric flask and make up to the mark. The titration can be carried out with $Pb(NO_3)_2$ solution (see p. 240). To do this, dilute 10.0 ml DTPA solution with 100 ml water and titrate with $Pb(NO_3)_2$ solution (0.1 mol/l) with XO (0.5% aqueous solution) as indicator until an intense red-violet coloration is obtained.

To prepare the $BaCl_2$ solution, 24.43 g $BaCl_2 \cdot 2\,H_2O$ is dissolved with water to make 1 l. Titration is carried out with DTPA solution.

Calculation: The consumption of $BaCl_2$ solution is the difference between the added volume and the consumption of DTPA solution:

1 ml $BaCl_2$ solution, $c = 0.1$ mol/l, corresponds to 9.606 mg SO_4^{2-}.

3.4.4.12 Carbamatometric titrations

This type of titration contains a preconcentration step with which a low limit of quantification can be achieved. A standard solution is used which contains the diethyldithiocarbamate (DDTC)-anion, a bidentate ligand:

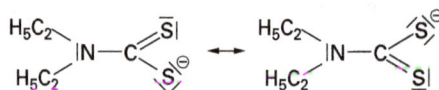

DDTC forms very stable complexes with a number of heavy metals. Although the chelate effect is not as strong with a bidentate ligand as with a hexadentate ligand, e.g. EDTA, the sulfur of DDTC as a soft base forms more stable bonds with large heavy metal ions than nitrogen or oxygen [140a]. The DDTC complexes contain just as many ligands as the charge number of the metal ion Z and are therefore electrically neutral. The structure is tetrahedral with two ligands:

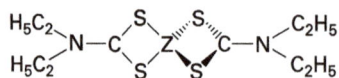

Due to the peripherally arranged organic residues, the complexes are nonpolar and can therefore be easily extracted with relatively nonpolar solvents, e.g. ethyl acetate, chloroform or trichloroethylene. A high enrichment factor and thus a low limit of quantification can be achieved by favorably selecting the volume ratio of aqueous sample solution and extraction agent.

As DDTC complexes are destroyed in acidic solutions, a pH range of 9–10 is set in the aqueous phase.

Especially Cu^{2+}, Zn^{2+}, Cd^{2+}, Hg^{2+} and Pb^{2+} can be determined well with this method, but other metals are also complexed by DDTC and can therefore interfere. It should be noted that the complexation constants decrease according to the following series:

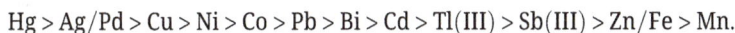

$$Hg > Ag/Pd > Cu > Ni > Co > Pb > Bi > Cd > Tl(III) > Sb(III) > Zn/Fe > Mn.$$

Dithizone is often used as an indicator, which sometimes forms colored complexes with many metal ions. It is also possible to use a metal ion as an indicator if its DDTC complex is colored and less stable than the DDTC complex of the analyte [24b]. Thus, copper(II) acetate, for example, is suitable as an indicator in the carbamatometric titration of mercury.

3.4.4.13 Determination of mercury

Procedure:
Standard solution: The trihydrate of the sodium salt can be used as a titer substance, $Na(C_2H_5)_2NCS_2 \cdot$ 3 H_2O. Depending on the amount of analyte, a concentration of 0.2 mol/l (45.06 g/l), 0.02 mol/l or 0.002 mol/l titer substance in NaOH, 0.01 mol/l is used. For example, it is adjusted against $AgNO_3$, $HgCl_2$ or $PbCl_2$.

Procedure: Pipette 5 ml of the sample solution into the titration flask, neutralize and make up to approximately 25 ml with water. Add 5.0 ml tartaric acid (1 mol/l), 0.5 ml $Cu(CH_3COO)_2$ solution (0.01 mol/l) and 2.0 ml ammonia (13.5 mol/l) and make up to about 50 ml with water. The pH should now be 9–10. After adding 10 ml of extraction agent (e.g. ethyl acetate or chloroform), titrate with 0.02 mol/l standard solution until a brown coloration appears in the organic phase:
   1 ml standard solution, c = 0.02 mol/l, corresponds to 2.0059 mg Hg [24b].

3.5 Two-phase titrations

3.5.1 Determination of anion and cation surfactants

Anion and cation surfactants play an important role in many areas of our lives. Representatives of these two groups can be found in detergents and cleaning agents, lubricants, as excipients in medicines or cosmetics, in disinfectants and antiseptics:

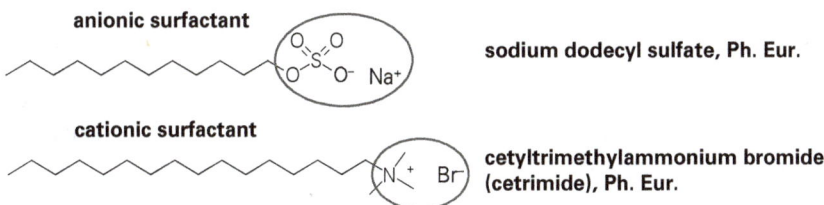

anionic surfactant

sodium dodecyl sulfate, Ph. Eur.

cationic surfactant

cetyltrimethylammonium bromide (cetrimide), Ph. Eur.

As suitable functional groups for a classic titration are often lacking, titration with a second surfactant to form ion pairs is an option. Anion surfactants can therefore be determined with cation surfactants, cation surfactants with anion surfactants. The ion pair of anion and cation surfactant formed during the titration is continuously shaken out into an organic solvent (usually chloroform or dichloromethane). This is therefore a two-phase titration (two-phase titration according to Epton, 1947). The indication is carried out using color indicators, which form less stable ion pairs with anion or cation surfactant. The indicator is shaken out into the organic phase and so the organic phase changes its color (often two indicators, cationic indicator and anionic indicator, are used so that before the equivalence point an ion pair of analyte and indicator I colors the organic phase and after the equivalence point an ion pair of standard solution and indicator II). The ion pair formed with the analyte must be more stable and easier to extract than the ion pair formed with the indicator. The end point is reached when a color change occurs in the organic phase.

Alternatively, the titration can also be indicated potentiometrically using a glass electrode (see Section 4) [24c]. When titrating anion surfactants in detergents, they can be titrated with benzethonium chloride alone (pH value 2) or in combination with soaps (pH value 11), depending on the pH value.

The surfactant titration (Epton titration) is not only suitable for classic surfactants, but also for numerous other anionic or cationic compounds that are sufficiently lipophilic and can form an extractable ion pair with a surfactant. Sodium dodecyl sulfate, sodium dodecyl benzene sulfonate or sodium tetraphenyl borate are used as standard solutions for cations, while benzethonium chloride (Hyamine® 1622), cetyltrimethylammonium chloride or cetylbenzyldimethylammonium chloride are used for anions.

Brilliant blue, bromophenol blue, dimidium bromide, sulfan blue, dimethyl yellow, disulfin blue, methylene blue, oracet blue or thymolphthalein can be used as indicators.

As an example, the titration of sodium dodecyl sulfate according to Ph. Eur. is listed [8a]:

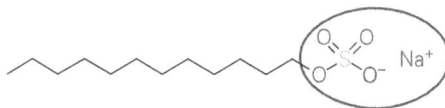

Procedure: Dissolve 1.15 g substance with water to 1000.0 ml, heating if necessary. Add 15 ml dichloromethane and 10 ml dimidium bromide-sulfan blue reagent to 20.0 ml of the above described solution. Titrate with benzethonium chloride solution (0.004 mol/l) while shaking vigorously, waiting for phase separation before each new addition. The end point is reached when the pink coloration of the dichloromethane phase has completely disappeared and a gray-blue color has developed:

1 ml benzethonium chloride solution, c = 0.004 mol/l, corresponds to 1.154 mg sodium dodecyl sulfate.

Preparation of the benzethonium chloride solution (0.004 mol/l): About 1.792 g benzethonium chloride (dried at 100–105 °C until constant mass) is dissolved in water to 1000.0 ml:

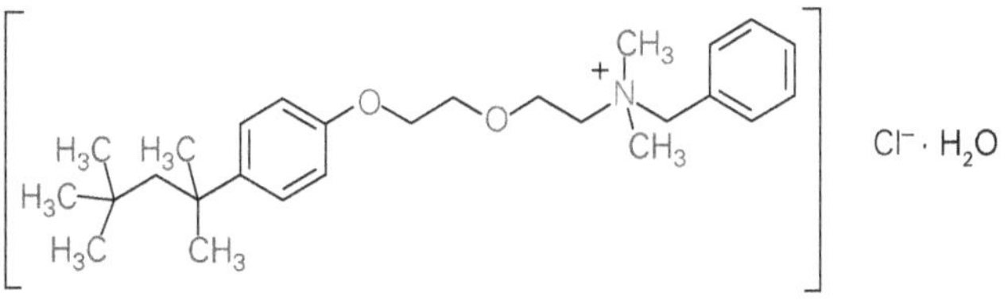

The standard solution is adjusted against an adjusted perchloric acid standard solution using the mercury acetate method (see Section 3.1.4):

1 ml perchloric acid, c = 0.1 mol/l, corresponds to 44.81 mg $C_{27}H_{42}ClNO_2$.

Preparation of the dimidium sulfan blue reagent: Separately, dissolve 0.5 g dimidium bromide and 0.25 g sulphane blue in 30 ml each of a hot mixture of one part anhydrous ethanol and nine parts water by volume. After stirring, the two solutions are mixed and diluted with the same solvent mixture to 250 ml. Add 20 ml of this solution to a dilution of 20 ml of a 14% solution (v/v) of sulfuric acid with approximately 250 ml of water. This solution is diluted with water to 500 ml:

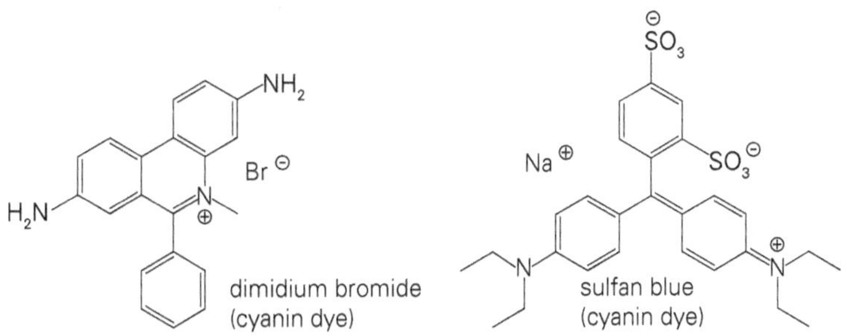

dimidium bromide
(cyanin dye)

sulfan blue
(cyanin dye)

Determination of cetrimide according to Ph. Eur.

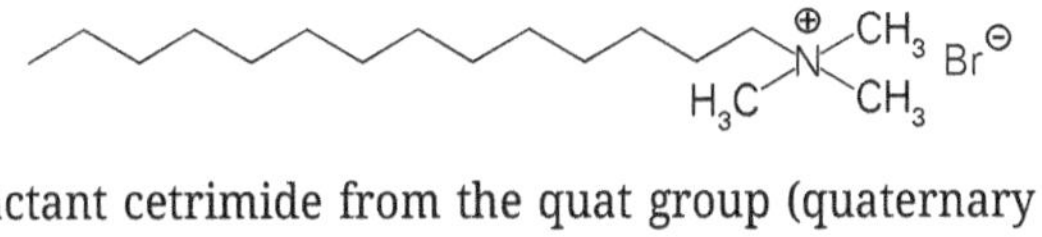

The cationic surfactant cetrimide from the quat group (quaternary ammonium salts) is determined indirectly iodatometrically according to Ph. Eur. First, the analyte is mixed with potassium iodide solution and then the ion pair formed from iodide and cetrimide is extracted with chloroform. After separation of the chloroform phase and renewed shaking with chloroform (quantitative extraction), the excess potassium iodide in the aqueous phase is titrated with potassium iodate solution using the iodine monochloride method.

The iodide first synproportions with iodate to form iodine, which is then oxidized to iodine monochloride in the strongly hydrochloric acid solution. During titration, some chloroform is added again. This first dissolves iodine, turning it purple. After all the iodine has been oxidized to iodine monochloride, the chloroform phase becomes colorless (equivalence point).

The exact content of the potassium iodide solution is determined by a blank titration:

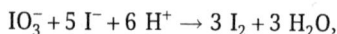

$$IO_3^- + 5\ I^- + 6\ H^+ \rightarrow 3\ I_2 + 3\ H_2O,$$

$$I_2 + HCl \rightarrow ICl + I^- + H^+.$$

Procedure: Dissolve 2.000 g of substance in 100.0 ml of water. Add 25 ml of chloroform, 10 ml of sodium hydroxide solution (0.1 mol/l) and 10.0 ml of a freshly prepared solution of potassium iodide (50 g/l) to 25.0 ml of the solution in a separatory funnel and shake. After separation of the layers, the lower chloroform phase is discarded. The aqueous phase is shaken thrice with 10 ml chloroform each time. The chloroform phases are discarded. The aqueous phase is mixed with 40 ml hydrochloric acid 36% and, after cooling, titrated with potassium iodate (0.05 mol/l) until the dark brown color has almost disappeared. After adding 2 ml chloroform, the titration is continued with vigorous shaking until the color of the chloroform phase no longer changes. A blank titration is carried out with a mixture of 10.0 ml of a freshly prepared solution of potassium iodide (50 g/l), 20 ml of water and 40 ml of hydrochloric acid:

1 ml potassium iodate solution, $c = 0.05$ mol/l, corresponds to 33.64 mg $C_{17}H_{38}BrN$.

4 Volumetric analysis with physical end point determination

4.1 Overview of the indication methods

Compared to gravimetric determinations, we have become familiar with titrations as a quick and generally simple method for quantitatively determining the content of a sample. Nevertheless, it is important to realize that in analytical chemistry, gravimetry, in particular, is an absolute method by which the accuracy of all alternative modern methods is measured. Only when one is convinced of the accuracy of a method by direct or indirect comparison with this absolute method, this will be used in routine operations.

With titrations, the problem lies in reliably recognizing the equivalence point. The previous chapters were dedicated to the classic possibilities of indication. At the same time as the equivalence point is reached, the color of the solution to be titrated should change in a characteristic manner – or, more rarely, the degree of turbidity. If the standard solution itself does not indicate the end point of the titration due to its own intense color, as is the case with permanganometry or iodometry (here, however, only in the presence of a starch solution), a color indicator must be added beforehand. The color change is then caused by a chemical process. Despite the simplicity and ease of use of chemically indexed titration methods, there is a great need for other types of indication. There are various reasons for this, and the most important of which are mentioned here:

- Depending on the chemical reaction on which the titration is based, a color indicator must meet a number of conditions that are not always easy to fulfill. As far as this is possible, this also means that a correspondingly wide range of different indicators must be available in an analytical laboratory if you want to be able to carry out many types of titration procedures at any time.
- A color indicator is always titrated by the standard solution in the same manner like the sample. The result can therefore only be correct if the amount of indicator required is negligible in relation to the amount of the substance to be determined. However, as soon as only very small quantities are to be determined (e.g. in trace analysis), the accuracy that can be achieved with indicators is usually no longer sufficient.
- Analysis samples that are inherently colored or even just cloudy cannot be titrated with the addition of a color indicator.
- The time at which the color change takes place during a titration is determined by the analyst's eye, whereby it is quite possible that individual errors come into play here. When developing new titration methods with color indicators, it must therefore always be checked whether such inadequacies in observation have only a negligible effect on the analytical result (e.g. the requirement of the steepest possible gradient of the titration curve in the area of the equivalence point).

https://doi.org/10.1515/9783111350127-004

Anyone who judges the course of chemical reactions by their color changes is, to put it more generally, already observing the physical properties of matter. Measuring them precisely is then often only a small step. The first question to ask is how certain physical properties change characteristically during a chemical reaction and to what extent these changes depend on the mass of the reactants. Only then can a suitable measuring method be developed. In the following section, some physically indicated methods will be discussed, as long as they are simple enough to be used in the routine operation of an analytical laboratory.

Let us return to the property of many substances being colored: While the eye is able to detect a rapid color change with some certainty, it is far less able to quantitatively assess the gradually changing color intensity. If a photometer is used instead, the relationship between the absorption of monochromatic light and the content of a solute can be precisely recorded, so that a titration in this case can be indicated **photometrically** (see Section 4.2). In addition to the advantage of being independent of the observer's eye, such a method offers above all the possibility of being able to quantitatively determine significantly smaller amounts of substance.

Ions are always involved in the chemical reactions on which the titration methods discussed so far are based. As it is known that the ions present in the solution conduct the electric current, the idea of measuring the electrical conductivity as a function of the course of the titration is obvious. The results are **conductivity titrations** or **conductometric titrations**. As electrodes must be present in the solution to measure conductivity, electrolysis would take place if a current were to flow through these electrodes. To prevent this disturbance, an alternating voltage must be applied, whereby the ions are then only excited to oscillate in time with the voltage frequency without discharging at the electrode surfaces. It is easy to imagine that, due to their greater inertia, heavy ion species can only follow such rapid voltage fluctuations incompletely, and that their contribution to conductivity can ultimately be suppressed completely by increasing the frequency. The application of such techniques is called **high-frequency titration**.

However, ions can also be exchanged between the solution and a solid phase. During this process, a phase-boundary potential occurs between the solid surface and the solution. If the solid phase is a metal electrode and the exchange process is associated with an oxidation or reduction process, this potential can be described by Nernst's equation, provided that the exchange equilibrium is set. The measurement of such potential changes in the course of a titration is the basis of **potentiometry**. It is desirable to have a large number of electrode types, at each of which only certain ion species build up a potential. In the case of metal electrodes, one is naturally limited by the voltage series of the elements, and the potential is always determined by the most easily oxidizable or reducible ion species. It was only with the use of ion exchangers with specific exchange properties as electrode materials that we learned how to overcome these limitations. These **ion-selective** electrodes (ISEs) allow the potentiometric indication of almost all elements.

While potentiometry requires equilibrium to be established in the vicinity of the electrode, under certain circumstances deliberately disturbing this state by applying a constant voltage or maintaining a constant current flow can also be advantageous for titration procedures (**voltammetry**, **amperometry** and **dead-stop method**).

Another way to quantitatively determine the substance conversion at electrodes is to apply Faraday's law. The reagent required for a titration can be generated in the sample solution by an electrode process, and the electric charge required to reach the equivalence point, which was used to generate the reagent, can be determined. However, one must be aware that these **coulometric titrations** replace the usual volume measurement with the burette, including the adjustment of the standard solution, while the indication problem must also be solved in each case.

Of course, many other physical properties dependent on the mass of a substance and their measurement can be used as an advantage for the indication of titration methods (for further information, see [141–144]).

Another instrumental indication method for titrations is **thermometry**. The theory of thermometric indication has been known since 1913, with the term first appearing in 1922. However, at that time, small temperature changes were difficult to record and required a great deal of equipment.

In thermometry, the heat of reaction released (rarely consumed) during a reaction is used for indication. The method can therefore be used for all types of titrations, as every reaction is accompanied by a change in enthalpy (ΔH).

In practice, numerous titrations can be indicated thermometrically, but the method fails if the heat tint is too low. This can still be partially compensated for by coupling with another exothermic reaction.

Numerous reactions, such as neutralizations, are exothermic. In exothermic reactions, the temperature rises up to the equivalence point, after which it remains constant or falls slightly due to the dilution of the sample solution.

If the heat of reaction released is too low, the indication can sometimes be facilitated by coupling with a second reaction. For example, the hydrolysis of paraformaldehyde can be partially utilized for titration with sodium hydroxide solution.

For each thermometrically indexed reaction, vigorous stirring is required to achieve a uniform temperature distribution in the sample.

Thermometric titration requires very sensitive temperature sensors called thermistors (temperature-sensitive resistors). A suitable titrator is available from Metrohm titrotherm, for example.

Applications of thermometric titration are:
- neutralization,
- precipitation titrations,
- redox titrations and
- complexometry.

4.2 Photometric titrations

Photometric titrations can in principle be carried out if the substance to be determined or the standard solution is colored or if the change of a color indicator is easier to recognize in this way. However, for all those titrations where the content to be determined is in the semi-micro range, the human eye is almost always sufficient to detect the color change and thus the equivalence point with sufficient clarity. There must therefore be special circumstances to justify the use of an instrumental method to measure the color change. But let us think, for example, of complexometric titrations with very low metal contents in the solution to be determined. A more dilute EDTA solution will be used in order to obtain a sufficiently accurate volume of the required standard solution, but then the user will be disappointed because the color change of the indicator is perceived as slow and the equivalence point is therefore only determined in a correspondingly blurred manner. Photometric titrations will be useful if the quantities of substances to be determined are in the micro range, if the color changes are only weak or in a spectral range that is not perceived by the human eye (ultraviolet (UV) and near-infrared (NIR)), if colorations of other origins interfere in the solution, or if the titration process is to be automated (see Further reading [145]).

4.2.1 Theoretical foundations

The interaction between electromagnetic radiation and matter results in an excitation of the electrons surrounding the atomic nuclei. Depending on the energy of the radiation used, the electrons close to the nucleus can be completely shot out of the atomic structure (X-rays) or the electrons of the valence shell can be excited by transitions to higher levels (UV/VIS range). During these processes, radiation is absorbed according to the energy required. In this context, photometry refers exclusively to interactions between the valence electrons and light in the VIS and UV range. While atoms have extremely sharp absorption lines, molecules are able to absorb light in a more or less broad range. The reason for this is the ability of molecules to absorb even lower energy IR radiation to excite molecular vibrations and rotational movements. Coupling these absorption lines with those of the valence electron spectrum results in the aforementioned broadening of the absorption lines to form bands.

If you want to quantitatively determine a substance by its ability to absorb light, you need monochromatic light. The spectral range of the incident light must always be narrower than the absorption band in question, because if it were otherwise, some of the incident light could not be absorbed, but would still excite the photocell used for the measurement to emit a signal. From what has been said so far, it should have become clear how difficult it is to fulfill these requirements for the monochromaticity of the light used in absorption measurements on atoms (atomic absorption spectrometry). However, the corresponding measurements on molecules (photometry) usually only require a relatively small amount of equipment, i.e. the necessary devices only need to have a simple monochromator or even inexpensive interference filters are sufficient, or even ordinary color filters are sufficient to block out a sufficiently narrow range from the spectrum of a filament lamp.

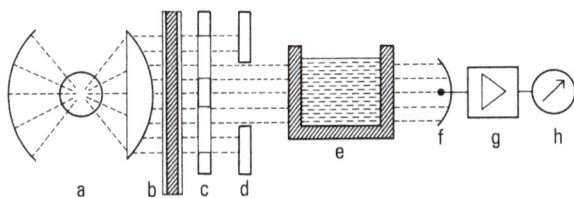

Fig. 4.1: Structure of a simple photometer (a) light source; (b) monochromator; (c) impeller; (d) slit; (e) cuvette; (f) photocell; (g) amplifier; (h) display.

A measuring device suitable for our purposes is shown schematically in Fig. 4.1. The light from a tungsten lamp is focused and passed in parallel through an interference filter. The luminous flux can be regulated by a slit whose width can be adjusted. A cuvette containing the substance to be determined in dissolved form is then shone through and the remaining luminous flux is measured using a photocell. As we intend to carry out photometric titrations, it is useful if an impeller (pinhole) rotates in the path of the light. The light is chopped into short pulses by this, so that the respective luminous flux can be displayed after amplification with an alternating current amplifier. This trick ensures that no light from the laboratory interferes with the measurement. Such unwanted light incidence can hardly be avoided during a titration; however, as this light is unpulsed, it is not amplified by the AC amplifier. The measurements are based on the well-known Lambert-Beer law:

$$A = \lg\frac{\Phi_0}{\Phi} = \varepsilon \cdot c \cdot d,$$

where A is the decadic absorbance, Φ_0 and Φ are the luminous fluxes before and after passing through a solution of layer thickness d, ε is the molar absorptivity and c is the concentration of the absorbing substance. At the start of a titration, the wavelength range absorbed by the substance to be analyzed is set on the monochromator (or a corresponding interference filter is placed in the beam path). The light that passes through the solvent-filled cuvette onto the photocell should bring the display device to full deflection (i.e. transmittance = 100% or absorbance = 0), otherwise the slit width must be changed accordingly (set Φ_0). The film thickness of the cuvette should be selected so that, under the given circumstances, the colored substance causes absorbances during the titration that are approximately between 0.1 and 1.1. The standard solution is best added from a microliter syringe in predetermined volume steps of 10–20 µl. As the cuvettes are made of optical glass that is not very resistant to mechanical stress, stir carefully (!) with a plastic rod after each addition of reagent. When the respective absorbance value has been reached, it is read off and noted. For a graphic evaluation of the titration result, the absorbance is then plotted against the volume of standard solution added.

No calibration with solutions of known concentration is required to carry out photometric titrations, nor is the constant and careful checking of the 100% and 0% transmittance marks necessary. The setting of Φ_0 to 100% transmittance described above serves exclusively to make sensible use of the full measuring range of the display unit. However, if photometric measurements are to be evaluated using calibration curves for the direct determination of concentrations, the checks mentioned must be carried out continuously and conscientiously.

4.2.2 Practical applications

4.2.2.1 Determination of calcium

Using the example of complexometric calcium determination with photometric indication, let us familiarize ourselves with the usual procedure. A volume of 5 ml of an aqueous solution, which should contain approximately 0.02–0.10 mg calcium, should be titrated with an EDTA solution, 0.01 mol/l, i.e. the standard solution should be added to the sample solution in volume increments of 10 μl using a microliter syringe. Calconcarboxylic acid is suitable as an indicator (see p. 236), which changes from wine red (calcium complex) to blue (free indicator) at the equivalence point. The indicator is generally triturated with NaCl in a ratio of 1:100. As there is a risk of using too much indicator with the small amounts of calcium to be determined here, it is perhaps better to work with a 1:150 salt trituration. After preparing by careful grinding in a mortar, add a very small spatula tip of this mixture to the sample solution. As the calcium-EDTA complex is relatively less stable, the solution to be titrated must be brought to pH 13 with 2 mol/l sodium hydroxide solution. Thereafter, the decision has to be made whether the calcium indicator complex or its destruction shortly before the equivalence point should be observed during the titration or rather the appearance of the free indicator after the equivalence point. In order to make the right choice, it is necessary to know the absorption spectra of both compounds, as it is important to adjust the monochromator so that only one of the two colored compounds is able to absorb light. A glance at Figs. 4.2 and 4.3 clearly shows that in the range of maximum absorption of the calcium indicator complex (560 nm), the free indicator also absorbs quite strongly. One could therefore at most use the smaller absorption maximum and set the monochromator to 460 nm. However, it is probably best to follow the formation of the free indicator at the equivalence point of the titration with EDTA solution in a spectral range around 640 nm.

If this last spectral range around 640 nm is selected, the titration procedure is as follows: The 5 ml of the calcium-containing solution (pH = 13) is in a cuvette with a path length of $d = 2$ cm and is titrated with EDTA as described after the addition of calconcarboxylic acid. The calcium initially present in excess forms the EDTA complex, and the

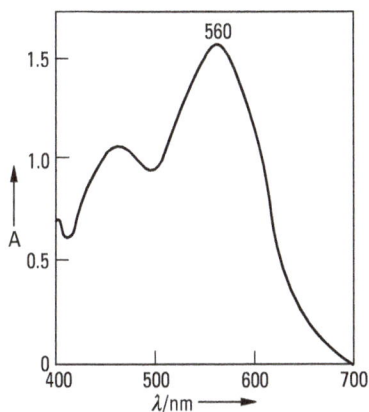

Fig. 4.2: Spectrum of the calcium-calconcarboxylic acid complex (A) absorbance; (λ) wavelength.

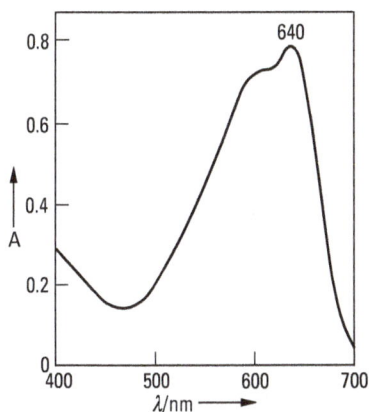

Fig. 4.3: Spectrum of the free calconcarboxylic acid indicator.

measured absorbance remains relatively constant at a low value as long as no free indicator is present in the solution. However, this is displaced from its less stable calcium complex shortly before the equivalence point when further EDTA is added, and the absorbance increases accordingly. As soon as the equivalence point is reached, the absorbance has reached its maximum value, because at this stage of the titration, the calcium-indicator complex has just been completely destroyed and therefore no further quantities of the free indicator absorbing here can be produced. Figure 4.4 shows the course of the absorbance as a function of the volume of EDTA solution added.

The limit of quantification of the photometrically indicated calcium titration is sufficiently low to determine the **calcium content in natural serum**. For this purpose, the serum separated by centrifugation is diluted in a ratio of 1:10 with sodium chloride solution of a concentration of 0.15 mol/l (so-called physiological saline solution). Bring 5 ml of this solution to pH 13 with sodium hydroxide solution and titrate as described above using EDTA and calconcarboxylic acid as an indicator.

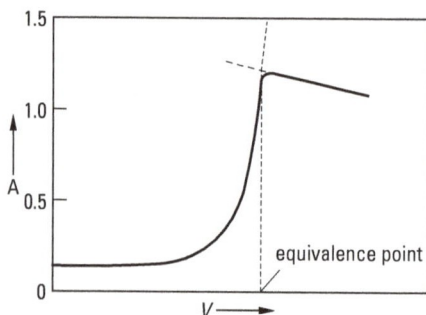

Fig. 4.4: Photometric titration of calcium with EDTA solution at 640 nm with calconcarboxylic acid as indicator (A) absorbance; (V) added volume of EDTA solution.

As a microdetermination, many metals can be titrated similarly complexometrically with simultaneous photometric indication. The procedure should be based on the calcium determination described above.

In addition to titrations using a suitable indicator, there are also so-called self-indexing methods. These involve measuring the light absorption of colored substances that are either contained in the standard or sample solution from the outset or are formed during the titration (see titrations with potassium permanganate). The titration curves are usually linear, i.e. the absorbance increases or decreases up to the equivalence point and then remains constant, or it is close to zero up to the equivalence point and then increases in a straight line.

4.3 Conductometric titrations

In **conductivity titration** or **conductometry**, the change in the conductivity of a solution is observed, which is caused by adding portions of the standard solution. The conductivity values κ – or proportions thereof – are displayed in a coordinate system as a function of the volume V of the standard solution added in each case. This results in curves as shown schematically in Fig. 4.5. The projection of, for example, the intersection point B of the reaction line AB with the line of the excess reagent BC onto the volume axis shows the reagent consumption up to the equivalence point. It should be noted that the conductivity is made up of the individual conductivities of all the ions present in the solution, regardless of whether they are involved in the reaction or not. If weak electrolytes are involved in the reaction, whose dissociation into ions repeatedly adapts to changing equilibrium conditions (law of mass action), this results in a more or less curved reaction curve A'B (e.g. the titration of monochloroacetic acid with sodium hydroxide solution). Favorable conditions for conductometry are therefore present when the titration reaction takes place between strong electrolytes and at the same time as few ions as possible are present that do not participate in the reaction. If the foreign electrolyte content is too high, the conductivity changes during the titration are often so small in relation to the total conductivity that it is difficult to recognize the end point of the reaction (see Further reading [146]).

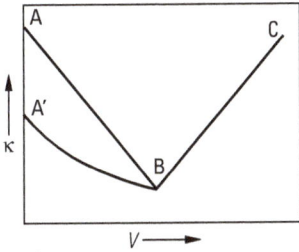

Fig. 4.5: Titration curve of conductometric titration (κ) conductivity; (V) added volume of the standard solution.

4.3.1 Theoretical foundations

Conductivity titration uses the property of aqueous electrolyte solutions to conduct an electric current. This conductivity is based on the electrolytic dissociation of the dissolved acids, bases and salts, i.e. the fact that these substances in aqueous solution have broken down into electrically charged particles, the ions. In the electric field, the ions migrate (the anions to the positively charged **anode**, the cations to the negatively charged **cathode**) and always transport the same electric charge per mole of equivalent particles, namely 96,494 coulombs, to the electrodes (Faraday's law). The conductivity of a diluted electrolyte solution depends on

 – the number of electricity carriers (ions) in the solution, i.e. on their concentration,
 – the number of elementary charges that each ion transports,
 – the **speed of migration** or **mobility** of the ions in the solvent in question,
 – the polarity of the solvent (the more polar the solvent is, the better the electrolyte is dissociated in it),
 – and the temperature (the conductivity increases by around 2.5% per degree increase in temperature).

The ion mobility u, measured in cm²/s V, is the speed at which ions move in the direction of the lines of force in the electric field (measured in V/cm). This mobility therefore depends on the nature of the ions, the size of their solvate shells, the viscosity of the solvent and the applied field strength. In an electric field, the movement of each type of ion can be regarded as independent of the others still present in the solution. Each one transports a certain proportion of the electricity, and the sum of all proportions determines the total measured **conductivity**. This quantity, formerly known as the specific conductivity of an electrolyte κ, is made up additively of the corresponding ion proportions according to

$$\kappa = \text{const.} \sum u_i \cdot z_i \cdot c_i$$

(u_i = ion mobility, z_i = charge number and c_i = concentration of ion type i). The SI unit of electrical conductivity is Siemens/meter (S/m). A Siemens (S) is defined as $1/\Omega$, where Ω (ohm) is the unit of electrical resistance. The conductivity of a completely dissociated electrolyte is a linear function of its concentration at constant temperature, as the valence and, in dilute aqueous solution, the mobility of its ions remain the same.

Every determination of electrical conductivity is based on a resistance measurement. The **conductance** G and the resistance R are linked to each other by the determination equation

$$G \cdot R = 1.$$

The resistance is directly proportional to the length of the conductor l and inversely proportional to its cross section q:

$$R = \rho \cdot \frac{l}{q}.$$

For the proportionality factor ρ, also known as the specific resistance, the following relationship applies accordingly

$$\rho \cdot \kappa = 1,$$

so that the desired conductivity can be calculated by

$$\kappa = \frac{1}{R} \cdot \frac{l}{q}.$$

If a direct current flows between the electrodes immersed in an electrolyte solution to measure the resistance, electrolysis occurs at these electrodes, and the contribution of the ohmic resistance of interest alone to the total resistance is so small that it is impossible to measure it. It is therefore necessary to measure the resistance with alternating current to determine the specific conductivity.

For conductometry, the concept of **equivalent conductivity** Λ is of particular importance. This results from the formula

$$\Lambda = \frac{1,000 \cdot \kappa}{c},$$

where c is the concentration of equivalents of the corresponding substance in mol/l. Λ is usually given in the unit S cm^2/mol. While the specific conductivity approaches zero with decreasing concentration, the equivalent conductivity tends toward a limit value Λ_0, which is the sum of the ion equivalent conductivities (these are proportional to the ionic mobilities) of the anion (l_A) and the cation (l_K): $\Lambda_0 = l_A + l_K$. In a way that cannot be discussed here, the equivalent conductivities of the individual ions have been compared with each other, and the values obtained are summarized in Tab. 4.1, which are valid for 25 °C.

How does the conductivity change in the course of a titration? The reaction of hydrochloric acid with potassium hydroxide solution, which can be formulated as an ionic equation as follows, may serve as an example:

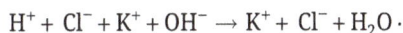

$$H^+ + Cl^- + K^+ + OH^- \rightarrow K^+ + Cl^- + H_2O \cdot$$

The hydroxide ions of the alkali combine with the hydrogen ions of the titrated acid to form practically undissociated water, while the potassium ions increasingly take the place of hydrogen ions. At the equivalence point, all hydrogen ions originally present in the solution have been replaced by potassium ions. Since, as given in Tab. 4.1, the potassium ions have a considerably lower equivalent conductivity (corresponding to a lower mobility) than the hydrogen ions, the total conductivity of the titrated solutions must decrease more and more in proportion to the progress of the neutralization. If alkali is added beyond the equivalence point, there is of course no further diminution, but rather an increase in conductivity, because the individual conductivities of the excess potassium and hydroxide ions are added to the conductivity at the equivalence point, which is caused only by the potassium chloride present. As shown in Fig. 4.6, the individual conductivities of the ions involved in the reaction add up to the total conductivity for each time point of the titration.

Tab. 4.1: Ion equivalent conductivities l in water at 25 °C.

Cation	l_K in $\frac{S\ cm^2}{mol}$	Cation	l_K in $\frac{S\ cm^2}{mol}$	Anion	l_A in $\frac{S\ cm^2}{mol}$	Anion	l_A in $\frac{S\ cm^2}{mol}$
H^+	349.6	½ Be^{2+}	45	OH^-	197	ClO_4^-	67
Li^+	38.7	½ Mg^{2+}	58	F^-	55	IO_4^-	55.6
Na^+	50.1	½ Ca^{2+}	59	Cl^-	76.4	MnO_4^-	61
K^+	73.5	½ Sr^{2+}	60	Br^-	78	$(HCOO)^-$	56
Rb^+	77	½ Ba^{2+}	63.2	I^-	77.1	$(CH_3COO)^-$	41.4
Cs^+	77.7	½ Zn^{2+}	54	CN^-	82	½ SO_4^{2-}	79
NH_4^+	74	½ Cd^{2+}	54	CNS^-	66	½ CrO_4^{2-}	83
Ag^+	62.2	½ Pb^{2+}	65	ClO_3^-	65.3	½ CO_3^{2-}	74
Tl^+	74	½ Mn^{2+}	50	BrO_3^-	56.0	½ $(C_2O_4)^{2-}$	63
		½ Cu^{2+}	55.5	IO_3^-	41.6	½ $(C_4H_4O_6)^{2-}$	55
		½ Ni^{2+}	49	NO_3^-	71.1	½ PO_4^{3-}	69

The titration curves are straightforward as long as the individual ion types present either do not react at all or react quantitatively. A great advantage of conductometry, as can be seen from this, is the fact that the equivalence point itself does not need to be determined during the titration steps, but can be found by graphic extrapolation. The respective curve character of a titration diagram is generally characterized by

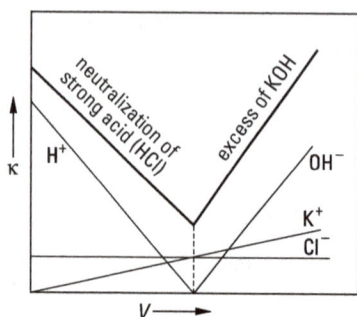

Fig. 4.6: Conductometric titration of a strong acid with potassium hydroxide solution (κ) conductivity; (V) added volume of potassium hydroxide.

the fact that instead of the disappearing ionic species of the sample solution provided, a new one appears from the standard solution – with greater or lesser conductivity. In the former case, an increase is obtained, and in the latter a decrease in the total conductivity up to the equivalence point. Once the equivalence point has been exceeded, an increase in conductivity is of course always observed if no further reactions follow.

Despite the temperature dependence of the conductivity, the use of a thermostat is only necessary in exceptional cases because most titrations are completed in a few minutes.

4.3.2 The titration device

In order to carry out a conductometric determination, suitable conductivity measuring vessels are required to hold the sample solution. These are usually glass vessels with platinized platinum electrodes. Examples of commercially available devices are shown in Fig. 4.7. You can opt for an immersion measuring cell fitted with electrodes, which is simply immersed in the beaker filled with the sample solution, or you can use a classic titration vessel with permanently installed electrodes. In general, provision is made for suitable stirring. The standard solution is added using the usual burettes or piston burettes.

The size and spacing of the electrodes in the conductivity measuring vessel depend on the resistance to be expected in the liquid to be titrated. The poorer the conductivity of the solution, the larger the electrodes should be and the smaller the distance between them. Care must be taken to ensure that the vessel resistance remains easily measurable, i.e. it is not below 30 and not above several thousand ohms. The purpose of platinizing the electrodes is to greatly increase their surface area. This effectively counteracts polarization of the electrodes, which would interfere with the conductivity measurement (see p. 304).

For platinization, the carefully cleaned vessel is filled with a solution of 3 g $H_2[PtCl_6]$ and 25 mg lead acetate in 100 ml deionized water. The two electrodes are conductively connected and a platinum auxiliary electrode is inserted as precisely as possible into the center of the gap separating them. A voltage of 4 V is applied to this as the anode and to

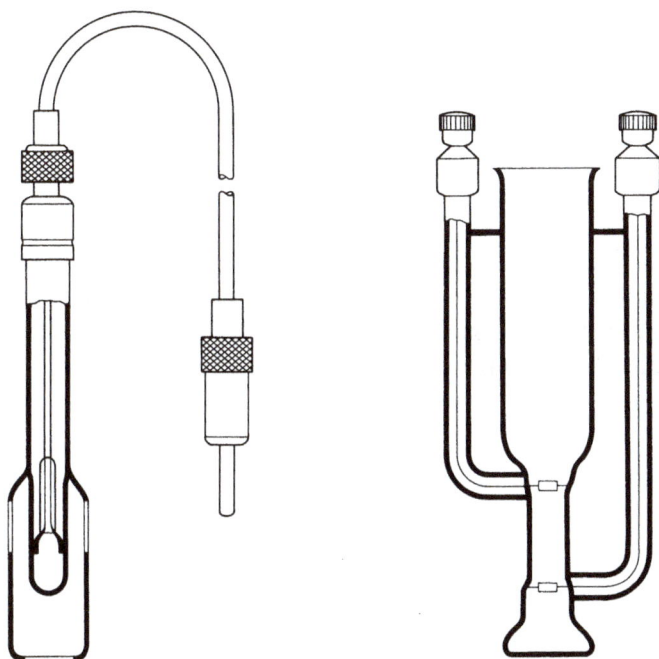

Fig. 4.7: Immersion measuring cell and conductivity measuring vessel.

the connected vessel electrodes as the cathode, whereupon the solution is electrolyzed with a current density of no more than 30 mA/cm² electrode surface (measured on one side) for about 10 min. The platinizing solution is then removed, the vessel is filled with diluted sulfuric acid and the residual $H_2[PtCl_6]$ still adhering to the electrodes is removed by briefly electrolyzing again. Finally, the conductivity measuring vessel is thoroughly cleaned with deionized water. Conductivity measuring vessels should never be left dry, but should always be filled with deionized water when not in use in order to maintain the effectiveness of the platinization.

Each conductivity measuring vessel has a cell constant Z that depends on the distance and cross section of its electrodes as well as on its filling level and other circumstances, which is

$$\kappa = \frac{1}{R} \cdot \frac{l}{q},$$

or, since l/q is not measurable here:

$$\kappa = \frac{1}{R} \cdot Z \quad \text{or} \quad Z = \kappa \cdot R.$$

Z is the resistance that a conductivity measuring vessel would have if it were filled with a liquid of conductivity 1. The cell constant can be determined using suitable calibration solutions of known conductivity (e.g. KCl solution, 1 mol/l: $\kappa_{25} = 0.11173$ S/cm). Manufacturers of conductivity measuring cells often specify the cell constant. Depending on the concentration of the solution to be titrated, the conductivity meter must be selected so that its cell constant has a value that allows a resistance measurement in a range that is favorable for the respective measuring device.

In order not to change the cell constant of the conductivity measuring vessels during a titration, the electrodes must not be placed too close to the surface of the liquid, and also the volume of the standard solution to be added must be kept low; a maximum of 5 ml of a correspondingly concentrated standard solution should be added to 50 ml of solution. It is advantageous to use smaller burettes, which are subdivided into 0.01 ml, so that the reading accuracy remains the same as with the usual titrations using burettes with a capacity of 50 ml subdivided into 0.1 ml.

4.3.3 Conductivity measurement

4.3.3.1 Wheatstone bridge circuit

As already mentioned, conductivity determinations are synonymous with resistance measurements. Anyone who demands the utmost precision and, above all, absolute accuracy in resistance measurement still use Wheatstone's bridge circuit today. The simplest method is the zero-point method. The circuit diagram is shown in Fig. 4.8.

The voltage U of an alternating current source is present at points A and B of the measuring bridge. This consists of four resistors R_x, R_1, R_2 and R_3, where R_x is the resistance to be measured of the conductivity measuring vessel and R_3 is an adjustable resistor. The circuit branches at A and B into two partial current paths ACB and ADB, i.e. the entire AC voltage applied drops completely at resistors R_2 and R_3 as well as at R_1 and R_x. If R_3 has been tuned so that points C and D are at the same potential (no voltage), the following relationship applies according to Kirchhoff's laws (see physics textbooks):

$$\frac{R_x}{R_3} = \frac{R_1}{R_2}, \quad \text{from which follows} \quad R_x = \frac{R_3 \cdot R_1}{R_2}.$$

Between C and D, the voltage (V) after amplification by the AC amplifier is measured at the galvanometer G. If R_3 has been set correctly, **no** voltage should be displayed at G (zero indication). After reading R_3, the cell resistance R_x can then be calculated using the above equation.

If very high accuracy is required, but especially if the electrodes of the conductivity measuring vessel are very close to each other due to the low conductivity of the electrolyte solution, it must be taken into account that a so-called capacitive resistance occurs at the conductivity measuring cell due to the use of alternating current. This can be compensated for by a controllable capacitor C connected in parallel with R_3. Accordingly, R_3 and C must be adjusted so that G does not display any voltage. However, this much more complex procedure can be simplified considerably by using an oscilloscope instead of the galvanometer G. In order to avoid polarization caused by electrolysis, an alternating voltage with a frequency between 50 Hz (concentrated solutions) and about 1,000 Hz (diluted solutions) should be applied to the well-platinized electrodes of the conductivity measuring cell (see p. 258). Instead of the complex voltage measuring device with amplifier and galvanometer, a headphone (telephone method) was connected between C and D in the past, and then had to tune to the tone minimum. However, working with such an acoustic indication was quite tiring in the long run.

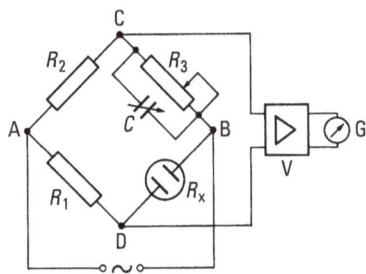

Fig. 4.8: Wheatstone's bridge circuit for measuring conductivity (R) resistors; (C) capacitor; (V) amplifier; (G) galvanometer.

To accurately determine the conductivity of an electrolyte solution, the method described above for measuring resistance using the Wheatstone bridge must be used. First, determine the cell constant Z (see p. 259) of the measuring vessel with an electrolyte solution of known conductivity (calibration). The measurements, which are then of course carried out with compensation of the cell capacitance, are very accurate, but also quite lengthy due to the discontinuous mode of operation. However, since conductometric titration only depends on relative measured values, a method with a direct display is often used today.

4.3.3.2 Direct reading methods

In routine direct reading methods, the alternating voltage U is dropped across the cell resistor R_x and a working resistor R. As shown schematically in Fig. 4.9, the voltage dropped across R controls an amplifier with a high-impedance input (the voltage is measured without current flowing) and thus regulates the current I of a secondary circuit operated with a constant DC voltage. As a result, the current I changes accordingly with the resistance R_x of the cell (voltage/current conversion). The more or less elaborately designed devices allow adaptation to different conductivities in the measuring cell with optimum utilization of the scale of the ammeter. The signal measured there is proportional to the conductivity in the cell and can be used directly to record the titration curves. There are commercially available device designs that allow calibration and therefore also absolute measurement of conductivity (e.g. conductometers from Metrohm).

The advantage of the described measuring arrangement is therefore that during a titration, after each addition of standard solution and after its mixing with the solution to be titrated (stirring), a measuring bridge does not have to be calibrated first, but the conductivity or a proportional thereof is read off immediately. The quality of the measurement depends on how linearly the amplifier operates in the required measuring range.

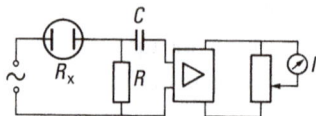

Fig. 4.9: Basic circuit of a conductometer (R) resistors; (C) capacitor; (I) ammeter.

4.3.4 Practical applications

4.3.4.1 Acid-base titrations

The titration of strong acids with strong bases has already been discussed on p. 87 and is shown graphically in Fig. 4.6. Strong acids and strong bases can also be determined precisely by conductometry down to very large dilutions. However, CO_2-free alkalis and CO_2-free water must then be used for dilution.

In the graphical representation of the neutralization of solutions of weak acids – hydrocyanic acid (hydrogen cyanide), boric acid, acetic acid, etc. – with a strong base, e.g. sodium hydroxide solution, 1 mol/l, a curve is obtained as shown schematically in curve 1 of Fig. 4.10. Initially, the solution has a relatively low conductivity due to the only slight dissociation of the weak (acetic) acid, which decreases even further as a result of further diminution in the H^+ ion concentration due to the formation of less dissociated water at the beginning of the titration because H^+ ions are replaced by Na^+ ions, which have a much lower equivalent conductivity than H^+ ions (see p. 257, Tab. 4.1), and at the same time the sodium acetate that is formed pushes back the dissociation of the acetic acid. Only in the course of the titration does so much strongly dissociated sodium acetate gradually form that the conductivity now caused by Na^+ and acetate ions can increase (AB). After the equivalence point is exceeded, the conductivity increases more strongly (BC) because the hydroxide ions of the base, which are good conductors, are no longer consumed. The reaction line and the line of excess alkali intersect at an obtuse angle. The weaker the acid to be titrated, the more obtuse is the angle. There is a curved transition piece near the equivalence point, which is due to the protolysis of the salt formed. The situation is very similar for the neutralization of weak bases – e.g. ammonia – by a strong acid, e.g. hydrochloric acid, 1 mol/l.

The shape of the curve obtained when neutralizing medium-strength acids or bases with strong bases or acids lies between the two extreme types discussed so far, depending on the dissociation and concentration conditions in the solution provided. If the buffer range is passed through during the titration of a medium-strength acid or base, the result is generally a strongly curved graph, making it very difficult to determine the exact equivalence point. In these cases, a measured volume of the strongly dissociated standard solution should be prepared and titrated with the medium-strength acid or base to be determined, because the large excess of the salt formed during neutralization can push back the dissociation of the acid or base after the equivalence point to such an extent that a straight titration curve is usually achieved. This makes it much easier to determine the equivalence point.

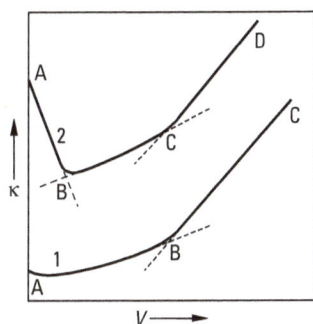

Fig. 4.10: Conductometric titration of acetic acid (1) or mix of hydrochloric acid and acetic acid (2) with sodium hydroxide solution (κ) conductivity; (V) added volume of sodium hydroxide.

In a solution containing a strong and a weak acid (e.g. hydrochloric acid and acetic acid) or a strong and a weak base next to each other, both components can be determined quantitatively in a single titration run. You then obtain curve shapes of the type shown in curve 2 in Fig. 4.10. AB indicates the decrease in conductivity of the solution caused by the neutralization of the strong acid, BC the increase in conductivity caused by the subsequent neutralization of the weak acid, CD the stronger increase in conductivity caused by the excess base. The projections of points B and C onto the abscissa indicate the volume of sodium hydroxide solution for the neutralization of the strong acid or the sum of strong and weak acids, respectively. This method is correct if the dissociation constants of the two acids differ sufficiently. Otherwise, the curved transition areas can become so large that the linear extrapolation of the intersection points gives incorrect results.

4.3.4.1.1 Displacement processes

In solutions of salts of weak bases with strong acids (e.g. ammonium chloride), the bound base can be determined conductometrically by **displacement titration** with strong bases, and in solutions of salts of weak acids with strong bases (e.g. sodium acetate and potassium cyanide), the bound weak acid can be determined by displacement titration with strong acid. The prerequisite for this is that the dissociation constants of the weak bases or acids whose salt solutions are to be titrated differ sufficiently from those of the strong bases or strong acids used for titration.

The shape of the curve for the displacement titration of salts of weak bases depends on the ratio of the equivalent conductivities of the cations, and for the displacement titration of salts of weak acids on the ratio of the equivalent conductivities of the anions. Curve 1 in Fig. 4.11 shows the titration of an ammonium chloride solution with potassium hydroxide solution, and curve 2 with sodium hydroxide solution:

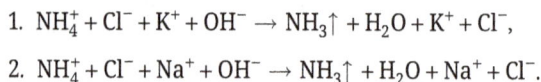

1. $NH_4^+ + Cl^- + K^+ + OH^- \rightarrow NH_3{\uparrow} + H_2O + K^+ + Cl^-$,

2. $NH_4^+ + Cl^- + Na^+ + OH^- \rightarrow NH_3{\uparrow} + H_2O + Na^+ + Cl^-$.

In the second case, the more conductive ammonium ion is replaced by the less conductive sodium ion, and in the first case by the equally conductive potassium ion (see Tab. 4.1). It is therefore possible to influence the shape of the curve by selecting a suitable standard solution and thus achieve the most suitable intersection angle for determining the equivalence point.

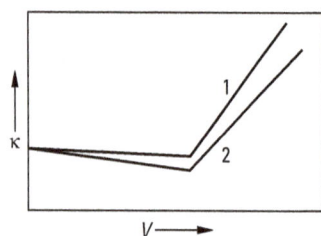

Fig. 4.11: Conductometric titration of ammonium chloride solution with potassium hydroxide solution (1) or sodium hydroxide solution (2) (κ) conductivity; (V) added volume of lye.

In this context, it should be pointed out once again that conductometric neutralization analysis should also be carried out with alkaline solutions ($Ba(OH)_2$ and $NaOH$) and other reagents that are as CO_2 free as possible; otherwise significant errors may occur. If a carbonate-containing alkali is titrated with a strong acid, its hydroxide ions are first neutralized, followed by the conversion of the carbonate into hydrogen carbonate and finally carbon dioxide is released and "displaced":

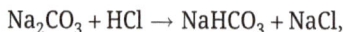

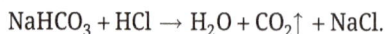

$$NaOH + HCl \rightarrow NaCl + H_2O,$$

$$Na_2CO_3 + HCl \rightarrow NaHCO_3 + NaCl,$$

$$NaHCO_3 + HCl \rightarrow H_2O + CO_2{\uparrow} + NaCl.$$

These processes can – when determining the equivalence point simply by extending the first larger section of the reaction line and the line of the excess acid up to the point of intersection – give rise to gross errors with a higher carbonate content. The situation is similar in the case of neutralization of acid with carbonate-containing alkali.

4.3.4.2 Precipitation titrations

Conductometric precipitation analysis is particularly important because there are numerous analytically usable precipitation reactions for whose end point detection a suitable indicator is missing. Their principles are illustrated using the example of the precipitation of bromide ions of a diluted sodium bromide solution by the silver ions of a relatively concentrated silver acetate solution:

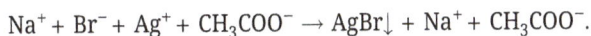

$$Na^+ + Br^- + Ag^+ + CH_3COO^- \rightarrow AgBr\downarrow + Na^+ + CH_3COO^-.$$

The resulting silver bromide is very poorly soluble and therefore does not contribute to the conductivity of the solution. The concentration of sodium ions remains practically constant during titration, while the more conductive bromide ions are replaced by the less conductive acetate ions. The conductivity therefore decreases until the end of the precipitation reaction. It then increases due to excess standard solution. The lowest possible solubility of the precipitate is important for the accuracy of the conductometric precipitation analysis. The lower the solubility, the more the curved transition section of the curve at the equivalence point disappears.

The nature of the respective precipitate must also be taken into account. It should have a defined composition, low adsorption capacity and not be involved in any subsequent reactions. The precipitation process itself should be fast and quantitative.

4.3.4.3 Conductivity titrations at elevated temperatures

It is often observed during titration that the conductivity only gradually reaches a constant value after each reagent addition. This can be the case, for example, if the precipitation of the precipitate takes place slowly. However, such processes generally take place much faster at elevated temperatures. It can therefore be advantageous to carry out conductometric titrations at higher but constant temperatures. In such cases, double-walled conductivity measuring vessels are used, which allow connection to the circulation of a liquid thermostat. The required standard solution should be added using a piston burette, whereby the tip of the burette, which is tapered to form a capillary, is immersed in the solution to be titrated in the conductivity measuring cell. Piston burettes are insensitive to pressure fluctuations that may occur due to increased temperature in the titration vessel.

If a thermostat is not available, an apparatus can be used in which the titration vessel is suspended in the vapor stream of a constantly boiling liquid (alcohol and water) (see Fig. 4.12). In this way, for example, the sulfate ion can be determined titrimetrically with barium acetate, though sulfate is otherwise difficult to titrate. The constancy of conductivity is achieved in the vicinity of 100 °C after a maximum of 1 min for each reagent addition. The entire determination does not take much longer than 10 min. The prerequisite is that the sulfate solution to be titrated, e.g. ammonium sulfate solution, reacts neutrally. This method can also be used to determine the sulfate content of drinking water.

The required ground joint apparatus is shown in Fig. 4.12. A suitable liquid boils in the flask; the vapor of which coats the actual conductivity measuring cell and is then condensed again in a reflux condenser. The liquid returns to the heated flask via a bypass and can be vaporized again. The conductivity measuring vessel suspended in

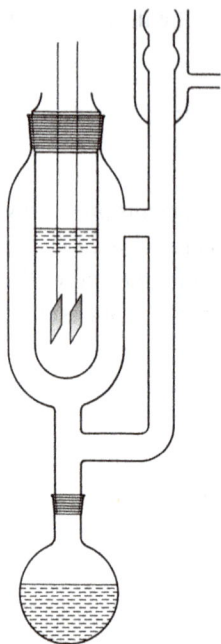

Fig. 4.12: Conductivity titration in the vapor stream of a boiling liquid.

the vapor flow has an external ground joint near its upper edge so that the vapor space remains securely sealed and no vapor can escape. The tip of the piston burette, the two electrodes and the motor-driven stirrer are inserted into the titration vessel from above. If you do not want to interrupt the steam flow after completing a titration in order to remove the conductivity measuring vessel, you should siphon off the titrated solution from above without opening the apparatus, then rinse several times and finally fill in a new sample solution.

4.3.5 High-frequency titration

In contrast to conductivity titration, in which the change in resistance is measured with electrodes immersed in the sample solution, high-frequency titration works with electrodes that are attached to the outside of the measuring vessel. This arrangement has the great advantage that a change in the electrodes due to chemical reactions and adsorption is excluded. Of course, the electrodes cannot be polarized either. One disadvantage of this method is the considerably greater measuring effort.

The measuring cell used is absolutely impermeable to direct current, as the glass wall between the electrodes and the solution has a very high ohmic resistance. However, the system behaves like a capacitor with respect to a high-frequency alternating current, the capacity of which depends on the composition of the sample solution with which the vessel is filled. As the composition of the filled solution changes during the titration, the resulting change in the AC resistance can be observed.

A basic measuring arrangement consists of connecting a capacitor, an induction coil and an ohmic resistor in a series circuit, as shown in Fig. 4.13. As a result of the phase shift between current and voltage, the so-called **reactance resistances** result at the coil and capacitor, whose magnitude depends on the AC frequency and the characteristics of the coil or capacitor. For each oscillating circuit of this type, there is a certain frequency at which the reactances of the coil and capacitor compensate each other, so that the current reaches a maximum value (**resonance**) due to the ohmic resistance remaining effective in the oscillating circuit alone. However, the more the resonant circuit is detuned from the resonance condition, e.g. by changing the capacitance of the capacitor, the more the current strength will decrease from its maximum value. If the measuring cell described above is connected to the AC circuit instead of the capacitor, the change in the composition of the solution caused by the titration will influence its capacity. The measured variable could be the current flowing in the oscillating circuit, but also, for example, the voltage drop across the ohmic resistor R. This voltage difference can be amplified electronically and a signal is obtained that assumes its highest value in the resonance case. In any case, different measurement variants are commonly used, but the aim is always to determine the AC resistance or its reciprocal value, the AC conductance of the titration cell. Due to the electrolyte inside the cell, the AC conductance can be made up of the aforementioned reactive component and an ohmic component, which depends on the conductivity of the solution. This component, also known as the **active component**, changes in a characteristic way during the titration. It is important to realize that the measuring cell cannot be regarded solely as a capacitor C, as initially shown in Fig. 4.13, but is rather described by the equivalent circuit diagram in Fig. 4.14. Here, C_1 represents the capacity component caused by the glass walls as a dielectric, while the electrolyte solution acts as both a capacitive (C_x) and an ohmic resistor (R_x), which must be imagined as being connected in parallel. If the AC conductance of the measuring cell is determined and plotted as a function of the logarithm of the specific conductivity of the electrolyte solution filled into the cell, a bell-shaped curve is obtained, which is also referred to as the **high-frequency characteristic curve** (Fig. 4.15). For this purpose, the frequency of the excitation voltage must be selected so that the conductance of the active component R_x is large compared to the blank conductance of C_x, so that the measured change in the AC conductance is practically caused by the change in ohmic conductivity (**active component method**). Figure 4.15 shows that measurements should only be taken in the area of the inflection points (shown in thick lines) because the titration curves then correspond to the usual conductometric curves. The further you move away from these zones, the more the linear relationship between the AC conductance and the concentration of the ions to be determined is lost, and more or less strongly curved graphs are obtained.

For the sake of completeness, it should be mentioned that it is also possible to detect the blank conductance of C_x as a function of the titration process by selecting different measurement parameters, which also allows an indication of the equivalence point under

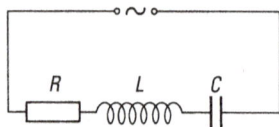

Fig. 4.13: Alternating current circuit with (R) an ohmic resistor, (L) an induction coil and (C) a capacitive resistor.

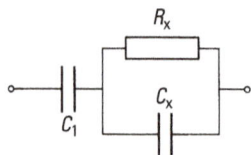

Fig. 4.14: Equivalent circuit diagram of a capacitance measuring cell (R) ohmic resistor; (C) capacitive resistor.

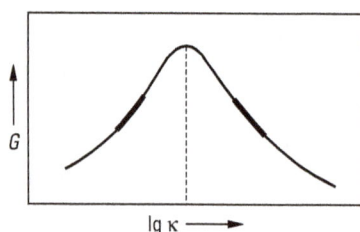

Fig. 4.15: High-frequency characteristic curve of the active component method (G) effective component of the AC conductance; (κ) conductivity.

certain conditions. Instead of capacitance cells, inductance cells are also in use. With these, the inductance of a coil is influenced by an electrolyte solution placed inside it.

High-frequency titrations are always used when the composition of an electrolyte solution interferes with the resistance measurement between immersed electrodes. Another important field of application is titrations in nonaqueous solvents (see Further reading [147]).

4.4 Potentiometric titrations

In potentiometric titrations, the potential changes that occur during a titration at an indicator electrode immersed in the concerning solution are measured. This electrode should only respond to those ions whose quantitative determination is the subject of the titration or should only be able to measure the redox potential (inert electrode). For example, a chloride solution can be titrated with a standard solution of $AgNO_3$, meanwhile the potential of Ag^+ ions is measured by an immersed silver plate as an electrode. By displaying the measurand and evaluation of this graph, the equivalence point can be determined. The equivalence point is the projection of the inflection point of the measured curve onto the abscissa (volume of the added standard solution). Figure 4.16 shows the situation described above.

It is not possible to measure the potential of a single electrode directly; it is only possible to determine the potential difference between two electrodes, which must be in a closed circuit. Of course, the second electrode must not be influenced by the titration process, as otherwise several influencing variables would be superimposed and the resulting curve would be difficult or impossible to interpret. The measurement of the potential difference itself cannot simply be carried out with any voltmeter, as the measuring instrument requires energy to display the voltage, which would have to come from the chemical conversion of substances at the electrodes. However, this influences the titration. It is therefore necessary to consider how such a measurement can be carried out without extracting energy from the chemical system. A number of questions need to be considered theoretically.

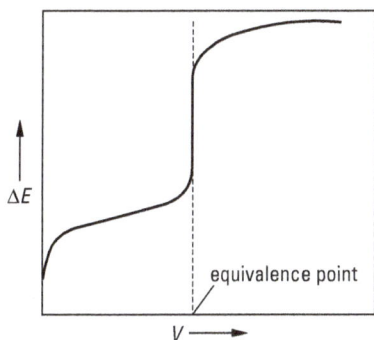

Fig. 4.16: Potentiometric titration curve (ΔE) electromotive force of the electrode; (V) added volume of the standard solution.

4.4.1 Theoretical foundations

First of all, we have to deal with oxidation and reduction reactions. If an Fe(II) salt solution is brought together with $Ce(SO_4)_2$ in a beaker, the Fe^{2+} ions are oxidized by Ce^{4+} to Fe^{3+}, while Ce^{4+} is simultaneously reduced to Ce^{3+}. We describe the reaction using the equation:

$$Ce^{4+} + Fe^{2+} \rightleftharpoons Ce^{3+} + Fe^{3+}.$$

Energy is released and the reaction solution heats up. A state of equilibrium is reached, and the position of which can be described by the law of mass action:

$$\frac{c_0(Ce^{3+}) \cdot c_0(Fe^{3+})}{c_0(Ce^{4+}) \cdot c_0(Fe^{2+})} = K.$$

The ion concentrations are exceptionally given the index 0 here to make it clear that the reaction has reached its equilibrium state. Some of the energy released can also be recovered in another form as heat. The maximum possible proportion is referred to as the free enthalpy G (second law of thermodynamics).

It is common practice to allocate the partial amount to each species involved in the reaction. The total conversion is calculated as ΔG by subtracting the energy amounts of the end products from those of the starting substances. However, if the system has already reached its equilibrium state, no further energy can be released voluntarily, and the equilibrium condition is then given as follows:

$$\Delta G = 0.$$

Depending on the freely selectable initial concentration c, the total free enthalpy is released in the course of a reaction up to the equilibrium state ΔG is released (ΔG is negative, because in chemistry an energy released by the system is given a negative sign). In a galvanic cell, this proportion of energy can be recovered in the form of electrical work (E_{El}):

$$-\Delta G = E_{El}.$$

Electrical work is defined as the product of the electric charge and the voltage. In a redox reaction, the electric charge $z \cdot F$ is exchanged between the partners (z is the number of electrons exchanged according to the reaction equation and F (96,494 A $\cdot$ s) is the amount of current of 1 mol of electrons exchanged, Faraday's constant). The voltage measured between two electrodes is ΔE (volt). This results in

$$-\Delta G = z \cdot F \cdot \Delta E.$$

In electrochemistry, this voltage ΔE is referred to as the potential difference.

If you want to generate electrical energy from the redox reaction described above, you must not allow the reactants to react with each other in a beaker. Instead, it must be ensured that the electron exchange takes place at electrodes. To do this, however, the Ce species and the Fe species must be spatially separated from each other. In practice, this means that Fe^{2+} and Fe^{3+} ions are dissolved in one glass, while Ce^{3+} and Ce^{4+} ions are dissolved in another. Both solutions must be ionically conductive, for example, connected by a salt bridge. Electron exchange can then take place at electrodes that are immersed in two glasses, with the oxidation $Fe^{2+} \rightarrow Fe^{3+} + e^-$ taking place at one electrode and the reduction $Ce^{4+} + e^- \rightarrow Ce^{3+}$ at the other. However, the reaction as a whole can only take place if the two electrodes are conductively connected to each other, thus extracting electrical energy from the system. What we have just constructed is a **galvanic element** as it could be used, in a modified form of course, to supply electrical energy. The conversion of chemical substances at the electrodes inevitably results in polarization effects (see p. 304), so that the effective voltage is lower than the maximum possible voltage stated above. Typical for the **galvanic** case is therefore the condition

$$\Delta G + z \cdot F \cdot \Delta E < 0.$$

If you want to reverse the voluntary reaction process, you must apply a countervolt-age to the electrodes that will be greater than that of the equilibrium condition due to the losses in the system, i.e.

$$\Delta G + z \cdot F \cdot \Delta E > 0$$

in **electrolytic** operation. The structure of a galvanic or electrolytic cell is shown schematically in Fig. 4.17.

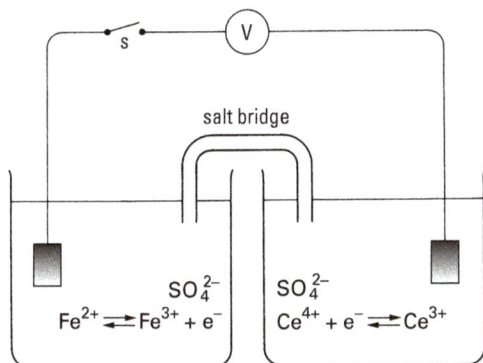

Fig. 4.17: Example of a galvanic or electrolytic cell (V) voltmeter.

The **electrochemical equilibrium** lies between the two forms of application de-scribed above with $\Delta G + z \cdot F \cdot \Delta E = 0$. Only in this case are the conditions defined, and the variables can be calculated or measured. However, energy may neither be with-drawn from the cell nor supplied to it, which means that no current may flow during a potential measurement ($I = 0$). Since this potential can be calculated theoretically from the concentrations of the species involved in the reaction in the respective solu-tions and their concentrations in the equilibrium state, exact potential measurements can be used to obtain information about the actual concentration ratios. For this pur-pose, we must first calculate the energy differences for each species separately be-tween the actual state given by the concentration c and that of the chemical equilib-rium with the concentration c_0. However, this is the reversible work A that is expended when changing the concentration from c to c_0:

$$A = RT \cdot \ln\frac{c}{c_0}.$$

According to the principle that the total free enthalpy converted during the reaction is equal to the sum of the proportions of the products ($\sum A_p$) minus the sum of the proportions of the starting materials ($\sum A_a$), the result is

$$\Delta G = RT \cdot \ln \frac{c(Ce^{3+})}{c_0(Ce^{3+})} + RT \cdot \ln \frac{c(Fe^{3+})}{c_0(Fe^{3+})} - RT \cdot \ln \frac{c(Ce^{4+})}{c_0(Ce^{4+})} - RT \cdot \ln \frac{c(Fe^{2+})}{c_0(Fe^{2+})}.$$

This expression is transformed in such a way that all equilibrium concentrations c_0 and all given concentrations c are summarized in one term each:

$$\Delta G = -RT \cdot \ln \frac{c_0(Ce^{3+}) \cdot c_0(Fe^{3+})}{c_0(Ce^{4+}) \cdot c_0(Fe^{2+})} + RT \cdot \ln \frac{c(Ce^{3+}) \cdot c(Fe^{3+})}{c(Ce^{4+}) \cdot c(Fe^{2+})}.$$

Since

$$\frac{c_0(Ce^{3+}) \cdot c_0(Fe^{3+})}{c_0(Ce^{4+}) \cdot c_0(Fe^{2+})} = K(\text{law of mass action}),$$

you get

$$\Delta G = -RT \cdot \ln K + RT \cdot \ln \frac{c(Ce^{3+}) \cdot c(Fe^{3+})}{c(Ce^{4+}) \cdot c(Fe^{2+})}.$$

From this follows with $\Delta G = -z \, F \, \Delta E$

$$\Delta E = \frac{RT}{zF} \cdot \ln K - \frac{RT}{zF} \cdot \ln \frac{c(Ce^{3+}) \cdot c(Fe^{3+})}{c(Ce^{4+}) \cdot c(Fe^{2+})}.$$

If you set $\frac{RT}{zF} \cdot \ln K = \Delta E^\circ$ (standard redox potential) for the constant term, you get Nernst's equation:

$$\Delta E = \Delta E^\circ - \frac{RT}{zF} \cdot \ln \frac{c(Ce^{3+}) \cdot c(Fe^{3+})}{c(Ce^{4+}) \cdot c(Fe^{2+})}.$$

It can be seen that if the specified concentrations are identical to the equilibrium concentrations, the galvanic cell no longer has a potential difference. Since the oxidation and reduction steps take place separately in their own reaction chambers, Nernst's equation is broken down accordingly:

$$\Delta E = E_{Ce} - E_{Fe},$$

providing

$$E_{Ce} = E_{Ce}^\circ + \frac{RT}{zF} \cdot \ln \frac{c(Ce^{4+})}{c(Ce^{3+})} \quad \text{or} \quad E_{Fe} = E_{Fe}^\circ + \frac{RT}{zF} \cdot \ln \frac{c(Fe^{3+})}{c(Fe^{2+})},$$

respectively.

When written in this form, care must be taken to ensure that the oxidized stage is always in the numerator and the reduced stage in the denominator in the concentration-dependent term:

$$E = E° + \frac{RT}{zF} \cdot \ln \frac{c(\text{Ox})}{c(\text{Red})}.$$

The so-called **standard potentials** $E°$ are not defined by themselves. Rather, for any reaction – and the oxidation of hydrogen has been agreed upon: $H_2 \rightleftharpoons 2\,H^+ + 2\,e^-$ – the associated $E°$ must be defined ($E_H° = 0$). This results in a relative scale for all other standard potentials (**voltage series**).

Let us now return to our actual problem of titrating an Fe^{2+} solution with a Ce^{4+} solution and following the progress of the titration potentiometrically. The foregoing has shown how the potential of a Pt electrode depends on the concentration ratio $c(\text{Ox})/c(\text{Red})$ of the redox partners in the solution. As potentiometry is intended to be an indication method, no current may flow during the potential measurement ($I = 0$), as otherwise a substance conversion would occur at the electrodes. The Pt electrode immersed in the titration vessel is used to measure the concentration ratio of only one redox pair at a time. When titrating with the Ce^{4+} solution, the Fe^{3+}/Fe^{2+} pair (Ce^{4+} is quantitatively reduced) is decisive for the potential setting before the equivalence point, while the Ce^{4+}/Ce^{3+} pair (Fe^{2+} is now completely consumed) is decisive after the equivalence point. The equilibrium setting between the reaction partners occurs immeasurably quickly after each addition of standard solution. Of course, potential differences can only be measured between two electrodes. However, it makes sense to use an electrode system as a second half-cell in which the potential remains constant, i.e. no other variable needs to be taken into account. An electrode system with such properties is referred to as a reference electrode (see p. 284). If the potential difference measured between the two electrodes is plotted against the volume of the Ce^{4+} solution, the titration curve is obtained (Fig. 4.18). It can be shown, that the standard potential $E_{Fe}°$ is reached with 50% of the reagent addition required up to the equivalence point and the standard potential $E_{Ce}°$ is reached with twice the addition. The inflection point (equivalence point) lies at the arithmetic mean $(E_{Fe}° + E_{Ce}°)/2$. The slope of the titration curve is greatest at the equivalence point. The potential jump does not depend on the concentration of the solution to be titrated. Titrations in low concentration ranges are therefore possible without loss of accuracy. It should be noted that this only applies to redox titrations, but not to neutralization, precipitation or complexation analyses.

The potentiometric indication of redox titrations is always carried out with precious metal electrodes (Pt, Pd and Au). As these are not directly involved in the potential-determining reaction, i.e. do not undergo any material changes themselves, they are referred to as **inert electrodes**.

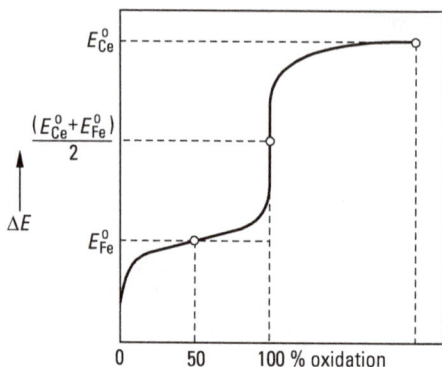

Fig. 4.18: Titration of Fe(II) ions with cerium(IV) sulfate solution, potentiometric indication (ΔE) EMF, electromotive force, of the measuring chain; ($E°$) standard potential.

4.4.2 Indicator electrodes

4.4.2.1 Metal electrodes

We have already seen the precious metal electrode for determining the concentration ratio of reversible redox pairs. However, equilibrium reactions between metals and their ions in the solution also lead to a charge on the surface of a corresponding metal electrode. The resulting potential can therefore be used to measure the relevant ion concentration. Examples of this are

$Ag \rightleftharpoons Ag^+ + e^-$	$E^°_{Ag} = 0.80$ V
$2\,Hg \rightleftharpoons Hg_2^{2+} + 2\,e^-$	$E^°_{Hg} = 0.79$ V
$Cu \rightleftharpoons Cu^{2+} + 2\,e^-$	$E^°_{Cu} = 0.34$ V
$Bi + H_2O \rightleftharpoons BiO^+ + 2\,H^+ + 3\,e^-$	$E^°_{Bi} = 0.23$ V

In all these cases, we are dealing with heterogeneous reactions in the sense of equilibrium theory. Since, according to the rules of thermodynamics, the solid phases involved in the reaction are not to be taken into account when establishing the law of mass action, the application of Nernst's equation leads to a particularly favorable form for potentiometric titration

$$E = E^°_{Me} + \frac{RT}{zF} \cdot \ln\, c(Me^{z+}).$$

The potential of the associated metal electrode depends solely on the concentration of its ions in the solution. The **silver electrode** can therefore be used for the potentiometric indication serving to argentometric titrations of halide ions.

For the **bismuth electrode**, the H^+ concentration of the solution must be taken into account when setting up Nernst's equation due to the protolysis of Bi^{3+} ions with the formation of bismuth oxide cations (BiO^+):

$$E = E^\circ_{Bi} + \frac{RT}{3F} \cdot \ln c(BiO^+) \cdot c^2(H^+).$$

In this case, an interference-free potentiometric indication requires that the H^+ concentration of the solution remains constant during the titration.

However, one disadvantage of these electrodes is obvious: the voltage series of the elements prevents any metal ions from being determined if ions of a more noble metal are simultaneously present in the solution to be titrated. If, as in the case of Cu^{2+} and BiO^+, for example, the standard potentials are close together, both types of ions determine the potential of the electrode. Such mixed potentials can be calculated in principle, but their occurrence prohibits the interference-free potentiometric indication of an ion species. The only solution is to completely mask one of the components with a suitable complex ligand. However, this presupposes that a very stable and at the same time specifically forming complex exists. In our case, the addition of potassium cyanide causes the Cu^{2+} ions to be reduced to Cu^+, and this forms a stable complex with the CN^-:

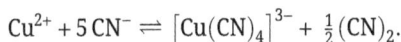

$$Cu^{2+} + 5\,CN^- \rightleftharpoons \left[Cu(CN)_4\right]^{3-} + \tfrac{1}{2}(CN)_2.$$

Bismuth can be titrated in the solution prepared in this way.

Less noble metals can only be considered as electrode materials for determining their ions in exceptional cases. This is mainly due to the fact that H^+ ions are reduced beforehand, i.e. they are more noble than these metal ions. On the other hand, the reaction

$$H_2 \rightleftharpoons 2\,H^+ + 2\,e^-$$

can be potential-determining at a Pt electrode. In principle, this opens up the possibility of a hydrogen or pH electrode. To avoid polarization effects, this electrode should be platinized to increase its surface area (see p. 258). It must be surrounded by H_2 gas. Its potential is determined by the equation

$$E = E^\circ_H + \frac{RT}{2F} \cdot \ln \frac{c^2(H^+)}{p(H_2)},$$

where $p(H_2)$ is the pressure referred to the standard pressure $p^\circ = 1.013$ bar, which is the pressure of the hydrogen flowing around the platinum electrode. In general, it will be 1.013 bar (1 atm) and should therefore be set to 1 in the equation. The standard potential E°_H is

$$E^\circ_H = 0\,V,$$

which has been defined internationally. As mentioned above, this defines the voltage series. The standard potential can be verified by providing an acid of activity $a(H^+) = 1$ mol/l and flushing the Pt electrode with hydrogen gas at a pressure of 1.013 bar. Since the logarithmic term becomes zero in this case, the voltage measured with respect to any other electrode can be used to determine the standard potential that applies to the reaction that determines the potential at the electrode in question. This standard hydrogen electrode is therefore an important reference electrode. However, it is no longer used for pH measurement or for monitoring potentiometric neutralization analyses, as its use is cumbersome and not interference-free due to a large number of possible side reactions.

Figure 4.19 shows a common form of hydrogen electrode. The hydrogen fed through the tube attached to the side emerges from a capillary ending below the electrode. It must first be carefully cleaned and the last traces of oxygen removed. This is done by washing with silver nitrate, alkaline permanganate and alkaline pyrogallol solution and by transferring it via platinum asbestos, which is located in a quartz tube heated to a weak red heat.

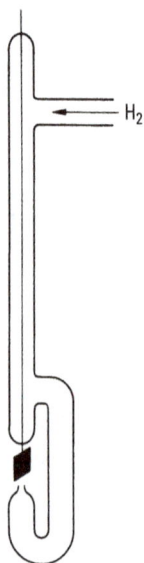

Fig. 4.19: Hydrogen electrode.

The titration vessel must be closed to the outside. The gas outlet is via a valve. If the hydrogen gas pressure is 1.013 bar (1 atm), the above potential equation is simplified to

$$E = 0 + \frac{RT}{F} \cdot 2.303 \cdot \lg c(\text{H}^+),$$

where the natural logarithm has been replaced by the decadic logarithm (conversion factor: 2.303). At 25 °C ($T = 298$ K), the value in front of the logarithm is calculated as 0.059 V (V = volt), and with the determination equation pH = $-\lg c(\text{H}^+)$, one finally obtains

$$E = -0.059 \text{ V} \cdot \text{pH}.$$

As far as the use of the discussed metal electrodes in analytical practice is concerned, the important requirement for specificity is only fulfilled in exceptional cases. Apart from the limitations resulting from the position of the relevant ion species in the voltage series, it is mainly reversible redox pairs whose presence in the titration solution prevents the desired potential setting. The adsorption of surface-active substances on the metal surface also causes undesirable interference.

However, metal electrodes have proved very successful in the manufacture of reference electrodes with a constant potential. Here, the potential-determining reaction on the metal surface is followed by the setting of a further chemical equilibrium (**electrodes of the second kind**). We will discuss this type of electrode in more detail (see p. 285).

4.4.2.2 Ion-selective electrodes (ISEs)

The charging of the metal electrodes that causes the potential can be understood as the transition of metal species sitting on the surface into the ionic state. The resulting ions pass into the adjacent liquid phase and leave their electrons behind on the metal surface. This type of potential adjustment should also take place when a solid ion exchanger changes some of the ions bound near its surface with the adjacent solution. The desired specificity depends on how specifically the exchanger can absorb only one ion species. However, the macromolecular ion exchangers known from organic chemistry with swellability in the surrounding solution are not suitable for our purpose, because the cavities are comparatively large and the ions, which compensate for the charges of the macromolecule inside, take their hydrate shells with them. However, this does not lead to sufficient specificity toward a certain type of ion. The suitability of such an ion exchanger as an electrode material requires electrical conductivity, which can only result from the mobility of ions bound in its interior. In the field of inorganic chemistry, on the other hand, there exist solids with ion conductivity for very specific ion species. These can be defined as crystals in which the anion or cation forming them is mobile, but also silicates with a spatial network structure in which lattice charges are generated as a result of regular intercalation of other valent elements (e.g. Al and B). The disadvantage of these compounds is their extraordinarily high ohmic resistance and thus the corresponding problems with the desired potential measurement. The only way to minimize these difficulties is to use very thin membranes made of the above-mentioned materials, but this again places demands on their mechanical stability.

A model for the formation of a phase-boundary potential is shown in Fig. 4.20. Ions that leave the exchanger and enter the dissolved phase leave uncompensated charges in the solid matrix. As these charges are mainly located near the surface, ions are attracted from the surrounding solution, but only certain species can be reabsorbed. This results in a phase-boundary potential that depends on the concentration of this type of ion in the surrounding solution:

$$E = a_0 + \frac{RT}{zF} \cdot \ln c.$$

Formally, this function corresponds to Nernst's equation, but the constant a_0 has a completely different meaning than the standard potential $E°$. It would now be ideal if, for example, a metal pin could be completely coated with the required exchanger material, and at the same time an electrically conductive coupling could be achieved between the metal and the ion exchanger. This can be achieved with a silver wire that is coated with a crystalline silver sulfide layer. The S^{2-} ions are densely packed in this crystal and the Ag^+ ions are small enough to be mobile in the cavities of the S^{2-} lattice. The exchanging material cannot accommodate other cation species, resulting in a very high specificity for Ag^+ ions. Compared to a simple silver sheet, this electrode has the advantage of not being an inert electrode with regard to reversible redox pairs. It should also be mentioned that the Ag_2S electrode also responds in principle to S^{2-} ions, as the solubility product of Ag_2S is established in the sample solution and the resulting Ag^+ equilibrium concentration can be measured. The results currently obtained with such simple electrodes are, with a few exceptions, unfortunately not very convincing. When people talk about ISEs today, they generally mean membrane electrodes, whose structure is shown in Fig. 4.21. A membrane (M) separates two solutions with different concentra-

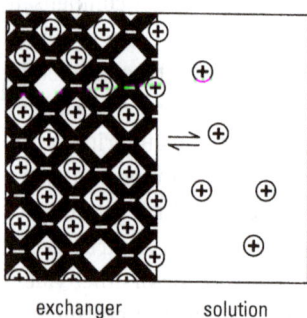

exchanger   solution

Fig. 4.20: Phase-boundary potential of an ion exchanger.

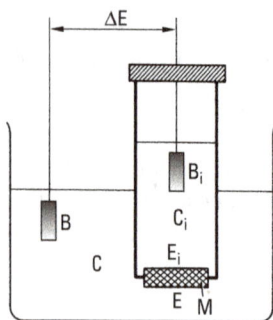

Fig. 4.21: Structure of an ion-selective membrane electrode (see text for explanation of the letters).

tions of exchanged ion species. One concentration (c_i) is known, and the other (c) is measured. A reference electrode (B) is immersed in each of the two solutions. (B) is immersed in each of the two solutions, but their potentials compensate each other if they are of the same structure. The measured potential difference ΔE then corresponds exactly to the difference between the two phase-boundary potentials E_i and E:

$$\Delta E = E - E_i = a_0 + \frac{RT}{zF} \cdot \ln c - a_0 - \frac{RT}{zF} \cdot \ln c_i.$$

As Fig. 4.21 shows, the known solution c_i is usually integrated together with the reference electrode B_i in the electrode body, which is sealed off from the outside. Since this reference solution retains its concentration c_i for all measurements, the concentration-dependent term of the potential of the inner phase boundary is therefore constant and is given by

$$\Phi_i = -\frac{RT}{zF} \cdot \ln c_i,$$

so that the measuring arrangement follows the simple relationship

$$\Delta E = \Phi_i + \frac{RT}{zF} \cdot \ln c.$$

Let us now consider the materials from which the exchanging membranes are made. On the one hand, there are inorganic single crystals with specific ionic conductivity and corresponding exchange properties, and on the other hand, the glass membranes are discussed later (see p. 281 ff.). Liquid membrane electrodes have recently become more and more common, a filter disk impregnated with liquid ion exchanger assumes the role of the exchanging membrane. This topic cannot be dealt with here in a detailed manner.

A pure fluoride ion conducting system is the LaF_3 crystal, which can be doped with a divalent cation such as Eu^{2+} to increase the ohmic conductivity. A membrane cut out of a single crystal is required to produce an electrode. The measuring range of the LaF_3 electrode is limited by the solubility of the membrane material LaF_3 in the sample solution. In practice, this means that F^- concentrations down to 10^{-7} mol/l can be measured. Below this, the fluoride content determined would largely originate from the electrode. To protect the electrode, the standard solutions should therefore be saturated with LaF_3 beforehand. Due to the adjustment of the solubility product of LaF_3 in the sample solution, La^{3+} ions can also be measured in principle, but bear in mind that the equilibrium position is disturbed by all those cations that also form fluorides that are poorly soluble with F^-. On the other hand, the LaF_3 electrode is suitable if, for example, Al^{3+} or Fe^{3+} solutions are to be titrated with a NaF standard solution, as the formation of the complexes $[FeF_6]^{3-}$ or $[AlF_6]^{3-}$ leads to potentiometrically well-indexable equivalence points.

The second widely used material for the production of crystalline membranes is silver sulfide. The electrode manufacturer benefits from the advantageous fact that no single crystals need to be grown; instead, it is sufficient to press the membrane from polycrystalline Ag_2S. An Ag_2S electrode produced according to the model in Fig. 4.21 is specific for Ag^+ and is also sensitive to S^{2-} ions due to the participation of these Ag^+ ions in the solution equilibrium of the membrane material.

The special feature of these Ag_2S compacts is the possibility of doping with other suitable substances, which on the one hand must be sufficiently sparingly soluble, but still more soluble than the host substance, and on the other hand, they must be made up of ions that influence the solubility equilibrium of Ag_2S in a defined way. First of all, the silver halides AgCl, AgBr and AgI should be mentioned in this context, whose solubility products in the sample solution determine the Ag^+ concentration, and the size of which depends on the given concentration of the respective counter ions (Cl^-, Br^- and I^-). This means that Cl^--, Br^-- or I^--sensitive electrodes can be produced, but it must be borne in mind that the electrode function is not interference-free with respect to the halide ions if the relevant solution equilibria are influenced by other factors.

Let us go one step further and consider how metal sulfides behave as additives in the Ag_2S matrix. Once the solubility equilibrium has been set, the metal sulfide incorporated will determine the position of the solubility product of Ag_2S via its corresponding S^{2-} concentration, and the Ag^+ concentration measured by the electrode will therefore be a measure of the content of the metal ions in question:

$$c^2(Ag^+) \cdot c(S^{2-}) = K_1,$$

$$c(Me^{2+}) \cdot c(S^{2-}) = K_2.$$

Since $c(S^{2-})$ is the same for both equilibria, dividing both equations yields

$$\frac{c^2(Ag^+)}{c(Me^{2+})} = \frac{K_1}{K_2}$$

respectively

$$c(Ag^+) = \sqrt{\frac{K_1}{K_2} \cdot c(Me^{2+})},$$

i.e. the measured potential difference of an Ag_2S electrode with MeS embedded in the membrane is determined by the Ag^+ concentration instead of the Ag concentration, and

$$\Delta E = \Phi_i + \frac{RT}{F} \cdot \ln c(Ag^+)$$

may depend on the Me^{2+} concentration according to the derived relationship:

$$\Delta E = \Phi_i + \frac{RT}{F} \cdot \ln \sqrt{\frac{K_1}{K_2} \cdot c(Me^{2+})}$$

respectively

$$\Delta E = \Phi_i + \frac{RT}{2F} \cdot \ln \frac{K_1}{K_2} + \frac{RT}{2F} \cdot \ln c(\text{Me}^{2+}).$$

As the first two summands are constants, the Ag^+ electrode has practically become an Me^{2+}-sensitive electrode. However, only those metal sulfides may be incorporated into the Ag_2S matrix, whose solubilities are greater than those of silver sulfide, but are soluble with sufficient difficulty to largely prevent leaching from the membrane material. The silver sulfide matrix itself cannot be dispensed with because there is no other sparingly soluble metal sulfide that has a sufficiently high ionic conductivity as a membrane material.

In practice, three metal sulfides, namely CuS, CdS and PbS, have proven to be suitable additives to Ag_2S membranes. According to what has been said so far, this means that the potential of the electrodes concerned depends on the respective concentration of Cu^{2+}, Cd^{2+} and Pb^{2+} ions. However, interference also occurs during the measurement if the setting of the relevant solubility products in the sample solution is influenced by a third party. This already applies to the pH value at which the measurement is carried out. Toward the alkaline side, the corresponding metal hydroxides are formed, whereas in a more acidic medium, the H^+ – with the S^{2-} – react to form HS^- ions. Using the example of a Pb electrode, Fig. 4.22 shows the ranges for different Pb^{2+} concentrations in which the potential setting is independent of the pH value. In general, before using any crystal membrane electrode, you must always consider whether there are substances in the sample solution that have an influence on the adjustment of the solution equilibria.

The longest known ISE is the **glass electrode** used for pH measurements. It has superseded all other electrodes previously used for H^+ determination, such as the quinhydrone electrode or the antimony oxide electrode . Here, the ion-exchanging membrane is made of a special glass which, for reasons of mechanical stability, is not left as a plane-parallel plate but is blown into a sphere. Such a glass has an approximate composition of 72% SiO_2, 22% Na_2O and 6% CaO, which can be produced by melting together the corresponding quantities of SiO_2, Na_2CO_3 and $CaCO_3$. This glass is characterized by a three-dimensional silicate structure with terminal $\equiv$ Si-$\overline{\text{O}}$|$^-$ groups. The negative charge is fixed to one oxygen at a time and has a correspondingly high anionic charge density. If such a membrane is placed in an aqueous solution, it can be seen that H^+ ions have an extremely higher affinity for the $\equiv$ Si-$\overline{\text{O}}$|$^-$ group than Na^+ ions. The outer glass layer swells and exchanges its Na^+ ions almost completely for H^+ ions. However, the hydrogen ions cannot penetrate any further into the interior of the membrane via the outer gel layer, as H^+ ions are absolutely immobile in the unswollen glass, which has been proven by diffusion experiments with the radioactive hydrogen isotope tritium. One must therefore imagine that although Na^+ ions migrate from the Na-silicate glass into the gel layer, the negative charge remaining on the scaffold is not compensated by H^+ ions, resulting in a so-called **diffusion potential**. This means that in the swollen layer immediately before the phase boundary to the unmodified part of the glass, a positive potential wall is initially created, but immediately behind it is a negative potential wall, a circumstance that ultimately prevents any further ion exchange across this boundary. This diffusion potential is also responsible for the fact that the Na-silicate glass cannot protolyze beyond an outer swelling layer.

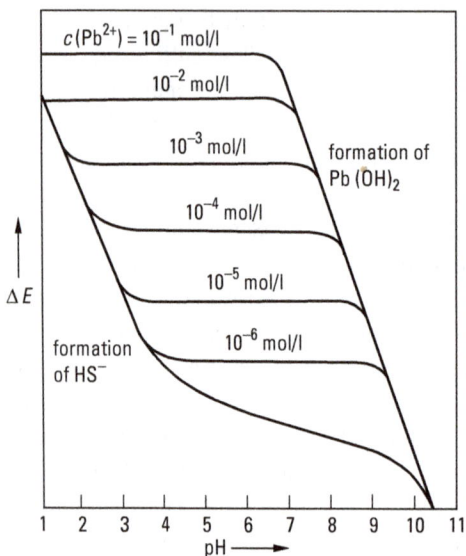

Fig. 4.22: Working range of a Pb electrode (ΔE) electrode potential.

A glass electrode constructed according to the scheme in Fig. 4.23 separates an inner buffer of pH 7 from an outer solution of a different pH. Since the inner phase-boundary potential E_i is constant, the electrode function is determined solely by the pH-dependent potential E. There is also a suitable reference electrode inside the glass electrode. The H^+-specificity of the glass electrode is extremely high. Only when the concentration of alkali metal ions in the sample solution exceeds H^+ ions by a factor of 10^{12}–10^{13} does the swelling layer noticeably absorb alkali metal ions, and the phase-boundary potential is gradually determined by their concentration. These are the alkali errors observed with glass electrodes which only become noticeable above pH = 12. Below this, the electrode works largely without interference and, above all, easily reproducible. It is important that H^+-selective glass electrodes are allowed to swell in a diluted NaCl or KCl solution for a longer period of time before use. After use, they must also be stored in such a solution.

More recently, it has been possible to produce membranes from such glasses that selectively exchange alkali or calcium ions with the solution. This is achieved by avoiding terminal $\equiv Si\text{-}\overline{O}|^-$ groups in the silicate framework and instead distributing the negative charge to the four oxygen atoms surrounding a heteroatom such as boron or aluminum. Such a heteroatom with the maximum oxidation number III can only be incorporated into the silicate framework if it formally accepts another electron. However, its charge is distributed among the four neighboring oxygen atoms. The much lower anionic charge density here compared to the H^+ electrode results in high affinity for alkali metal ions. In order to produce glasses that are capable of exchanging divalent cations with the solution, an atom with coordination number 6 and oxidation number IV is occasionally required in the silicate framework instead of silicon. This results in a structure with oxygen atoms arranged approximately octahedrally around this atom, with two negative elementary charges distributed between them. A Ca-selective glass electrode can be produced according to this principle if zirconium is chosen as the heteroatom, and an appropriate amount of $CaCO_3$ is also added to the molten glass so that, after cooling, cavities are created in the glass that are adapted to the Ca^{2+} ion, and it can also move optimally.

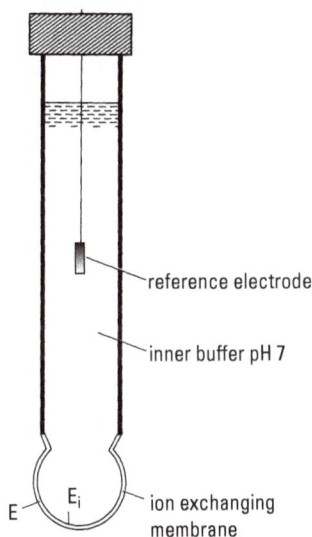

reference electrode

inner buffer pH 7

E E_i ion exchanging membrane

Fig. 4.23: Glass electrode for pH measurement (E) pH-dependent potential; (E_i) inner phase-boundary potential.

Glass electrodes are constructed from the glasses assembled as described according to the construction principle of Fig. 4.23, except that a salt solution is filled inside with the cation for which the electrode is selective. Although the melting of the glasses results in a charge distribution in the silicate framework that is adapted to the ion to be exchanged, and optimum cavities for the absorption and mobility of this particular ion are created after cooling, the selectivity of the respective electrode does not reach the level that we have seen with the pH electrode. However, selectivities are still achieved, which allow only one type of alkali metal ion to be detected by potential measurement in solutions that have an approximately equal concentration of several alkali metal ions. Against that, if the interfering ion species is about a hundred times more concentrated than the one to be determined, this will usually result in a phase-boundary potential whose magnitude is determined by the concentrations of both ion species. In this range, the interpolation of results can be based on the equation proposed by Nikolsky:

$$E = \Phi_i + 2.303 \frac{RT}{F} \cdot \lg(c_a + S \cdot c_n),$$

where c_a is the concentration of the ion type a for which the electrode is selective, and S is the selectivity toward the interfering ions whose concentration should be c_n. If you set $S = 10^{-2}$, as is realistic for alkali electrodes, you obtain a representation corresponding to Fig. 4.24. The concentration of the interfering ion is arbitrarily set to $c_n = 0.1$ mol/l. As soon as the curve is no longer linear, it is helpful to apply the Nikolsky equation; on the left before this, the electrode works without interference because the concentration c_a is much greater than the product $S \cdot c_n$ of the selectivity coefficient and the concentration of the interfering ion. However, if the concentration c_a becomes very small, the phase-boundary potential is ultimately only dependent on the "interfering component" in the solution. It should be noted, however, that Nikolsky's equation is only partially correct in representing the potential curve, as it neglects the existence of a diffusion potential inside the glass membrane. This inaccuracy is eliminated by an equation from Eisenman [148], but its application is considerably more complicated (for more details, see Cammann [149]).

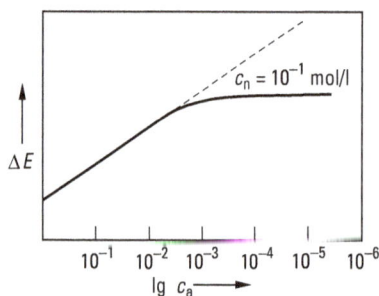

Fig. 4.24. Disturbance of the potential setting by foreign ions (c_a) concentration of the alkali ions to be determined; (c_n) concentration of the interfering ions.

More recently, membranes have been produced from porous hydrophobic solids impregnated with a "liquid ion exchanger". This is an anion or cation exchanger that is only soluble in a water-immiscible (organic) solvent. A membrane impregnated with such a solution can separate two aqueous phases from each other without water penetrating the pores of the hydrophobic material. Alternatively, ion exchangers can also be immobilized in polyvinyl chloride, resulting in membranes that are easy to process and tough. This expands the possibilities for producing ISEs for a wide variety of requirements.

4.4.3 Reference electrodes

We have already pointed out several times that it is necessary to measure the potentials of the indicator electrodes against a reference electrode with a constant potential. It has also been said that these electrodes are exclusively metal electrodes. Compared to what we have already learned, however, it is also possible that the metal ion concentration is determined by the solubility product of a sparingly soluble salt that is in equilibrium with the solution as a solid phase. As the actual electrode function is followed by a chemical equilibrium, this is also referred to as a **second kind of electrode**.

The **calomel electrode** is a very frequently used electrode that you can easily make yourself in the laboratory. To do this, mix mercury and Hg_2Cl_2 together in a porcelain mortar until a homogeneous-looking paste is formed. This is used as a solid phase in a saturated KCl solution, whereby a current flow to the Hg/Hg_2Cl_2 mixture must be created, isolated from the electrolyte solution. The Cl^- concentration specified by the KCl solution allows as much Hg_2Cl_2 to be dissolved until the solubility product is set:

$$c(Hg_2^{2+}) \cdot c^2(Cl^-) = K_{sp}.$$

However, the corresponding Hg_2^{2+} concentration determines the potential of the mercury electrode:

$$E = E^\circ + \frac{RT}{2F} \cdot \ln c(Hg_2^{2+}).$$

Replace $c\left(Hg_2^{2+}\right)$ with the solubility product:

$$E = E^\circ + \frac{RT}{2F} \cdot \ln \frac{K_{sp}}{c^2(Cl^-)}$$

respectively

$$E = E^\circ + \frac{RT}{2F} \cdot \ln K_{sp} - \frac{RT}{F} \cdot \ln c(Cl^-),$$

and with

$$E' = E^\circ + \frac{RT}{2F} \cdot \ln K_{sp},$$

$$E = E' - \frac{RT}{F} \cdot \ln c(Cl^-).$$

The Cl^- concentration is fixed in a saturated KCl solution. At a temperature of 25 °C, this results in a constant potential of 0.246 V (compared to the standard hydrogen electrode). If a KCl solution with $c = 1.0$ mol/l is selected instead of the saturated KCl solution, the potential is 0.2801 V at 25 °C (room temperature). Whenever the concentration of Hg_2^{2+} changes as a result of a current passing through the electrode, a corresponding amount of Hg_2Cl_2 precipitates or dissolves until the original state is reached again. The ionically conductive connection to the titration solution is a current key filled with electrolyte or a diaphragm. Figure 4.25 shows the modern design of a calomel electrode. It is immersed in the solution to be titrated with its diaphragm melted at the lower end. The spent KCl solution can be added as required via a filler opening. Care must be taken to ensure that the liquid level in the reference electrode is always higher than in the titration solution, as this is the only way to prevent it from entering the calomel electrode via the diaphragm. Conversely, if some of the KCl solution flows into the solution to be titrated as a result of excess pressure, the diffusion potential occurring at the diaphragm is also largely suppressed. Of course, KCl must not interfere with the titration being carried out. In such a case, a reference electrode with a different electrolyte must be used. The calomel electrode no longer works reliably at higher temperatures. This is due to the disproportionation of Hg(I) into Hg(II) and Hg.

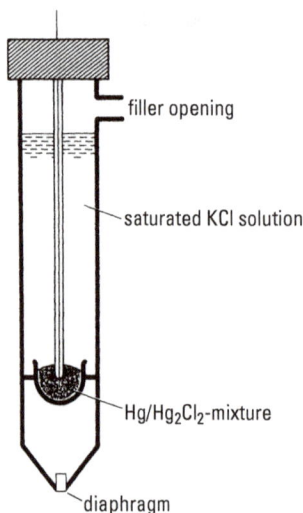

filler opening

saturated KCl solution

Hg/Hg$_2$Cl$_2$-mixture

diaphragm

Fig. 4.25: Calomel electrode.

For most electrochemical measurements, the very reliable **Ag/AgCl electrode (silver/ silver chloride electrode)** is preferred today with AgCl as the solid phase. Those who value good measurement results at higher temperatures should use the **thalamide electrode**. It consists of thallium amalgam as an electrode material and a TlCl base in a KCl solution with a concentration of 3 mol/l. If a titration is disturbed by Cl ions, a **mercury(I) sulfate electrode** can be used as an alternative, which also has a constant potential.

4.4.4 Measuring chains

In potentiometric titrations, the potential of an indicator electrode responding to the characteristic changes in concentration is measured against a reference electrode. We must now consider how to establish an ionically conductive connection between the electrolyte solutions in the titration vessel and in the reference electrode. In principle, there are two possibilities for this, namely the use of a diaphragm or a current key filled with an electrolyte. Whichever is chosen, the test setup ready for measurement is called a measuring chain.

The titration device shown in Fig. 4.26 uses a current key. The silver nitrate solution is to be titrated with NaCl solution. A simple silver plate can be used as the measuring or indicator electrode. If the reference electrode is a calomel or Ag/AgCl electrode, the problem arises that Cl⁻ ions flow through the diaphragm into the sample solution. For this reason, the reference electrode is immersed in a beaker filled with KNO_3 solution, for example, and this electrolyte is connected to the titration vessel using a current key also filled with KNO_3 solution. In the simplest case, this is a U-tube containing agar gel in addition to the electrolyte solution, which prevents leakage. The intermediate electrolyte KNO_3 does not interfere with the titration of the silver ions with NaCl. If the current key is filled with an electrolyte, whose anion and cation have approximately the same mobility, no diffusion potential occurs. However, this circumstance is only of secondary importance as long as the titration is to be indicated potentiometrically.

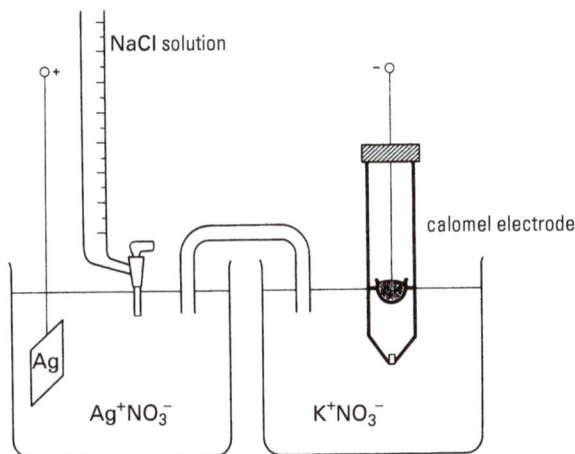

Fig. 4.26: Measuring chain with current key.

The most convenient way of ionically connecting two electrolyte solutions is to use a diaphragm. Such a diaphragm can be made of clay with a suitable pore size, but can also consist of an asbestos thread or many thin platinum wires twisted against each other, which, when melted in glass, allow small amounts of electrolyte to be exchanged between two solutions. We have already seen this with the calomel electrode (see Fig. 4.25), which can be immersed directly in the titration vessel, provided that the KCl solution it contains does not interfere. However, it must also be said that diaphragms are the main source of error in potential measurements with high accuracy requirements. If electrolyte solutions with different compositions come into contact with each other via a diaphragm, their ions migrate in the potential gradient according to their mobility. In the diaphragm, however, different mobilities of the ion species involved in the conduction process lead to the buildup of a diffusion potential, which is also recorded as an additive element when measuring the potential difference between two electrodes. As long as this potential occurring in the diaphragm remains unchanged during a series of measurements, it can also be taken into account. However, in the neutralization analysis, for example, very mobile H^+ or OH^- ions are neutralized, and the diffusion potential then assumes very different values. Nevertheless, when carrying out a potentiometrically indexed titration, the absolute values of the potential are less important than the shape of the curve, as only the inflection point needs to be determined correctly. In this case, we can therefore disregard the changes in the diffusion potential in the diaphragm, but should still ensure that the liquid level in the reference electrode is higher than in the titration vessel. To suppress diffusion potentials practically completely, increase the hydrostatic pressure in the reference electrode by extending the liquid column to such an extent that a small but constant flow of electrolyte from the electrode reservoir through the diaphragm into the solution under investigation takes place. However, this will only be necessary if ion concentrations are to be determined directly with an ISE.

Figure 4.27 shows a so-called **single-rod measuring chain** for pH measurement in aqueous solutions. The special feature of this setup is the combination of indicator electrode and reference electrode into a single rod. The actual glass electrode with its inner buffer and the inner reference electrode is surrounded by a further glass sheath in the shaft area fused above the exchanging spherical membrane, which provides space for the second reference electrode. Its electrolyte filling is in ionically conductive contact with the solution to be examined via a diaphragm attached to the side. Electrolyte solution for the reference electrode can be added as required via the filling that opens at the top. Single-rod measuring chains for pH determination must also have stood in KCl solution for a sufficiently long time before the first use so that the glass membrane is activated by the swelling process for the H^+ ion exchange. During the measurement, it is important to ensure that the active glass membrane is immersed not only in the solution but also the diaphragm. For many types of ISEs, single-rod measuring chains are now commercially available, which greatly benefit the simplicity of the intended measurements.

4.4.5 Currentless potential measurement

The dependency of the electrode potentials on the concentration of the exchanged ions in the solution is only then described correctly by the equations of Nernst and Nikolsky if there is no material turnover at the electrodes. In practice, however, this can only be realized in such a way that no energy is taken from the electrode even during a potential measurement. However, any voltage measuring device connected directly between the electrodes would consume energy for the measurement and there would be a current flow, albeit small.

When measuring the potential difference between two metal electrodes (e.g. a Pt plate against a reference electrode in redox titrations), it is possible to fulfill the no-

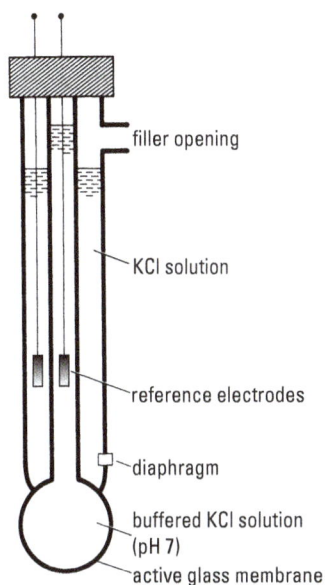

filler opening

KCl solution

reference electrodes

diaphragm

buffered KCl solution
(pH 7)

active glass membrane

Fig. 4.27: Single-rod measuring chain for pH measurement.

current condition with a simple setup. A voltage divider is used to generate a correspondingly lower voltage from the constant voltage emitted by a battery, which corresponds exactly to the potential difference to be measured. To achieve this, this voltage is applied to the electrodes of the measuring chain, and the resistance is regulated until both voltages are equal and opposite. A sensitive ammeter in the circuit then indicates the absence of current. This **compensation method** originates from Poggendorf and is usually referred to by this name in the literature. Figure 4.28 shows the circuit diagram. The position of the sliding resistor is proportional to the voltage output. However, in order to be able to assign a voltage in mV to each position of the sliding resistor, the measuring device has to be calibrated by means of a standard element (this is a galvanic cell with a known potential, e.g. the Weston element) that is placed instead of the electrode. Unfortunately, this compensation method cannot be used if the electrode contains a glass electrode as the indicator electrode. This is due to the extremely high resistance of a glass membrane, which is in the order of 10^{10} Ω. Under these circumstances, firstly, the sliding resistance would practically short-circuit the electrode, which would cause it to lose its state of equilibrium, and secondly, it would be impossible to measure such small currents with a conventional ammeter so that it would not be possible to calibrate with the voltage divider.

Without wishing to discuss metrological problems, it is only necessary to say that an **electronic measuring amplifier** is required here, whose input resistance must be at least a factor of 1,000 higher than the resistance of the glass electrode if the measurement signal is to be recorded with an error of no more than 0.1% and displayed correctly after amplification. The insulation of the shielded cable used to feed the sig-

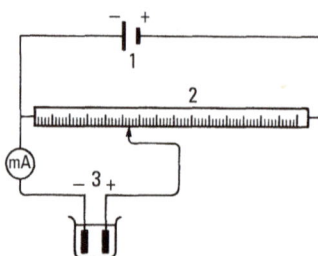

Fig. 4.28: Poggendorf's compensation method for potential measurement: (1) DC voltage source, (2) sliding resistor (calibrated in mV) and (3) measuring cell.

nal to the measuring amplifier must also be of the same quality if signal distortion is to be avoided. The measuring devices available on the market, mostly pH or ion meters allow almost currentless potential measurement at an input resistance of approximately 10^{13} Ω with an accuracy of up to ±0.2 mV (precision devices). However, this means that with an analog display device, the available measuring range must be divided into 10 consecutive subranges, which requires precise and time-constant compensation of the individual voltage levels. Such precision devices are not required for potentiometrically indicated titrations; it can be regarded as sufficient if they allow a potential measurement with an accuracy of ±2 mV.

In order to be able to carry out more than just relatively correct measurements with these devices, especially on measuring chains that contain ISEs, they must allow calibration. For pH measurements, for example, two buffer solutions with precisely known but sufficiently far apart pH values must be measured in order to bring the display of the pH meter into agreement according to the device manufacturer's specifications. Many devices also have an input for a temperature sensor. If this is immersed in the standard solution, the device corrects for temperature fluctuations. If the device does not have such an option, the temperature in the measuring vessel should be kept constant using a thermostat.

Anyone using an ion meter with a measuring accuracy of ±0.2 mV for direct concentration determination should bear in mind that diffusion potentials at the diaphragm of the reference electrode can reach a magnitude of more than 1 mV. They must therefore always be suppressed. With such precise measurements, activities must be calculated instead of concentrations.

In order to be able to correctly assess the mode of action of the various color indicators when determining the equivalence point of classic titration methods and the quality of the statements made, we discussed the course of the corresponding titration curves for the various types of volumetric analysis. Basically, they are obtained by logarithmically plotting the concentration of the ion type that is produced or consumed as a result of a chemical reaction during the titration. Since, as we have seen, the potential of an indicator electrode also depends on the logarithm of this concentration, potentiometric titration is a method for recording these titration curves. At 25 °C, the determination function is

$$\Delta E = \Phi_i + \frac{0.059\,\text{V}}{z} \cdot \lg c.$$

In this context, we would like to come back to the calibration of a pH meter: With the first buffer solution, we bring the known pH value into agreement with the corresponding display of the device by regulating a counter voltage. In order to be able to do the same with the second buffer solution, the slope of the electronic amplifier must be changed until agreement is also achieved for the second pH value. The set slope can now be read off the device. When measuring a monovalent ion ($z = 1$) and at 25 °C, it should of course have the value 0.059 V. Practice has shown that the quality of the approximation to this theoretical slope is a criterion for the correct function of an ISE. If a smaller value is found, the functionality of the electrode is at least reduced. It must then first be checked for cleanliness (this also applies to the reference electrode in the diaphragm area) and reactivated if necessary, which in the case of a glass electrode may involve etching it briefly (!!) with hydrofluoric acid (caution! highly toxic!), for example. It must then be left in an electrolyte solution for a longer period of time to swell. If the slope of the electrode function still does not come close to the theoretical value, the electrode must be replaced with a new one.

While all pH or ion meters require the analyst to carry out the titration manually and read off the corresponding measured values in order to create a graphical representation of the titration curve, there are also so-called **potentiographs** that can automatically register the required titration curves. Their design corresponds to the devices described above, but they also have a writing device (compensation recorder) and a motorized burette (see Section 4.5). In order to ensure a correct correlation between the measured value and the reagent addition, there must be a precise mechanical or even better electronic coupling of the paper feed and the piston position of the motor burette. With such a device, it is then possible to titrate more slowly in the area of the equivalence point and, above all, to follow a predetermined and then automatically reproduced schedule from measurement to measurement. To facilitate curve evaluation, the measurement curve can also be differentiated; the maximum of a peak then has to be determined instead of the inflection point. The reproducibility of the titrations carried out can be significantly increased with these devices, which is particularly beneficial for routine operation.

Today, modern electronics allow the construction of playing card-sized, battery-operated potentiometers, which, after setting the potential value expected at the equivalence point, digitally display the degree of titration until the end point is reached. This allows the advantages of potentiometric indication to be utilized, leading to very accurate titration results.

4.4.6 Practical applications

The following examples will show how versatile potentiometric titrations can be used to indicate precipitation, complexation and neutralization reactions as well as oxidation and reduction processes. The main advantages of potentiometry are as follows:

- The possibility of determining several substances in the course of a single titration (simultaneous determinations).
- The possibility of determining the quantity of a substance in the presence of accompanying substances that would interfere with the analysis using other methods (selective determinations). The possibility of titrating turbid or strongly colored solutions should also be mentioned here.
- The possibility of a very significant extension of the volumetric analysis methods by the fact that standard solutions can now also be used for which no useful indicator is known.
- The possibility of carrying out microdeterminations, as the accuracy of many potentiometric titrations exceeds that of the corresponding classical methods.

4.4.6.1 Precipitation and complexation titrations

Determination of halides and of silver: The argentometric determination of halides is carried out in a weak sulfuric acid solution. A silver plate is used as the indicator electrode and a mercury(I) sulfate electrode as the reference electrode, both of which are immersed in the titration vessel. It is important not to titrate too quickly and to wait for the somewhat hesitant potential setting, especially near the equivalence point. This applies in particular to the bromide and iodide determination, less so for the determination of chloride. The potential-determining process is

$$Ag^+ + Hal^- \rightleftharpoons AgHal.$$

The observed potential jumps therefore depend on the solubility products of the silver halides: K_{sp} (AgI) = 10^{-16}, K_{sp} (AgBr) = 10^{-12} and K_{sp} (AgCl) = 10^{-10}, thus decreasing in magnitude in the order mentioned. Iodide and bromide solutions can be determined down to concentrations of 10^{-5} mol/l, but chloride solutions can only be determined down to concentrations of 10^{-3} mol/l.

The same reactions can of course also be used to determine silver ions. If the Ag^+ concentration in the solution to be titrated is above $c(Ag^+) = 10^{-3}$ mol/l, a chloride solution is used as the standard solution, which offers the advantage of faster potential adjustment. More dilute silver solutions are titrated with bromide solutions, as these allow determination down to concentrations of $c(Ag^+) = 10^{-5}$ mol/l with sufficient accuracy.

Determination of halides side by side: It will be discussed using a specific example: 100 ml of a solution with iodide and chloride concentration of 0.01 mol/l each is to be titrated with a silver nitrate solution of concentration $c(Ag^+) = 1$ mol/l. How does the Ag^+ concentration change during the titration? At the beginning, the concentration is 10^{-14} mol/l due to the solubility product of AgI, at the equivalence point it is 10^{-8} mol/l and after the addition of another ml of the $AgNO_3$ solution it would be 10^{-2} mol/l if the chloride were not present (curve 1 in Fig. 4.29). However, as this has a concentration of 0.01 mol/l, the precipitation of AgCl begins at an Ag^+ concentration of 10^{-8} mol/l (see the solubility product of AgCl). As long as this continues, the Ag^+ concentration increases only slightly until the equivalence point of the chloride determination is reached at an Ag^+ concentration of 10^{-5} mol/l (inflection point of curve 2 in Fig. 4.29). The measured curve thus shows two characteristic concentration jumps: the first of which indicates the end of silver iodide precipitation, while the second is observed after precipitation of the chloride ions. Strictly speaking, however, the actual turning point of the iodide precipitation is no longer reached because the curve bends slightly beforehand due to the onset of AgCl precipitation. The turning point, which is actually determined a little earlier, is only in the right place if the curve is very steep in the equivalence range. To achieve this, however, the values of the solubility products must be small and sufficiently different from each other, too.

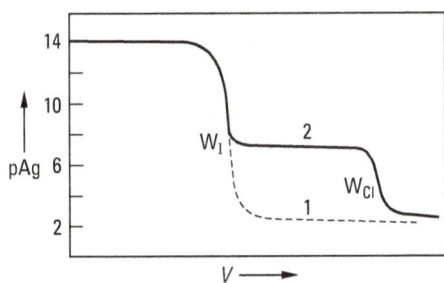

Fig. 4.29: Joint titration of chloride and iodide with $AgNO_3$ solution (pAg = –lg $c(Ag^+)$, (V) added volume of $AgNO_3$ solution.

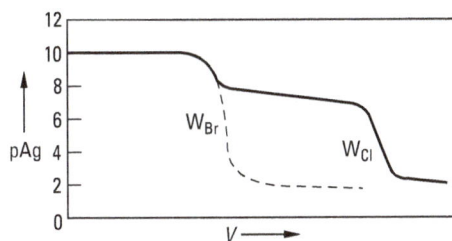

Fig. 4.30: Joint titration of chloride and bromide with $AgNO_3$ solution (pAg = –lg $c(Ag^+)$, (V) added volume of $AgNO_3$ solution.

It is comparatively more difficult to determine two ions that can be precipitated by the same reagent in one titration if the differences in solubility of the precipitates are small. This is shown clearly in Figure 4.30. Here, the theoretical course of the change in the silver ion concentration of a solution of chloride and bromide ions titrated with the same silver nitrate solution is shown, the concentration of each being 0.01 mol/l. The termination of the bromide precipitation is only reflected by a small and blurred jump in concentration or potential, while the second inflection point, which corresponds to the sum of both precipitable ions, is easy to determine. In fact, a further problem arises during the determination: AgBr and AgCl form solid solutions with each other, even if only in a narrow range of existence. This phenomenon shifts the first equivalence point a good 2% too far to the right. In such a case, the titration curve has to be evaluated according to the Gran method (see p. 300 f.).

Determination of zinc: The determination of zinc as potassium zinc hexacyanoferrate(II) can be indicated potentiometrically if the standard solution contains small amounts of $K_3[Fe(CN)_6]$ in addition to the $K_4[Fe(CN)_6]$ required for precipitation. At a platinum electrode, this process determines the potential:

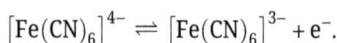

$$[Fe(CN)_6]^{4-} \rightleftharpoons [Fe(CN)_6]^{3-} + e^-.$$

The concentration ratio of the two ions changes during the titration because $[Fe(CN)_6]^{4-}$ is consumed precipitating Zn^{2+} according to the equation:

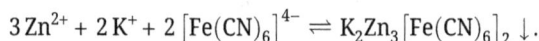

$$3\,Zn^{2+} + 2\,K^+ + 2\,[Fe(CN)_6]^{4-} \rightleftharpoons K_2Zn_3[Fe(CN)_6]_2 \downarrow.$$

The silver chloride electrode is used as the reference electrode.

Work is carried out, preferably with the addition of potassium sulfate, in a very weakly hydrochloric acid solution at a temperature of 60–70 °C. The $K_4[Fe(CN)_6]$ solution with a concentration of 0.1 mol/l, which also contains 1 g/l $K_3[Fe(CN)_6]$, serves as the standard solution. Sodium, magnesium, calcium and aluminum salts interfere with the titration if they are present in large quantities. Iron(III) ions, which also interfere, can be bound by adding ammonium fluoride and a small amount of acid to form complex $[FeF_6]^{3-}$ ions so that the analytical result is not distorted. It is very important to stir vigorously during the entire titration so that the initially precipitating $Zn_2[Fe(CN)_6]$ precipitate can be quantitatively converted into $K_2Zn_3[Fe(CN)_6]$. The method can also be used to determine $[Fe(CN)_6]^{4-}$ ions.

Determination of fluoride: It is best carried out in an almost neutral solution using a fluoride-selective LaF_3 electrode by titration with $La(NO_3)_3$ solution. The fluoride ions are precipitated according to the equation

$$3\,F^- + La^{3+} \rightleftharpoons LaF_3\downarrow,$$

and the phase-boundary potential caused by the fluoride ions is measured as a function of $La(NO_3)_3$ addition.

A solution, which should contain between 1 and 50 mg fluoride ions, is titrated slowly after adding 10 ml acetate buffer pH 6. A lanthanum nitrate solution of concentration $c(\frac{1}{3}\,La^{3+}) = 0.1$ mol/l (1 ml corresponds to 1.90 mg F^-) serves as standard solution. The acetate buffer pH 6 contains approximately 85 g CH_3COONa dissolved in 1 l of water, to which glacial acetic acid was then added until the pH value of exactly 6 was reached. A standard solution of concentration $c(\frac{1}{3}\ La^{3+}) = 0.1$ mol/l is prepared by dissolving 14.44 g $La(NO_3)_3 \cdot 6\ H_2O$ in 1 l of deionized water. The titer is determined by potentiometric titration of NaF solution prepared by dissolving a weighed quantity of solid NaF (with high degree of purity).

Metal ISEs can be used for the potentiometric indication of complexometric titrations with EDTA.

4.4.6.2 Acid-base titrations

Using the pH glass electrode, all determinations of acids and bases can be conveniently indicated potentiometrically. The procedure corresponds to what has already been said about acid-base titrations using color indicators. The advantages of potentiometry only really come into play when the quantities to be determined are very small or the solutions are colored. As long as a neutralization process causes an inflection point in the titration curve, the equivalence point can also be determined by potential measurements. Therefore, unlike when using color indicators, it is not necessary for the curve to be so steep at the equivalence point that the changeover range of the color indicator causes only a negligible volume error. The steepness of the titration curve is known to depend on the concentration and strength (dissociation constant) of the acid or base used. Mixtures of strong to moderately strong acids or bases only produce a potential jump, and the position of which depends on the total H^+ or OH^- concentration. In contrast, weak acids or bases can be determined simultaneously if their dissociation constants are sufficiently different from each other. However, the dissociation constant must never fall below the value 10^{-9}, as otherwise the dissociation of water will produce so many H^+ or OH^- ions that the potential jump in the titration curve will be suppressed. If the glass electrode has been calibrated beforehand, the pK value $(pK = -\lg K)$ of the acid or base just titrated can be taken from the titration curves by determining the pH or pOH value exactly at $\tau = 0.5$. The simple relationship $pH = pK$ applies here, which can be easily derived from the law of mass action.

For laboratories that have to carry out acid or base determinations in routine operation, potentiometric titration will be irreplaceable insofar as it can be easily automated in contrast to titrations using color indicators. Since the second derivative of the potential curve indicates an equivalence point as zero, the titration can be automatically terminated when the zero point is reached and the corresponding consumption can be transmitted to a data acquisition device.

Determination of phosphoric acid in Coca-Cola® (based on [149a]):

Procedure: First, transfer approximately 200 ml of Coca-Cola® to a 300 ml Erlenmeyer flask and heat for approximately 10 min. A volume of 40.00 ml of the cooled Coca-Cola® sample solution is placed in a 100 ml beaker and titrated with NaOH standard solution ($c = 0.1$ mol/l) with constant stirring until a constant pH value (approx. 10.5) is achieved.

Near the equivalence point (steep rise in pH value), wait a few seconds before taking a reading, as the pH value usually rises very slowly here.

4.4.6.3 Oxidation and reduction titrations

Simultaneous determination of iron and manganese with permanganate: In this determination, iron must be present as iron(II) and manganese as manganese(II). Titration of iron(II) ions in sulfuric acid solution results in a sharp jump in potential after the reaction is complete:

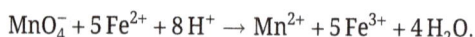

$$MnO_4^- + 5\,Fe^{2+} + 8\,H^+ \rightarrow Mn^{2+} + 5\,Fe^{3+} + 4\,H_2O.$$

If the solution also contains an excess of potassium fluoride (caution! highly toxic!), this reaction is followed by a second reaction, which consists of the oxidation of manganese(II) ions formed during the first process to manganese(III) ions (the F^- ions complex the resulting Mn^{3+}):

$$MnO_4^- + 4\,Mn^{2+} + 8\,H^+ \rightarrow 5\,Mn^{3+} + 4\,H_2O.$$

A second jump therefore appears in the potential curve when a quarter larger volume of the permanganate solution is added than was required to achieve the first jump. If the volume of permanganate solution required to oxidize the iron(II) ions is denoted by a and the volume required for the second jump is denoted by b, then $b = (a + a/4)$ if the solution to be titrated contains only iron(II) ions. However, if manganese(II) ions are already present from the outset, the volume of the permanganate solution b required up to the second jump is greater than $(a + a/4)$, and the volume of the permanganate solution used to titrate the manganese(II) ions originally present is $x = b - (a + a/4)$ or $x = b - \tfrac{5}{4} \cdot a$.

When carrying out the determination in practice, it should be noted that a fluoride-containing iron(II) salt solution is sensitive to air and that an acidic fluoride solution precludes the use of a glass titration vessel. A platinum dish or a polytetrafluoroethylene vessel containing the initially fluoride-free, sulfuric acid solution (5 ml conc. H_2SO_4 per 100 ml) is suitable as a titration vessel; the indicator electrode is a platinum plate. First, titrate at room temperature. As soon as the first equivalence point is reached, 7 g of potassium fluoride per 100 ml of solution is added. The solution is then titrated further at 80 °C up to the second equivalence point corresponding to manganese(II) concentration.

Determination of tin and antimony with dichromate: Tin(II) or antimonate(III) can be oxidized to tin(IV) or antimonate(V) by potassium dichromate in a strongly hydrochloric acid solution:

$$Cr_2O_7^{2-} + 14\,H^+ + 3\,Sn^{2+} \rightarrow 2\,Cr^{3+} + 3\,Sn^{4+} + 7\,H_2O,$$

$$Cr_2O_7^{2-} + 8\,H^+ + 3\,SbO_2^- + 5\,H_2O \rightarrow 2\,Cr^{3+} + 3\,[Sb(OH)_6]^-.$$

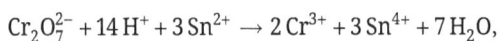

The end point is detected potentiometrically using a platinum electrode. The titration must be carried out in solutions with $c(HCl) \approx 2$ mol/l. The potential jump for tin is about 10 times greater than for antimony.

Selective determination of antimony in the presence of tin: If Sn(II) and Sb(III) ions are present together in the solution, the potentiometric titration only results in a potential jump corresponding to the sum of both components. However, if an excess of mercury(II) ions is added to a second sample of the solution, tin(II) is oxidized ($2\,Hg^{2+} + Sn^{2+} \rightarrow Hg_2^{2+} + Sn^{4+}$), while antimony(III) does not react. In the potentiometric titration, a potential jump is now obtained, which corresponds only to the amount of antimony. The sparingly soluble mercury(I) chloride is not oxidized by the dichromate solution.

Determination of vanadium with cerium(IV): Cerium(IV) sulfate is a strong oxidizing agent that is often preferable to potassium permanganate. The reaction always takes place according to the equation: $Ce^{4+} + e^- \rightleftharpoons Ce^{3+}$. Cerium(IV) sulfate solutions are titer stable for a long time and are neither light- nor temperature-sensitive. In contrast to permanganate solutions, they can also be used for titration in strongly hydrochloric acid solutions. They can therefore be used for a wide range of analytical measurements. When titrating with cerium(IV) sulfate solution, the equivalence point can be indicated not only potentiometrically but also using suitable redox indicators (see p. 179 f.).

Preparation of cerium(IV) sulfate solution: The preparation of the solution is described in p. 179 f. The titer determination is carried out potentiometrically at 70 °C with a weighed quantity of sodium oxalate, which is dissolved in 10 ml concentrated hydrochloric acid and diluted to 100 ml. (The reaction proceeds faster in the presence of HCl than in the sulfuric acid solution.)

Determination of vanadium: It is carried out according to the equation

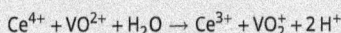

$$Ce^{4+} + VO^{2+} + H_2O \rightarrow Ce^{3+} + VO_2^+ + 2\,H^+$$

in a hot, strongly mineral acidic solution. If vanadium with the oxidation number V is present, it must first be reduced in an acidic solution with SO_2 gas. Excess sulfur dioxide must then be completely removed by passing CO_2 through a closed apparatus for a longer period of time.

Joint determination of vanadium and iron: The acidic solution containing vanadyl(IV) and iron(II) ions is first titrated in the cold to the first potential jump, which corresponds to the iron content, and the solution is heated to 50–60 °C and titrated further to the second turning point, which indicates the vanadium content. This method is used to analyze vanadium-containing steel grades.

Determination of copper and iron with chromium(II): The standard potential of the redox system $Cr^{3+} + e^- \rightleftharpoons Cr^{2+}$ is -0.408 V. Therefore, aqueous chromium(II) salt solutions have a strong reducing effect. They outperform titanium(III) chloride, which is also used in potentiometric titrations. Both reducing agents are used to quickly and accurately analyze various binary and ternary alloys.

> **Preparation of chromium(II) sulfate solution:** Boil the purest potassium dichromate with concentrated hydrochloric acid until the chlorine stops developing (be careful!). The solution is cooled and reduced in a flask with a Bunsen valve attached for a few hours with the purest zinc. When the solution is pure blue in color, it is pressed through a siphon tube fitted with a glass wool filter with hydrogen gas into excess sodium acetate solution that has previously been boiled. Poorly soluble chromium(II) acetate precipitates here, which is washed about 10 times in a hydrogen atmosphere by decanting with boiled water until the chloride reaction disappears and then dissolved in boiled, diluted sulfuric acid. After this solution has been allowed to settle, it is also lifted into the storage bottle under hydrogen gas and diluted with boiled water.

Chromium(II) salt solutions, like titanium(III) chloride solutions, are extremely sensitive to air. They must therefore be carefully stored away from air. An automatic burette, as shown in Fig. 4.31, is suitable for this purpose. The standard solution is stored under hydrogen gas in the storage bottle c, on which a burette a is placed. To fill this, the taps f and b are closed and tap e is opened. The standard solution then rises through the tube g into the burette by suction at e and by the pressure of Kipp's hydrogen developer connected at h via a wash bottle. When the burette is full, e is closed and f is opened. A particular advantage of this arrangement is the fact that the solution does not come into contact with greased taps when it enters the burette. The purpose of the Bunsen valve d is to prevent air from entering during suction.

A reducing burette with amalgamated zinc as the reducing agent can also be used for titration with chromium(II) sulfate solution. In this case, $KCr(SO_4)_2$ dissolved in hydrochloric or sulfuric acid ($c(H^+) = 0.1$ mol/l) serves as the standard solution [154].

The chromium(II) sulfate solution is best adjusted by potentiometric titration of a copper(II) sulfate solution of known content.

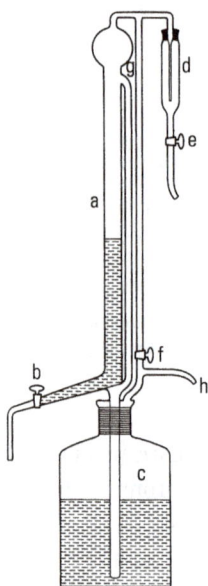

Fig. 4.31: Automatic burette for titration with air-sensitive solutions (see explanation of the letters in the text).

Determination of copper: The titration is carried out at 80 °C in a chloride-free, sulfuric acid solution. Naturally, titration must be carried out in the absence of air. In a sealed titration vessel, the sample solution is carefully rinsed with oxygen-free nitrogen to remove most of the dissolved oxygen. Residual oxygen is reduced by adding 1 ml of the chromium(II) sulfate solution. A reduction of copper(II) ions is reversed by adding a stronger oxidizing agent, e.g. a few milliliters of potassium bromate solution. Only now does the actual titration with the chromium(II) sulfate solution begin: a first potential jump indicates the complete reduction of the added oxidizing agent (in this case, potassium bromate) and a second the complete reduction of copper(II) ions to metallic copper:

$$2\,Cr^{2+} + Cu^{2+} \rightarrow 2\,Cr^{3+} + Cu.$$

A potential jump corresponding to the copper(I) stage only occurs to a limited extent in sulfuric acid solution, as the primary copper(I) ions formed according to the equation

$$2\,Cu^{+} \rightleftharpoons Cu + Cu^{2+}$$

disproportionate to metallic copper and copper(II) ions. In the presence of chloride ions, however, the copper(II) ions are only reduced to the copper(I) level. Therefore, hydrochloric acid interferes with the titration of copper(II) ions with chromium(II) sulfate solution. Nitric acid must also not be present due to its oxidizing properties. The copper(II) ion content of the solution under investigation can be calculated from the volume of chromium(II) sulfate solution consumed between the first and second potential jumps.

Simultaneous determination of copper and iron: The sulfuric acid solution containing iron(III) and copper(II) ions is pre-reduced at 80 °C and titrated with chromium(II) sulfate solution after the addition of a small amount of potassium bromate solution. Three potential jumps occur: the first of which indicates the reduction of the excess bromate, the second the reduction of Fe^{3+} ($Fe^{3+} + e^{-} \rightarrow Fe^{2+}$) and the third the reduction of Cu^{2+} ($Cu^{2+} + 2\,e^{-} \rightarrow Cu$). Any arsenic present is oxidized to arsenic acid by the bromate, but antimony prevents the third potential jump if more than 10 mg/l are present. The specified method allows the determination of iron even in the presence of 2,000 times the amount of copper.

If a chalcopyrite is to be examined, it is first digested with boiling concentrated sulfuric acid with the addition of potassium peroxodisulfate.

4.4.7 Evaluation

Normally, the standard solution is added to the solution to be titrated in small, precisely measured portions, and the deflection of the galvanometer (or the bridge sections in the compensation method) is read and noted each time. It should be noted that the adjustment of the equilibrium potentials requires a certain amount of time for some reactions. It is advisable to add the standard solution in smaller, but always equally large volume steps near the equivalence point in order to be able to determine the end point as accurately as possible. The potential values are entered in a coordinate system as a function of the volume of the standard solution added in each case. The task is then to determine the inflection point of the titration curve in the area of the potential jump. Its projection onto the abscissa corresponds to the consumption of standard solution up to the equivalence point.

A simple geometric method for finding the inflection point of a titration curve is the **tangent method**. To do this, draw two parallel tangents to the curve as shown in Fig. 4.32 so that their points of contact enclose the equivalence point. The point of intersection of the central parallel line with the titration curve is then the inflection point W. The result is always correct if the titration curve is symmetrical.

If you have many measuring points in the area of the equivalence point, it can be advantageous to plot the potential differences observed per volume portion of the added standard solution ($\Delta E/\Delta V$) against the total added volume of the standard solution (see Fig. 4.33). The individual points are connected to each other, and the two connecting lines on both sides of the equivalence point are extrapolated. The projection of their point of intersection onto the volume axis indicates the titration end point. The difference quotient reaches its greatest value when the original titration curve has the greatest gradient.

The smaller the volume step between two potential measurements, the closer the resulting representation approximates the first derivative of the titration curve. With modern potentiographs, this type of differentiation can be carried out electronically.

Despite the relatively high computational effort, the linearization of titration curves according to Gran is a very useful method (**Gran method,** Gran plot [155]). It is based on the potential equation

$$\Delta E = \Phi_i + \frac{s}{z} \cdot \lg c \qquad \left(\text{with } s = \frac{RT \cdot 2.303}{F} \right)$$

and forms them into

$$\frac{z \cdot \Delta E}{s} = \frac{z \cdot \Phi_i}{s} + \lg c$$

around. This forms the powers in base 10

$$10^{z \cdot \Delta E/s} = 10^{(z \cdot \Phi_i/s + \lg c)}$$

and applies the rules of arithmetic with powers:

$$10^{z \cdot \Delta E/s} = 10^{z \cdot \Phi_i/s} \cdot 10^{\lg c},$$

respectively

$$10^{z \cdot \Delta E/s} = m \cdot c$$

if you set $10^{z \cdot \Phi_i/s} = m$ for the constant value. If the variable $10^{z \cdot \Delta E/s}$ calculated from the measured potential values is now plotted against the concentration c, a straight line with the gradient m is obtained, which runs through the origin of the coordinate system.

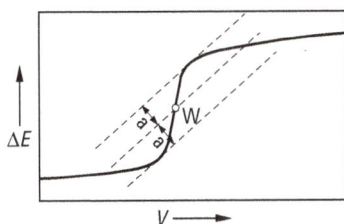

Fig. 4.32: Tangent method to find the turning point W (ΔE) EMF, electromotive force of the electrode; (V) added volume of the standard solution; (a) distance.

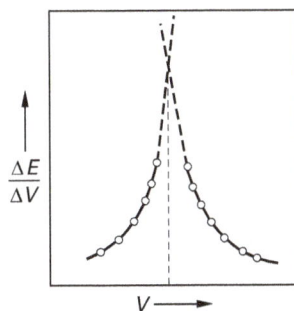

Fig. 4.33: Plot of the difference quotient $\Delta E/\Delta V$ to find the equivalence point (ΔE) EMF, electromotive force of the electrode; (V) added volume of the standard solution.

Our task is to meaningfully evaluate titration curves obtained potentiometrically. It should be borne in mind that the concentration of the ion to be determined is decreased proportionally with the addition of the standard solution in the titration vessel, which of course presupposes a quantitative reaction. Therefore, the variable $10^{z \cdot \Delta E/s}$, plotted against the volume of the added standard solution, must also result in a straight line, which, however, has a negative slope corresponding to the decrease and intersects the abscissa at the equivalence point. The same applies to the excess solution added. Its concentration in the titration vessel increases constantly after reaching the equivalence point and, according to the Gran method, an increasing straight line must be obtained starting at the equivalence point. An important criterion for accuracy is the common intersection of both lines with the abscissa (see Fig. 4.34).

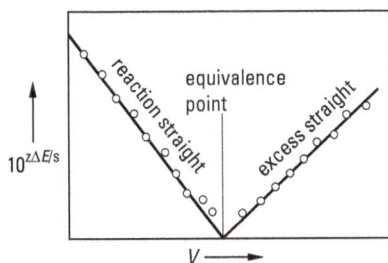

Fig. 4.34: Titration curve using the Gran method (z) equivalent number; (ΔE) EMF, electromotive force of the electrode, $s = 2,303\ RT/F$; (V) added volume of the standard solution.

Before we come to the practical implementation of the Gran evaluation, it is necessary to give some thought to the function on which the method is based. Formally, the exponential function is the inverse of the logarithmic function, and in English-speaking countries, for example, the equivalent notation $10^x = \text{antilg } x$ is usually found. This means that in our calculation above for $10^{\lg c}$, the identity $10^{\lg c} = c$ results because the two operations to be performed are inverses of each other.

The easiest thing to do now is to consider how errors in the calculation of the variable $10^{z \cdot \Delta E/s}$ affect the accuracy of the method. If the quotient $z \cdot \Delta E/s$ is affected by a proportional error f, so that $(10^{f \cdot z\Delta E/s})$ is actually calculated, the result would be $(10^{z \cdot \Delta E/s})^f$ according to the rules of power calculation. Since only $10^{z \cdot \Delta E/s}$ is proportional to the concentration c, the more f differs from one, the more likely its power with the exponent f will result in a recognizably curved graph instead of the expected straight line. In other words, to calculate the Gran variable, the ratio $\Delta E/s$ must be largely error-free, because z only takes into account the charge of the potential-determining ion and is therefore an integer. Although the so-called Nernst factor s can be calculated for the prevailing temperature (e.g. $s = 0.059$ V for 25 °C), this must be determined experimentally on a dilution series in order to use the Gran method, which also corrects for any possible incorrect behavior of the indicator electrode.

However, the quotient $z \cdot \Delta E/s$ can also be incorrect by one additive element. If you follow this influence, it is easy to show that only the slope m of the equalization line will be different, without affecting the position of its intersection with the abscissa. This property is utilized when applying the Gran method.

The potentiometric titration is carried out in the usual way after the relevant Nernst factor s for the indicator electrode used has been determined by calibration. From the measurement data obtained during the titration, a series of values is initially compiled, the absolute value of which decreases as the titration progresses toward the equivalence point. It is only important that the potential differences measured after each addition of standard solution correspond exactly to those of the original series of measurements. For the area after the equivalence point, a similar series of values is required, which is now created according to the principle that, if the measured potential differences are taken into account exactly, their absolute values increase again with the addition of further standard solution. If a fixed amount is added to or subtracted from all the original measured values when calculating these data series, the result is not affected. The original sign of the measured values is also not taken into account. Next, the Gran variables are calculated, and the results are plotted against the added volume of standard solution, as shown in Fig. 4.34. Incidentally, it does not matter if a different scale is selected for the ordinate to represent the excess straight line.

However, if during a titration the original titration volume V_0 is significantly increased by the addition of the standard solution, the values to be applied using the Gran method must be corrected. If necessary, proceed as follows: In the solution to be titrated, the amount of substance n present in each case is independent of the dilution. Thus, its concentration is calculated according to the dilution currently present according to

$$n = c_0 \cdot V_0 = c \cdot (V_0 + V_t)$$

to

$$c_0 = c \cdot \frac{V_0 + V_t}{V_0},$$

where c is the real concentration, c_0 is the theoretical concentration and V_t is the volume of the standard solution added at that time. Accordingly

$$\frac{V_0 + V_t}{V_0} \cdot 10^{z \cdot \Delta E / s} = \frac{V_0 + V_t}{V_0} \cdot m \cdot c = m \cdot c_0.$$

All Gran variables must first be multiplied by the applicable factor $(V_0 + V_t)/V_0$ before the graph is displayed.

If all these calculations are too complicated for you, you can purchase special **Gran plot paper (a kind of logarithmic paper)** from a specialist retailer and enter your measurement data directly into it. On the other hand, since programmable computers or laboratory computers are increasingly available in laboratories, it is easy to create programs that convert the measured potential values into the pairs of values required for the Gran method on the basis of the mathematical relationships given. This procedure is applied very quickly and, in the long run, more cost-effectively.

The Gran method for evaluating potentiometrically obtained titration curves would certainly be too complex if it did not also offer special advantages. This will be briefly explained using two examples. The simultaneous determination of bromide and chloride by titration with silver nitrate solution yields two potential jumps: the first of which corresponds to the bromide content and the second to the chloride content. The Gran representation of the titration curve is shown in Fig. 4.35. It has already been said (see p. 293 f.) that the first equivalence point is found too late when the inflection point method is used, since the solid solution formation between AgBr and AgCl plays a role under the conditions given there. However, the equalization line according to Gran is mainly based on the data that apply to the beginning of the titration, i.e. when no solid solution formation is observed. (The measuring points near the equivalence point can be taken less into account.) The result is more accurate than that determined using the inflection point method.

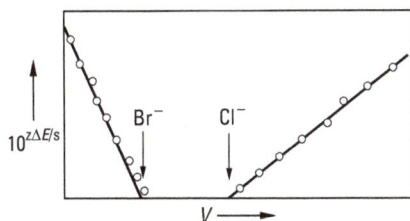

Fig. 4.35: Titration of bromide next to chloride with AgNO$_3$ solution; evaluation according to Gran (z) equivalent number; (ΔE) EMF electromotive force of the electrode, $s = 2.203\ RT/F$; (V) added volume of the standard solution.

It often happens that an inflection point is observed in a titration curve, but after applying the Gran method it is found that the intersection points of the two equalization lines do not coincide as expected, but leave a small gap. If the measuring points are otherwise well balanced by the respective straight lines, such a finding is an indication that, in addition to the substance to be determined, small amounts of a second substance are present, which reacts with the standard solution. This will be taken as an opportunity to systematically search for the impurity.

It should be noted in this context that the undisputed value of the Gran method only becomes clear when evaluating direct ion measurements with an ion-sensitive electrode in the trace range and when multiple additions of standard solutions are applied to take into account matrix effects of the solution investigated (standard addition method).

Of course, the ability to determine ion concentrations directly with ISEs, even in the trace range, is what makes these electrodes so valuable. Such ISEs are constantly coming onto the market in new variants and specifications, but it should be borne in mind that such use is only economically viable in routine operation, as it must always be preceded by a complex and time-consuming calibration procedure. First of all, the measuring devices must be correctly adjusted and then, above all, the influences of the respective matrix (matrix effects), if a correct result is to be obtained at the end. However, it is obvious that it is an advantage if, after such complex preparations, series analyses can be carried out automatically and with fast sample throughput. To meet these requirements, the device manufacturers also offer the corresponding electronic and technical equipment with which the required analyses can be carried out directly on the screen on the basis of a **Gran plot**. Experience shows that, despite all the automation and sophistication of the equipment used, the respective operator must have sufficient knowledge of how ISEs work and the limits of their applicability.

All this effort can be saved if ISEs are only used to determine the end point of titrations, because here, as usual, the consumption of correctly adjusted standard solution is the measured value required to calculate the analysis result. However, trace analysis cannot be carried out in this way, as the actual performance of ISEs is not optimally utilized.

4.5 Titrations with polarized electrodes

4.5.1 Polarization of electrodes

Conductometry and potentiometry can only be used as indication methods if polarization of the electrodes in the solution is avoided. In contrast, **polarization titration methods** deliberately use polarized electrodes. They use the sudden change in voltage or current caused by polarization or depolarization of the electrodes at the equivalence point to indicate the titration end point.

If the potential of a metal electrode shall be described by Nernst's equation, the condition of currentlessness must be fulfilled for the respective measuring method.

However, if current flows through an electrode, deviations occur, for which the term **polarization** has been chosen. Accordingly, the difference between the Nernst voltage (open-circuit voltage) and the voltage measured under current flow is called the **polarization voltage**.

Polarization is caused by a variety of inhibitions that occur in the case of current flow between the electrode and the solution as well as within the solution. Transport problems play a very important role here. Ions migrate in the electric field, and their different mobilities cause differences in voltage and concentration within the solution. Even if these effects can still be eliminated by stirring, layers of liquid (boundary layers) remain on the surfaces of the electrodes, which cannot be influenced by stirring. However, the ions must also migrate through these layers, possibly taking part in chemical reactions, exchanging their charge in order to finally be incorporated into the crystal lattice of the electrode material. Such processes take time and can only follow the constraints of the external current flow to a greater or lesser extent. The system comes out of a state of equilibrium; one also speaks of irreversible processes when more than just concentration effects occur.

It is obvious that an electrode is particularly easily polarized if its surface area is very small. Under certain conditions, its polarization can become so large that a current flow is completely blocked despite the voltage being applied. The indication with such electrodes is based, for example, on the fact that ions occur at the equivalence point of a titration, which considerably reduce or even cancel out an existing polarization (**depolarizers**). The occurrence of such ions can be recognized by the fact that current can flow again.

Of course, polarizable electrodes can only be operated in relation to a second electrode. In most cases, however, it is important to have only one such polarizable electrode in the respective measuring circuit. Your counter electrode will then be a nonpolarizable electrode. This is an electrode on which the electrochemical processes can take place uninhibited. Both the complete polarizability and the nonpolarizability of an electrode are ideal concepts that can at best be approached. Compared to a metal electrode with a very small surface area, an electrode of the second kind, as we have already seen in the calomel or Ag/AgCl electrode, behaves in many cases as unpolarizable. Only when the current load increases, it is advisable to use a metal sheet with a sufficiently large surface area as a counter electrode. The potential of the polarizable electrode is then measured without current against a reference electrode (third electrode) using a compensation circuit. The counter electrode takes over the current exchange with the solution.

Sometimes, two polarizable electrodes are used, namely for **biamperometric or dead-stop titrations (see Section 4.5.4)**.

Using the example of the reduction of Fe^{3+} ions to Fe^{2+} ions, we want to clarify the mode of action of a polarizable electrode, a platinum wire, connected as a cathode. The Fe^{3+} ions diffuse through the boundary layer in front of the electrode and are reduced there, provided a sufficient voltage U is applied. The result is a measurable current I. If the voltage is increased, the current increases until all Fe^{3+} ions that reach the surface

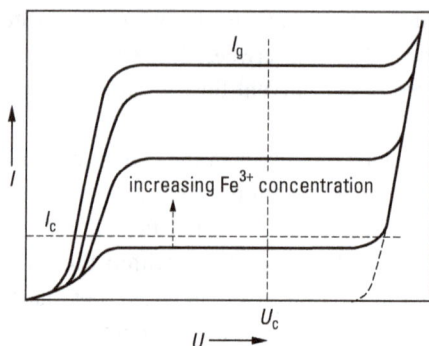

Fig. 4.36: Current-voltage curves for different Fe(III) concentrations during titration in the presence of a polarized electrode (I) current; (I_g) limiting current; (I_c) for voltametric titrations specified constant current; (U) voltage; (U_c) for amperometric titrations specified constant voltage.

by diffusion are immediately reduced. A limiting current I_g is reached that cannot be exceeded even with further increases in voltage, although its magnitude depends on the concentration of Fe^{3+} ions in the solution. If the Fe^{3+} concentration were to be increased, more ions could diffuse through the boundary layer and the limiting current would be correspondingly higher. The occurrence of a limiting current is therefore caused by the fact that fewer ions reach the surface than can be recharged there due to slow diffusion, i.e. an inhibition. The reduction $Fe^{3+} \rightarrow Fe^{2+} + e^-$ nevertheless represents a clear system because under the given conditions no further inhibitions occur, which would cause polarization effects. Current-voltage curves determined for different Fe^{3+} concentrations are shown in Fig. 4.36. It can also be seen that the current strength only increases further at very high voltages, namely when the latter is high enough to reduce other ions present in the solution, for example, hydrogen ions.

4.5.2 Voltametric titrations

We want to take the above considerations further in order to find out how Fe^{3+} ions will behave during the voltametrically indicated titration. Figure 4.37 schematically shows the equipment required for this. An electronically stabilized voltage source supplies direct currents with voltages between 10 and 100 V as required. Resistors with values between 10 and 100 MΩ, which can be switched into the circuit and considerably exceed the resistance of the measuring cell, make it possible to maintain constant currents in the range between 0.1 and 20 μA. The voltage between the electrodes is read on a voltmeter. A calomel electrode could serve as a counter electrode whose active surface is large enough to withstand the specified current flow.

With this measuring device, we set a constant current I_c and titrate a solution of Fe^{3+} ions in the measuring cell with EDTA standard solution to which some NaCl has been added to suppress migration effects within the solution. The processes are best illustrated in Fig. 4.36. At the beginning of the titration, the specified current I_c causes a low polarization voltage U due to the still high Fe^{3+} concentration. As the Fe^{3+} con-

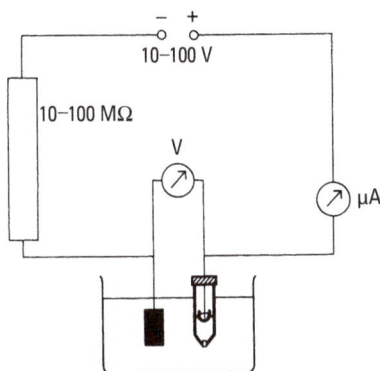

Fig. 4.37: Measuring arrangement for voltametric indication.

centration is decreased during the titration, U gradually increases slightly until, at the end, the Fe^{3+} concentration has become so low that the limiting current caused by it is less than I_c. The voltage U rises sharply here, and the current flow is increasingly caused by the reduction of H^+ ions. However, the observed potential jump by no means indicates the equivalence point. Rather, it occurs when the limiting current of the Fe^{3+} reduction has reached the arbitrarily specified value I_c. I_c must therefore be selected so that the Fe^{3+} concentration present in the solution at the potential jump has become negligibly small compared to the concentration present at the start; only then can the equivalence point be considered to have been reached. These uncertainties naturally severely restrict the use of voltametrically indicated titrations, although the potential jump is much sharper than with a potentiometric indication. However, ions whose oxidation or reduction is irreversible can be made accessible to an electrometric indication using this method. If you wanted to work potentiometrically in such a case, very long measurement times would have to be provided for due to the slowly developing equilibria. Figure 4.38 shows the voltametrically indicated procedure for the titration of Fe^{3+} ions with EDTA standard solution. If I_c was selected sufficiently small, the potential jump practically coincides with the equivalence point. In principle, voltametric methods can be used for the indication of complex formation and redox analyses, but also in argentometry down to low concentrations.

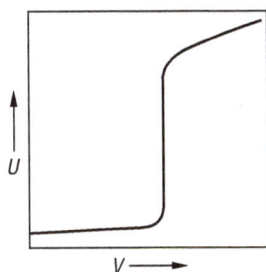

Fig. 4.38: Voltametric indication of Fe(III) determination with EDTA solution (U) voltage; (V) added volume of EDTA solution.

4.5.3 Amperometric titrations

Amperometric titrations can be carried out using a measuring arrangement similar to the opposite schematic diagram (Fig. 4.39). A voltage divider is used to apply a constant voltage U_c to the two electrodes, and the resulting current I is measured with a microammeter. If U_c is in the limiting current range of an oxidizable or reducible ion species, a current proportional to its concentration is obtained. Depending on whether the ion detected by the polarizable electrode is consumed during the titration or only enters the solution after the equivalence point, the resulting titration curves correspond in shape to the left or right curve in Fig. 4.40. An electrode based on mercury drops is a popular polarizable cathode for amperometric titrations. Two important kinds of mercury electrodes are distinguished: the dropping mercury electrode and the hanging mercury drop electrode. Due to the high overvoltage of hydrogen, mercury is an electrode material that is suitable for the reduction of many metal ions that are less noble than hydrogen ions. Most titrations down to concentration ranges of 10^{-5} mol/l can also be indicated amperometrically. This is the reason why these methods, also known as **limiting current titrations**, are used much more frequently than the voltametric methods discussed above. Even kinetic processes with time-varying concentrations of certain ion species can be recorded at the mercury drop electrode by continuously measuring the limiting current. It is then only necessary to amplify the measured current signal and record it automatically with a *compensation recorder.*

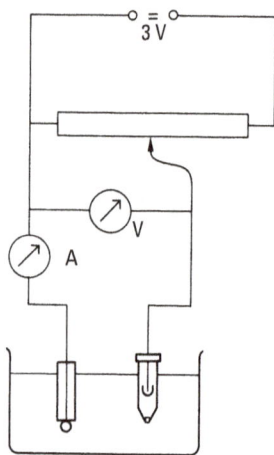

Fig. 4.39: Measuring arrangement for amperometric titration.

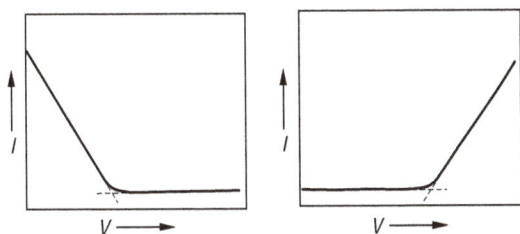

Fig. 4.40: Titration curves with amperometric titrations (*I*) current strength; (*V*) added volume of the standard solution.

4.5.4 Biamperometric or dead-stop titrations

The amperometric titration with **two polarizable** electrodes, also known as the **dead-stop method**, is of particular interest because of the sharpness of the equivalence points displayed and because of its very low equipment requirements. The apparatus required for the measurement corresponds in its structure to Fig. 4.39, except that now two polarizable electrodes are immersed in the solution, usually two platinum wires. A voltage between 10 and a maximum of 100 mV is selected so that the two electrodes completely block any current flow as long as no reversible redox pairs are present in the solution due to the structure of a boundary layer acting as a capacitor. For example, if you titrate an Fe^{2+} solution with an oxidizing agent such as $K_2Cr_2O_7$, the sample solution contains both Fe^{2+} and Fe^{3+} ions up to the equivalence point, which together represent a reversible redox couple. Accordingly, a current flow is observed on the microammeter. However, at the moment when all Fe^{2+} ions are consumed (equivalence point), there is no longer a reversible redox couple present – the dichromate is irreversibly reduced – and the current practically drops to zero (zero point, dead stop).

The same applies when titrating an iodine solution with sodium thiosulfate:

$$2\,S_2O_3^{2-} + I_2 \rightarrow S_2O_6^{2-} + 2\,I^-.$$

As long as iodine is present in the solution, the reversible equilibrium can be established at the electrodes

$$I_2 + 2\,e^- \rightleftharpoons 2\,I^-.$$

At the cathode, this process runs from left to right, while conversely at the anode I_2 is formed. The currents in the microampere range show that the material conversion at the two platinum wires is practically negligible. Once all the iodine has been consumed, the irreversible reaction

$$2\,S_2O_3^{2-} \rightarrow S_4O_6^{2-} + 2\,e^-$$

takes the place of the reversible I_2/I^- redox couple, and the current flow is interrupted due to the immediate onset of polarization (dead stop). As in the case of Fe^{2+} oxidation with potassium dichromate, the measured titration curve corresponds to type 1 in Fig. 4.41. If, in the opposite case, a chromate solution is prepared and titrated with a standard solution containing Fe^{2+} ions, the two platinum electrodes remain in the polarized state up to the equivalence point. Only when the reversible Fe^{2+}/Fe^{3+} redox pair occurs, the electrodes are depolarized and current flows according to curve type 2. A third possibility is to replace one reversible redox pair with another: the titration curve would then correspond to type 3 in Fig. 4.41. An example of this is the titration of Fe^{2+} with Ce^{4+} ions. In addition to the pairs mentioned so far, $Br_2/2Br^-$, Ti^{4+}/Ti^{3+} and VO_2^+/VO^{2+} are further reversible redox pairs that have a depolarizing effect on polarized platinum wires and thus enable an indication according to the dead-stop method.

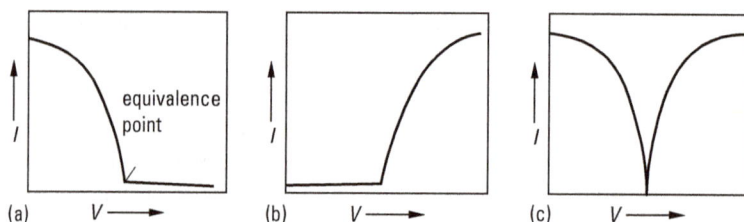

Fig. 4.41: Titration curves with indication using the dead-stop method (I) current intensity; (V) added volume of standard solution); polarization of the electrodes takes place (a) from the equivalence point, (b) up to the equivalence point and (c) only at the equivalence point.

4.5.4.1 Water determination according to Karl Fischer

The Karl Fischer method has become generally accepted as an important routine method for determining water in organic solvents. The special feature of this method is that the standard solution used contains oxidizing agent (I_2) and reducing agent (SO_2) together. However, both can only react with each other in the presence of water. The following equation is sometimes formulated for this purpose:

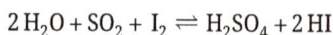

$$2\,H_2O + SO_2 + I_2 \rightleftharpoons H_2SO_4 + 2\,HI$$

(Bunsen reaction). The deviations are reported below.

When all the water has been consumed, I_2 appears next to I^- for the first time, and the equivalence point can be indicated by the presence of a reversible redox couple using the dead-stop method. (When titrating in an organic medium, the excess I_2 after the equivalence point has been exceeded cannot be detected by the blue coloration of the starch complex, as is usually the case in iodometry.) For the simple performance of water determinations, many well-known equipment manufacturers offer more or less automated apparatus that all use this iodometric method.

The Karl Fischer reagent used for the titrations is a solution of elemental iodine and sulfur dioxide in a mixture of pyridine and methanol. Iodine dissolves in methanol without losing its oxidizing effect, while pyridine in particular, and also methanol, is able to bind the starting and end products. The presence of pyridine has a stabilizing effect on the shelf life of the Karl Fischer solution. Naturally, the organic solvent containing water is then titrated in a sealed vessel with absolute exclusion of moisture. When the two platinum electrodes are increasingly depolarized after reaching the equivalence point, the current begins to rise.

In practice, however, it turns out that the reduction of an I_2 molecule does not consume two molecules of water, as the reaction equation given above would suggest, but under optimal conditions only a ratio of $I_2/H_2O = 1{:}1$ is obtained. The methanol-pyridine ratio plays a role here and, as was later discovered when replacing pyridine with other bases, the pH value reached in the reaction solution. The starting point was the realization that methanol is able to compete with water to a certain extent due to its similarity to water. This also explains why the titer of a Karl Fischer solution constantly decreases when all components involved are stored together in a stock solution. Johansson [156] therefore suggested using two solutions, namely a solution of iodine in methanol and a second solution of SO_2 in a pyridine-methanol mixture. In this way, a fresh reaction mixture can be prepared by mixing both solutions before titration.

Scholz [150] summarizes the latest investigations and findings on the course of the Karl Fischer reaction with the following equations:

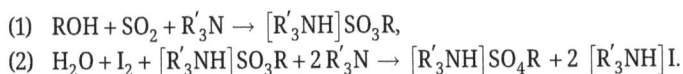

(1) $\quad ROH + SO_2 + R'_3N \rightarrow [R'_3NH]SO_3R,$
(2) $\quad H_2O + I_2 + [R'_3NH]SO_3R + 2\,R'_3N \rightarrow [R'_3NH]SO_4R + 2\,[R'_3NH]I.$

ROH is a "reactive" alcohol and R'_3N is a "suitable" base. The equations show that the higher the pH value of the solution or the more basic the R'_3N used reacts, the better the acidic components are bound. In practice, it is best to work at pH values between 5 and 7, as the reaction takes place both stoichiometrically and at a constant reaction rate in this range.

Of the numerous attempts to use the classic solvents methanol and pyridine by others, only two are highlighted here. In 1955, Jungnickel and Peters [157] suggested using 2-methoxyethanol instead of methanol in some cases, and Scholz [158a] pointed out the advantages of the more basic imidazole over the otherwise common pyridine. The common goal of such efforts is above all not only improved titer stability, but also greater speed and thus better accuracy of the titration procedure itself. Anyone who needs to make the most of the versatility of Karl Fischer titration should refer to the monographs by Scholz [150] and P. Bruttel et al. [150a].

With its ready-made solutions, the trade adapts to the respective level of knowledge. The division of the required components into two solutions, which must be mixed before titration, is preferred.

1. Procedure:
1. Preparation of the solutions
Even if pure substances are used, they must be dried and all operations must be carried out in the absence of moisture.

Preparation of the chemicals
Pyridine is dehydrated by distillation with benzene. The ternary mixture of water-benzene-pyridine distils over at 67 °C. After driving off the excess benzene (Bp. 80.1 °C), pyridine is converted at 115.5 °C. The water content is then still about 0.05–0.08%.

Methanol is boiled with magnesium powder at the reflux condenser (5 g/l Mg) until the water content has fallen below 0.05% and then distilled (Bp. 64.6 °C).

2-Methoxyethanol is dried by simple distillation (Bp. 124.3 °C). All water is contained in the first 5% of the feed.

Dioxane is dehydrated by distillation over sodium (Bp. 101.5 °C).

Iodine is sublimated twice in a closed grinding apparatus in the absence of moisture.

Sulfur dioxide is taken from a pressurized gas cylinder and introduced into ice-cooled, dry pyridine to saturation (1 mol SO_2/mol pyridine) via a drying tube filled with P_4O_{10} while excluding moisture.

Karl Fischer solution
To prepare a stock solution according to Mitchell and Smith [151], dissolve 762 g iodine in 2.4 l pyridine. After dissolving (iodine is relatively poorly soluble in cold pyridine!), 6 l of methanol is added. For use, 4 l of this indefinitely stable solution is gradually mixed with 192 g liquid sulfur dioxide (cold mixture: methanol-dry ice (solid carbon dioxide)) under ice cooling and swirling in a burette storage bottle. After a few days, the solution is ready for use. Theoretically, 1 ml of solution corresponds to 6 mg of water, but in practice – for the reasons already mentioned – only 3.5–5 mg of water.

For smaller quantities, according to Eberius [152], 530 g methanol, 265 g pyridine and 100 ml of SO_2 solution in pyridine are mixed together in a storage bottle. If the solution still contains too much water, add bromine drop by drop until a faint brown coloration remains (stir!). Now slowly add 45 ml of liquid sulfur dioxide during moderate cooling, excluding moisture. Dissolve 84 g of iodine in the still warm solution.

Solution according to A. Johansson
Solution A is prepared by dissolving 100 g of sulfur dioxide in a mixture of 500 ml of anhydrous methanol and 500 ml of anhydrous pyridine.

Solution B consists of a solution of 30 g iodine in 1 l of anhydrous methanol. A volume of 1 ml of the solution corresponds to 2 mg of water.

Solution according to Peters and Jungnickel
Dissolve 133 g iodine in 425 ml pyridine. Then add 425 ml of 2-methoxyethanol and finally 70 ml of liquid sulfur dioxide. The solution is ready for use after 8 h and is stable over a longer period of time. A volume of 1 ml of the solution corresponds to about 6 mg of water.

Solution according to Scholz
First, dissolve 136 g of imidazole ($C_3N_2H_4$) and then 45 ml of liquid sulfur dioxide in anhydrous methanol. Add more methanol to bring the volume to 1 l (solvent).

Titration is carried out with a solution of iodine in anhydrous methanol.

2. Water determination
To carry out the titrations, you need a vessel that can be tightly closed with a lid. There are several ground joint openings in the upper part, which allow the burette tip, electrodes and drying tube to be

fed in gastight. A magnetic stirring device must be provided from below. The titration solutions are dosed with a piston burette, whereby the storage vessel must be protected against the penetration of moisture with a drying tube. Two platinum wires (polarizable electrodes) fitted with a ground joint holder are inserted into the titration vessel from above. Titration is then carried out at a voltage of 15–20 mV until the microampere meter indicates an increase in current (reversible redox couple). The best way to do this is to first titrate anhydrous solvent (methanol or dioxane) with Karl Fischer solution until the current has increased by a small amount that is maintained in all subsequent titrations (this also removes traces of moisture). The substance to be analyzed for water is then added, and titration is repeated up to the selected current mark. If the two-component method is to be used, the solution containing the base is prepared in the quantity subsequently required and titrated with the iodine solution. The substance to be tested is then added and titrated again with iodine solution, making sure that the base (pyridine or imidazole) present in the solution exceeds the theoretically required quantity by at least 50%. For example, 10 ml of the solvent mixture specified by Scholz is required in the template for the determination of 70–80 mg of water. Always stir sufficiently during titrations (magnetic stirrer).

The **titer determination of** the Karl Fischer reagents is carried out on a standard methanol solution containing a defined amount of water. Taking appropriate precautions, a weighed portion of water can be added to the anhydrous methanol. Another possibility is to weigh out storage-stable salts with a defined crystal water content and dissolve them in methanol or dioxane. Suitable compounds are oxalic acid (2 H_2O), sodium tartrate (2 H_2O), citric acid (1 H_2O) and others.

Water determination according to Karl Fischer has a wide range of applications. Many organic solvents can be titrated directly, and other substances must first be dissolved in suitable anhydrous solvents. In this context, many inorganic salts with a water of crystallization content should be mentioned, but also fats, oils, foodstuffs, etc. Water determination is disturbed by simple or halogenated hydrocarbons, aldehydes, ketones and mercaptans.

4.5.4.2 Determination of primary aromatic amines

Primary aromatic amines (anilines) can be easily determined by titration with a sodium nitrite solution in an acidic solution (nitritometry). The diazonium ion formed in the process is sufficiently stable under titration conditions. Redox indicators such as ferrocyphen or metanil yellow can be used for indication. In most cases, electrochemical indication by means of dead-stop titration (biamperometry) is used. The depolarization of the double platinum electrode takes place after the equivalence point by excess sodium nitrite, which is oxidized as a redox amphoter at the anode and reduced at the cathode. Bromide is added to stabilize the intermediate nitrosyl cation, and ice cooling is often recommended to stabilize the diazonium ion:

$$NaNO_2 + 2\ H^+ \longrightarrow NO^+ + Na^+ + H_2O$$

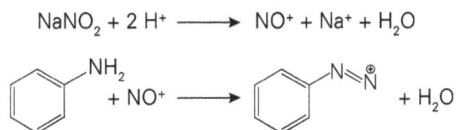

The method can be found in many pharmacopoeias, e.g. determination of sulfonamide antibiotics or local anesthetics such as procaine or benzocaine. Carboxylic acid hydrazides (USP: determination of isoniazid) can also be determined with sodium nitrite, as can some secondary aromatic amines (e.g. tetracaine), which form stable nitrosamines.

4.5.4.3 Determination of procaine according to the European Pharmacopoeia

Procedure: 0.400 g of procaine hydrochloride is dissolved in 50 ml of dilute hydrochloric acid. After adding 3 g potassium bromide, the solution is cooled in an ice-water mixture and slowly titrated with sodium nitrite solution, 0.1 mol/l, with continuous stirring. The indication is carried out using biamperometry [8a].
   1 ml sodium nitrite solution, $c = 0.1$ mol/l, corresponds to 27.28 mg $C_{13}H_{21}ClN_2O_2$.

4.6 Coulometric titrations

Whereas with all the methods we have become familiar with so far, we first had to prepare a suitable standard solution and determine its titer before we could start the actual titration, coulometric titration provides us with a method of generating the reagent directly in the solution to be tested by means of an electrolytic process. The measured variable is the electric charge Q exchanged with the solution at the electrodes, from which we can calculate the amount of substance of the reagent produced according to Faraday's laws. The resulting possibilities extend to all areas of volumetric analysis, whereby the inaccuracy achieved does not exceed 1% down to the trace range. The determination of the end point is independent of this type of reagent generation. All suitable indication methods are used here, although electrometric methods are preferred due to the already relatively high equipment requirements (see further literature [153]).

4.6.1 Theoretical principles

According to Faraday's laws, the electrical equivalent of 1 mol of exchanged electrons is $F = 96{,}493$ C/mol; this quantity is the already known Faraday constant. If a substance with the molar mass M exchanges electrons at an electrode z and requires the amount of electricity Q to do so, the converted amount of substance n results in

$$n = \frac{Q}{z \cdot F}.$$

Since the amount of substance is $n = \dfrac{m}{M}$ (m = mass of the substance), you get

$$\frac{m}{M} = n = \frac{Q}{z \cdot F} \quad \text{or} \quad m = \frac{M \cdot Q}{z \cdot F},$$

so that the mass m of the substance to be determined can be calculated from the measured electric charge Q. The prerequisite for accuracy is a current yield of 100%, i.e. only the substance of interest may accept or release electrons, or in other words, side reactions due to the conversion of other species must be excluded.

Now we know from electrolysis that a sufficient conversion of substances within a reasonable time can only be achieved by applying a voltage, whose magnitude exceeds the equilibrium voltage calculated according to Nernst. As long as the electrodes are polarized while the current is flowing, only the desired conversion of substances generally takes place at them, i.e. the applied **overvoltage** serves exclusively to overcome polarization effects. Toward the end of electrolysis at the latest, the concentration of the substance to be converted becomes so low that the current drops and the polarization of the electrodes also decreases. However, the voltage applied is then sufficient to convert species that are difficult to oxidize or reduce, and a side reaction takes place. For this reason, the voltage is limited during electrolysis so that the secondary reaction consists at most of the decomposition of the solvent to gaseous products (e.g. water to H_2 or O_2). However, this does not interfere with the electrolytic deposition of a metal, for example, because it is the increase in mass of the electrode that is determined and not the amount of electricity.

In coulometric analysis, the problem described also exists if the voltage applied is greater than the equilibrium conditions for the conversion of the desired substance. As a secondary reaction, however, a process should now take place that yields a product that is itself capable of reacting with the substance to be determined in the same way as it would directly with it at the electrode. For example, arsenite could be oxidized to arsenate at an anode:

$$AsO_3^{3-} + H_2O \rightleftharpoons AsO_4^{3-} + 2H^+ + 2e^-.$$

However, this reaction is not reversible, so that the electrode would quickly become polarized and the conversion would soon come to a complete standstill. However, if the same reaction were to take place in a solution with a high concentration of additional NaBr, bromine would be formed anodically from bromide at a correspondingly higher voltage:

$$2Br^- \rightleftharpoons Br_2 + 2e^-,$$

which in turn would oxidize the arsenite to arsenate:

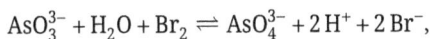

$$AsO_3^{3-} + H_2O + Br_2 \rightleftharpoons AsO_4^{3-} + 2H^+ + 2Br^-,$$

whereby the anodically generated Br_2 would be quantitatively converted back to the starting material Br^-. When balancing the amount of current converted, it does not matter in principle whether the arsenite was oxidized directly at the electrode or whether there was a detour via bromine. It is only necessary to ensure that no gaseous or other products that do not oxidize arsenite are produced at the anode as a result of secondary reactions. However, since Br_2/Br^- is a reversible redox couple, no polarization occurs at the anode. Therefore, a reaction requiring even greater voltage, such as the oxidation of water to oxygen, cannot take place. This means that we have already familiarized ourselves with the principle of coulometric titration using an example.

Since, under the given conditions, an electrolyte that is present in excess and at the same time reversibly oxidizable – such as sodium bromide in the above case – does not allow further side reactions at the anode, a very simple method can be used to determine the electric charge Q converted. According to its definition, $Q = I \cdot t$ (t = time; $1\,C = 1\,A \cdot s$), you need a device that emits an electronically stabilized and therefore constant current I. If it takes t seconds to reach the equivalence point, you can use the knowledge of I to calculate the electric charge Q and thus the mass of the converted substance. The coulometric titration therefore amounts to a time measurement. It is not necessary to know how such a constant current device works in detail, but it is only important to understand how it affects the chemical system. It sets a voltage and checks whether I corresponds to the set value. If this is not the case, the voltage is changed until I has reached its set value. If the resistance increases during an electrolytic reaction due to a decrease in concentration, the device will increase the voltage in accordance with Ohm's law until I has reached the specified value again. If, as in the case discussed above, a reversibly oxidizable electrolyte (NaBr) does not keep the overall resistance of the solution low and constant, unpleasant side reactions will occur. This should have made clear the limits within which coulometric titration can be used with a constant current.

So far, we have only ever considered the electrode at which the reaction of interest takes place. As with all electrochemical processes, a second electrode is of course required, and it is clear that an equal electric charge is exchanged at this electrode and that an electrochemical reaction takes place accordingly. As long as the reaction products do not interfere with the actual titration, no problems arise. However, if this is no longer the case, the electrolyte space of the counter electrode must be separated from the sample solution by a diaphragm. Such a diaphragm must be large enough so that its resistance in the circuit remains sufficiently low. It must be said, however, that in view of the electric charge transported through the solution, such a diaphragm is not always unproblematic, so that one will try to avoid its use if possible. In the electrolyte solution used in our arsenic determination, elemental hydrogen is generated at the cathode from the H^+ ions of the water. In statu nascendi, this is able to reduce arsenate again if it gets into the solution to be titrated. For this reason, the cathode compartment must be delimited. As the solution surrounding the cathode is quickly saturated, the hydrogen escapes, and no elaborate precautions are required. Instead of a diaphragm, a reagent glass with a small hole at its lower end immersed in the solution is sufficient to delimit the cathode chamber, at least for small quantities of arsenite to be determined.

An apparatus for carrying out coulometric titrations therefore consists of two generator electrodes, one of which is used for the electrochemical generation of the reagent, a device that emits a constant current and a clock that automatically determines the titration time (see Fig. 4.42). When the switch S is pressed, current is supplied to the two generator electrodes and the electrical clock is started at the same time. When the titration is finished, which is indicated by a suitable indicator system, the clock stops again when the switch is pressed again and the current flows back to the device via a resistor R instead of through the titration vessel.

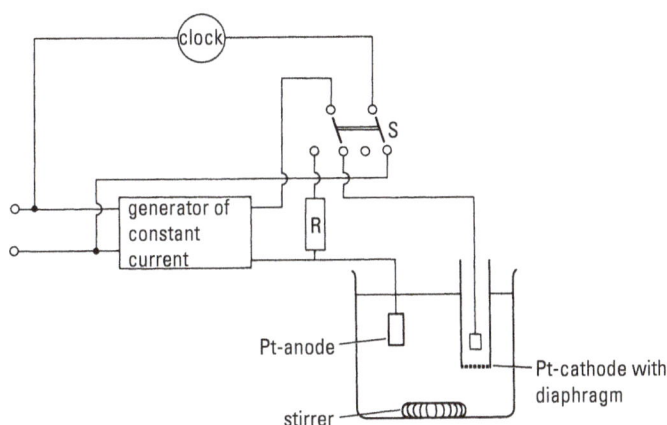

Fig. 4.42: Principle of coulometric titrations at constant current (S) switch; (R) resistor.

4.6.2 Practical applications

4.6.2.1 Determination of arsenic with dead-stop indication

The electrochemical principles of the process have already been discussed in detail above. A dead-stop system is particularly suitable for determining the end point (see p. 309 f.). Since the bromine produced at the anode of the generator is consumed as long as the solution contains unoxidized arsenic(III) species, the two platinum wires immersed for indication remain polarized. The equivalence point of the titration is characterized by the occurrence of the reversible Br_2/Br^- redox couple, which causes depolarization of the indicator electrodes and allows a current to flow via these electrodes.

The titration vessel (Fig. 4.42) is filled with 30 ml of a solution prepared from 20.6 g NaBr and 5.5 ml concentrated sulfuric acid by making up to 1 l with water. A voltage of 50 mV is applied to the platinum wires (indicator system), and an electrolysis is carried out to remove any oxidizable substances until the indicator current begins to rise. Once this has reached a selected value, e.g. 2 µA, the electrolysis is stopped. Now add the solution to be tested, which should contain around 15–30 mg As (III), to the titration vessel. The indicator current drops to zero. Then electrolyze at a constant current strength of 10 mA (generator) until the indicator current strength has reached the selected level (2 μA) again. The arsenic content of the sample is calculated from the measured time and the set generator current. The NaBr solution in the titration vessel can be used for three more arsenic determinations before it is replaced by a fresh one.

4.6.2.2 Determination of thiosulfate

Similar to the above procedure, add 30 ml of KI solution (0.5 mol/l) to the titration vessel and then 3–5 drops of diluted H_2SO_4 solution (caution: if the acid concentration is too high, thiosulfate could decompose).

After adding the analysis solution, I_2 is generated electrolytically at a generator current of 10 mA. This oxidizes the thiosulfate present to tetrathionate:

$$2\,S_2O_3^{2-} + I_2 \rightarrow S_4O_6^{2-} + 2\,I^-.$$

The end of the titration is recognized by an increase in the indicator current strength (dead-stop system) as soon as I_2 is no longer consumed by reduction. The thiosulfate content of the sample is calculated from the measured time and the set generator current.

4.6.2.3 Alkalimetric titrations

For the determination of acids, a sodium chloride solution with a concentration of 1 mol/l is used as the fundamental electrolyte. OH^- ions are formed at the platinum cathode of the generator according to the equation

$$2\,e^- + 2\,H_2O \rightarrow H_2 + 2\,OH^-.$$

Under no circumstances should a platinum sheet be used as an anode in a solution containing halide ions, as this will be oxidized (Cl_2 evolution). Instead, a silver plate has proven to be suitable because the Ag^+ ions formed during oxidation are immediately precipitated as AgCl by chloride. In this experimental setup, it is not necessary to separate the anode and cathode compartments from each other using a diaphragm. In contrast, the use of a fundamental electrolyte such as Na_2SO_4 requires very careful separation of the electrode compartments, as oxygen and H^+ ions are formed at the anode in this case. The equivalence point can be determined either visually using a color indicator or potentiometrically using a glass electrode.

4.6.2.4 Complexometric titrations

A solution of Hg-EDTA (0.02 mol/l) and NH_4NO_3 (0.05 mol/l) is suitable as the fundamental electrolyte. The mercury bound in the complex is electrolytically deposited on a cathode made of metallic mercury, releasing a corresponding amount of EDTA, which is then available for the titrations. Base metals such as magnesium and calcium can be determined in this way.

4.6.2.5 Argentometric titrations

A precipitation titration of halides using coulometrically generated silver ions is called argentometric titration. This plays an important role in the summary determination of organohalogen compounds in water and provides a meaningful sum parameter which, along with others, is used as a substance parameter in the quality control of water and wastewater in environmental analysis. The organohalogen compounds in question can be removed from the aqueous phase by blowing them out of a heated sample, by extraction or by adsorption. These three fractions are digested separately in an oxygen stream at approximately 950 °C in a combustion furnace, producing gaseous hydrogen halides (HX), which in turn are absorbed in an aqueous phase. After transferring to a closed titration cell, the halides are precipitated as silver salts with coulometrically generated Ag^+. As can be seen, three specific material parameters are determined in this way: **POX** (purgeable), **EOX** (extractable) and **AOX** (adsorbable), each as organically bound halogen content. As the substance class is questionable from a toxicological point of view, there are relevant DIN regulations (German standards) for determining such sum parameters.

The hydrogen halides are also absorbed in an acetic acid electrolyte in accordance with DIN, as the silver salts to be precipitated are less soluble in this medium than in water. The measuring cell used is a setup as shown schematically in Fig. 4.42. Instead of the Pt anode, a silver anode should of course be used, as the Ag^+ ions required for the titration are generated at this anode. The end point is best determined potentiometrically at a silver electrode (Section 4.4.6). Please note that a mercury(I) sulfate electrode, for example, should be used as a reference electrode because the analysis sample must not be contaminated by traces of chloride that could flow out of the diaphragm of a reference electrode. As the electric charge Q required to calculate the halide content is the product of the two very precisely measurable quantities of current (I) and time (t), the method described here is very sensitive and, above all, does not require any additional titer determination, as coulometry is an absolute method. The result is always given for $X = Cl$. If it is necessary to differentiate between Cl, Br and I, ion chromatic methods must be used for detection. The tolerable limit values of the halogen components are regulated by the wastewater legislation of the country concerned. In order to achieve a high sample throughput, special machines compatible with the DIN regulations are available.

4.6.2.6 Redox titrations

Many redox titrations can be carried out with coulometric reagent generation. This applies in particular to redox-sensitive reagents such as Ti(III) ions, as this eliminates the need to regularly check the titer. Whenever reversible redox pairs are involved in a reaction, the dead-stop method can be used to indicate the equivalence point. Recently, this method has also been used with great success in organic solvents.

4.6.2.7 Coulometric water determination according to Karl Fischer

The iodine required for the titration is generated coulometrically at the anode. The electric charge required to generate it is directly proportional to the amount of iodine formed in situ. At a current of 100 mA, exactly 1.315 mg of iodine is produced in 10 s, as long as the interfering side reactions can be avoided in the titration cell that influence the current yield and stoichiometry of the Karl Fischer reaction. One molecule of formed I_2 then consumes exactly one molecule of H_2O. A solution consisting of anhydrous methanol, sulfur dioxide, an organic base, usually imidazole, and iodide is filled into the titration cell. The greatest risk of contamination is the absorption of water from moisture in the air. The titration cell must therefore be absolutely airtight, so all outlets must be carefully protected with drying tubes. The method is much more sensitive than volumetric water determination and is therefore particularly suitable for determining small quantities of water, which should be in the range of 10 μg to a maximum of 100 mg (the European Pharmacopoeia (Ph. Eur.): microdetermination of water). Since coulometry is an absolute method, no titer determinations are required; only the exact time measurement at a precisely known electrolysis current is sufficient for the accuracy of the analysis results. K. Schöffski [158h] states that more than half a million Karl Fischer titrations are carried out worldwide every day. The range of applications extends from petrochemical products to pharmaceuticals and food industry products; the water content of crude oil is just as economically relevant as that of gummy bears in food quality assurance.

The titration cell again corresponds to the model shown in Fig. 4.42. The required I_2 is generated at the Pt anode, while gaseous hydrogen is formed at the cathode. As long as the solution to be titrated does not contain any other substances that can be cathodically reduced under the given conditions, a diaphragm can be dispensed with. Nevertheless, the cathode should remain separated from the rest of the solution by a glass tube to allow the H_2 to escape unhindered. However, the diaphragm is always required as soon as other reducible substances are present in the solution. In such a case, the cathode chamber must be filled with a suitable catholyte. Care must be taken to ensure that its filling level matches that in the cell in order to avoid cross-contamination. Catholytes usually contain alkylenediamines and alicyclic or aromatic amines and are commercially available as ready-made solutions.

The end point is best indicated using the dead-stop method. For this purpose, as described above, another electrode is immersed in the titration solution, which has two platinum tips between which a constant voltage of 10–30 mV is applied. Under these conditions, a current can only flow when a reversible redox couple enters the solution, i.e. in the given case when, after the equivalence point is exceeded, the no longer consumed I_2 forms such a reversible redox couple with the I^- (Section 4.5.4) and cancels the polarization of the electrodes. The currents to be measured are in the μA range and therefore do not cause any further material conversion. Follow the instructions for water determination (p. 310 ff.). As already mentioned, the working range for water determinations is between 10 μg and 100 mg H_2O per analysis sample.

As a rule, commercial devices are used to carry out coulometric titrations, and the practical use of which is described in detail in the operating instructions of the various manufacturers. To get an idea of the many possible applications of the Karl Fischer method, we recommend reading a monograph by Metrohm [150a], for example. Here you will also find detailed information on how to avoid interferences that occur with certain substance classes when determining their water content. There are binding regulations for the analysis of pharmaceutical products, which can be found summarized in Ph. Eur. 10.0 [8a]. Bear in mind that considerable effort must always be made to keep humidity from the laboratory away from the analysis samples.

The high sample throughput required in modern testing laboratories has led to the production of automated analyzers with integrated data processing. Such devices offer the possibility of optimally adapting to the requirements of the analysis. For example, it should be mentioned that anodic I_2 generation with pulsed currents. is realized. This means that the product of current and pulse length can be compared with a "drop" of a set I_2 solution. This allows slower titration near the equivalence point, as recommended in various work instructions.

5 Instrumental volumetric analysis

5.1 Apparative development

The development of volumetric analysis progressed from the visual determination of the titration endpoint using color indicators to the use of physical indication methods with initial step-by-step recording and then continuous recording of the measured values to electronically controlled titration devices. The objective was characterized by the pursuit of higher precision, greater reliability, an extended range of applications and extensive automation.

Initially, this path led via devices that could be attached to conventional burettes and eliminated the need for manual operation of the burette tap. They are to be understood as titration aids. Two examples may serve to illustrate this. With the **Tri-Stop device** (Gebrüder Klees, Düsseldorf), the burette outlet was connected by a silicone tube to a solenoid valve, which could be fitted with glass tips of varying fineness to produce different drop sizes. The solenoid valve was operated by a controller controlled by three buttons. Pressing one button started the titration. A second button was used to switch from a continuous flow of the standard solution to dropwise addition, the speed of which could be continuously adjusted using a rotary knob. Pressing the third button stopped the titration. Another option was to use a universal controller in conjunction with a pointer device. In addition to a solenoid valve for regulating the addition of the standard solution, this required a control unit with a transistor photo amplifier and a stick-on microphoto barrier. The latter was attached to the glass pane of the instrument scale (e.g. a millivoltmeter) at a preselected position. When the pointer of the measuring device passed under the barrier, the solenoid valve was switched (universal controller **auto-tit**, Gebrüder Klees, Düsseldorf).

However, the development toward the largely automated titrators outlined at the beginning was only made possible by the introduction of the piston burette in combination with improved sensors to indicate the end point and the use of electronic components for control and evaluation. As explained on p. 22, the piston burette, which operates according to the displacement principle, allows a more precise volume measurement than the conventional outlet burette due to its higher resolution and less falsification by systematic deviations due to the elimination of the lag and parallax error. After equipping the piston burette, which was initially intended for manual operation, with a motor drive digital volume display and controllable valve switching for filling and dosing, a precise dosing device was created for use in modern automated titration systems (e.g. Mettler Toledo, Greifensee, Switzerland; Metrohm, Herisau, Switzerland; Dr. Bruno Lange Radiometer Analytik, Düsseldorf; Schott-Glass, Mainz). The piston movement, which is proportional to the ejected volume, can be converted into electrical signals by inductive pulse generation or by stepper motor drive [159]. Flexibility is provided by the range of interchangeable units with different cylinder sizes (volume: 1, 5, 10, 20 or 50 ml),

https://doi.org/10.1515/9783111350127-005

which can be easily exchanged. This means that the standard solution or cylinder volume can be changed quickly using just one drive and display unit. The connections of the interchangeable attachments fit onto standard glass or plastic bottles containing ready-to-use standard solutions.

The first titrators to appear on the market were devices with which potentiometrically indicated titrations could be recorded (**potentiographs**). Initially, the titration curves had to be evaluated manually. The second step was the development of semiautomatic devices with which it was possible to titrate to a preselected end point. This was later followed by digital potentiometric titration with a linked computer system.

In this context, it should be mentioned that the term **automatic or automated titration** has undergone a fundamental change in the course of development. Initially, a potentiograph with a limit switch was called an automatic titrator [160, 161], while later an automatic analyzer was understood to be a device system in which at least the sample transport and the dosing of reagents and standard solutions are carried out automatically [144]. The degree of automation can vary. In semiautomated systems, certain steps of the analysis process run automatically, while others must be performed manually, such as sample preparation and evaluation. With fully automated device, on the other hand, no manual intervention is required. With their help, samples can be taken and analyzed directly and automatically discontinuously or continuously (**online working method**). The necessary optimal adaptation of the equipment to the analytical problem requires a high level of effort and low flexibility. Devices of this type are used to monitor and control processes or operating streams (**process titrators**). They are often part of a closed control loop. More or less fully automated titrators are used to analyze samples taken discontinuously and manually in the laboratory (**offline working method**). They have a wide range of possible applications and can be used in conjunction with sample changers and can also be used for serial analyses. The sample changers work according to the conveyor belt or carousel principle and are used to mechanically feed samples from transport containers. The sample containers are called up one after the other by a control system and fed to the titration stand, which contains stirrers, electrodes, a rinsing system and devices for dosing the standard solution and, if necessary, auxiliary solutions. Further stations for sample preparation measures can be connected. The titration takes place in the transport container and the sample must be quantified before the automated titration process begins. In other embodiments, a defined volume of the sample is removed from the transport container and placed in a stationary titration vessel containing the aforementioned devices. The titration vessel is automatically emptied and rinsed after each titration. If the titrator is linked to a corresponding computer system, sample changers can be used to achieve an automated operating sequence – with the exception of sampling. With electronic digital scales, the sample weighing can also be integrated into the overall system.

It seems appropriate to categorize titration devices according to the type of volume dosing to be carried out. The addition of the standard solution can be carried out in a continuous or discontinuous manner. In the latter case, this is done in portions in the form of volume steps or volume increments. Both with the continuous as well as with the incremental method, the speed of dosing can be controlled and adapted to the course of the titration curve. A further subdivision option results from the fact that the volume steps for discontinuous addition of the standard solution can be selected to be the same size (**monotonic titration**) on the one hand, and variable on the other. It makes sense to choose larger volume steps at the beginning and reduce them later. The size of the increments can be controlled via a signal comparison with a preselected signal value for the titration end point or via the slope of the titration curve (**dynamic titration**). Figure 5.1 provides an overview of the possibilities (according to [193]).

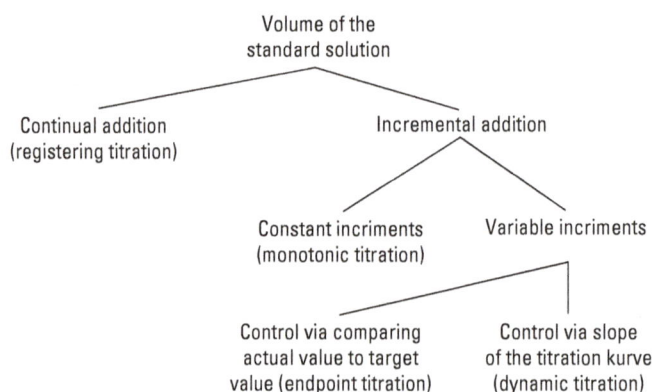

Fig. 5.1: Systematic classification of the types of instrument-controlled titrations.

5.2 Registering titrators

Recording the titration curve with an analog device (e.g. potentiograph) is achieved by continuously adding the standard solution from a motor-driven piston burette synchronously with the paper feed of the recorder, with which the analog measuring signals are recorded as a function of the reagent volume. The evaluation can be performed manually using a graphical method, e.g. the tangent method described on p. 300 ff, the circular method according to Tubbs [162] or the intersection method according to Ebel [163] can be used. An overview of the graphical evaluation of analog registered titration curves can be found in [141, 144] and in detail in [164]. There and in [165], the problem of the mismatch of equivalence and inflection points, especially with asymmetric titration curves, and the resulting deviations are also discussed.

Since the mid-1960s, the recording titrators have been equipped in such a way that the addition of the standard solution can be automatically adapted to the change in the potential difference of the electrode ΔE. The rate of addition is controlled depending on the slope of the curve. At the beginning of the titration, when the measurement signal changes only slightly, the dosing is carried out at a higher speed than in the area of the signal increase; as the equivalence point is approached, it is greatly reduced. In this way, distortion of the curve in the rise range can be avoided and the risk of over-titration can be minimized. It also saves time when carrying out the titration.

At around the same time, the devices were also equipped with a differentiation unit, which made it possible to record the first derivative of the titration curve. In the older devices, the measurement signal ΔE was derived according to time as $d\Delta E/dt = f(V)$. As a result, the signal height also decreased when the titration speed was reduced. This disadvantage was overcome as the development of the device progressed. Since the introduction of electronically controlled digital burettes, it has been possible to build potentiographs in which the measurement signal is differentiated according to the volume of the standard solution added. If the titration speed is now greatly reduced in the rise range, the curves show a high degree of selectivity. The advantage of this representation becomes apparent in difficult titration problems, e.g. when a mixture of similar components is to be titrated simultaneously (Fig. 5.2) or when the simultaneous determination of components that are present in very different concentrations is to be carried out (Fig. 5.3). In such cases, evaluation is sometimes only possible using the differentiated titration curve.

The graphical evaluation of the differentiated titration curves is less difficult, as the volume of the standard solution used can be read at the peak and the inflection point does not have to be determined first. Potentiographs can also be equipped with an interface (analog-to-digital converter) that enables online evaluation of normal and derived titration curves by a computer [193].

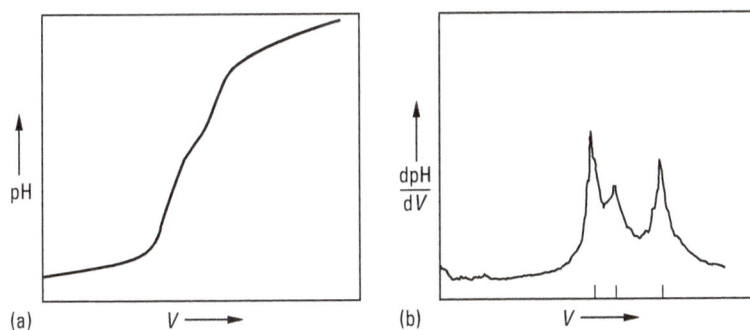

Fig. 5.2: Titration curves of an alkalimetric simultaneous determination: HCl 0.01 mol/l, acetic acid 0.033 mol/l and 4-nitrophenol 0.017 mol/l with NaOH 0.1 mol/l; potentiograph Metrohm E536, combined glass electrode; (a) normal and (b) differentiated titration curves (according to [166]).

In combination with appropriate peripheral devices, titration with a recording titrator can not only be indicated potentiometrically with glass and metal electrodes as well as with ion-sensitive electrodes, but can also be carried out using conductometric, ampero-

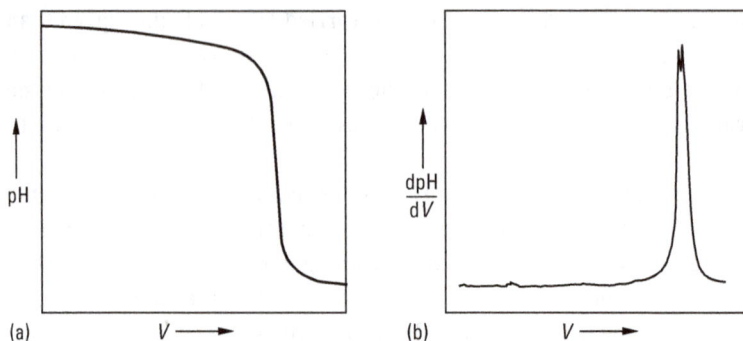

Fig. 5.3: Titration curves of a carbonated sodium hydroxide solution containing carbonate, 0.1 mol/l, with HCl, 0.1 mol/l; potentiograph Metrohm E536, combined glass electrode; (a) normal and (b) differentiated titration curves. Both curves with adjustment of the dosing rate to the potential difference (according to [166]).

metric, voltametric and photometric indication. Further expansion, e.g. with a sample changer, is possible. Digitally operating devices are dealt with in Section 5.4.

5.3 End point titrators

In end point titration, a specific potential difference of the electrode ΔE is specified as the potential value of the end point. This value is referred to as the target value It must be determined before the titration and must be sufficiently reproducible. During the titration, the device compares the actual potential value of the sample solution (actual value) with the preselected target value after each volume step and, depending on the size of the deviation, sends a more or less large dosing pulse to the motor of the burette, which drives the piston forward. When the target value is reached, the device switches off the burette control after a constant or variably selectable delay time, thereby ending the titration. The volume of the standard solution used is displayed by the dosing device and can also be printed out if a printer is connected. The titration curve is generally not recorded.

The titrator works according to the principle of a controller, it doses the standard solution discontinuously in variable volume steps. Feedback of the volume added by the dosing device to the titrator – as with the recording devices – is not required.

As the target value is known, it is not necessary to titrate slowly over the entire potential range, so that short titration times can be achieved. This is particularly desirable when large sample series are to be analyzed using sample changers. However, as the titration time is reduced, the risk of over-titration increases especially with very steep titration curves. It is therefore advisable to work with a double control dynamic and to adapt not only the size of the volume steps but also the titration speed to the course of the titration curve. This can be done by using the difference between the actual and target values to control the pulse duration and also the pulse frequency for the motor burette or by using the analog-formed first derivative of the titration curve to control it. Other influencing variables that can lead to deviations in end point titrations are discussed in [164, 166]. If the continuous addition of reagent is used for end point titration, the standard solution is added at high speed until a defined control band of the potential is reached. From this point onward, the volume increments are automatically and continuously reduced by the device so that the end point of the titration is reached as quickly as possible, but not exceeded. The titrators work according to the fuzzy logic control algorithm (DL 5x titrators, Mettler Toledo). The control band and the smallest increment must be specified.

Commercially, there are compact devices for end point titration, in which the measuring and control sections are housed in one device, for processing series of similar samples in routine analysis, as well as modular systems that can be assembled from a millivoltmeter, pulse controller, dosing device and printer, as well as a recorder and computer if required. This allows adaptation to different tasks in the laboratory. If millivoltmeters with a built-in polarization current source are used, a device for Karl Fischer titration (see p. 310 ff.) can be set up. However, special single-purpose devices are available for this purpose.

5.4 Digital titration systems

With the digital titrators, the measurement data (dosed volume of the standard solution and associated signal value) are recorded and printed out in a data sequence as exact numerical values in numerical form. The evaluation must be carried out by calculation. It can be carried out using a programmable pocket calculator, but automated titration systems will use a desktop calculator in online mode.

The most important components are the stepper motor-driven piston burette for precise dosing of the volume increments and the control unit for monitoring the system and controlling the time at which the measured value is taken. The latter is necessary because a constant measurement signal is only obtained with a certain time delay. The reasons for this are the time required for complete mixing, the kinetics of the underlying titration reaction (particularly in the case of precipitation and nonreversible redox reactions) and the finite response speed of the indicator electrode.

The first digital titration device was commercially available in 1968 (**Titroprint**, Metrohm, Herisau). It works as follows: At the beginning of the titration, a preselectable volume of the standard solution is added continuously in a rapid dosing step and then – after switching to stepwise addition – the range of interest around the equivalence point is run through with a constant, selectable increment size (0.1–0.4 ml). Each addition is followed by an adjustable constant waiting time (0–300 s) adapted to the titration problem, after which the measured data is transferred and printed out (**time-controlled transfer of measured values**). After adding a selectable stop volume, the titration is terminated.

The equilibrium titrator works according to the same principle of monotonic titration but with a different transfer of measured values (modular system DK/DV, Mettler Toledo, Greifensee). After each addition of volume, the device waits until the potential change of the electrode has fallen below a preselectable small threshold value in mV/min and then takes over the measured value (**drift-controlled or equilibrium-controlled transfer of measured value**).

The type of measured value transfer must be specified for each titration method in the titration systems currently available on the market. This means that the device knows when the potential must be measured and the next volume increment dosed. If time intervals are selected as the measurement type, the device waits for a specified time before adding the next increment. This type of measurement is used for unstable conditions (nonaqueous titrations). With equilibrium-controlled measured value transfer, the measured value is only transferred after each increment addition when the measurement signal has become stable, i.e. when the potential does not change by more than a specified amount ΔE within a specified time Δt. If no equilibrium is reached, a maximum waiting time must be specified. The additional specification of a minimum waiting time prevents the excessively rapid addition of further standard solution. The equilibrium-controlled transfer of measured values has proven to be the superior measuring method.

Numerous methods have been described for the mathematical evaluation of digital titrations. Compilations can be found in [141, 144, 164, 166]. The development of sample changers that can be coupled with digital titrators resulted in largely automated analyzers in the early 1970s.

The use of microprocessors has made it possible to carry out computer-controlled volume dosing of the standard solution. This has allowed the previously used principle of monotonic (incremental) titration with addition in constant volume steps could be further developed into a titration principle that works with variable volume increments. This procedure, known as dynamic titration [193] aims to dose the volume in such a way that constant potential steps are achieved with potentiometric indexing of the titration. For this purpose, the size of the next volume step must be calculated based on the slope of the titration curve, which requires the curve to be memorized. Figure 5.4 illustrates the difference between the two types of titration [193]. The curves are shown in the figure as they are documented by the recorder: The ordinate (direction of paper feed) is the volume axis, and the measured values are plotted on the abscissa.

In monotonic titration, there is an accumulation of measured values at the start and end ranges of the titration. In the equivalence range, which is important for evaluating the curve, there are only a few measured values. However, if the measured value recording is adapted to the potential change by computer-controlled modification of the volume steps, considerably more measured values are available for evaluation.

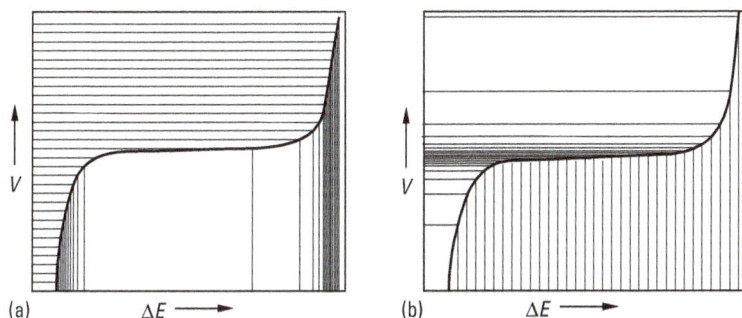

Fig. 5.4: Potentiometrically indexed titration curves: (a) monotonic titration with constant volume steps and (b) dynamic titration with constant potential steps (idealized representation).

The traditional procedure with incremental addition of the standard solution is used for titrations whose titration curves show steep jumps or where the measurement signal is not completely stable. Typical areas of application are redox and complexometric titrations as well as certain acid-base titrations in nonaqueous solvents.

Since titration systems with dynamic titrant addition have been available since the mid-1980s (1985 DL compact titrator, Mettler Toledo), this method has become increasingly popular. To prevent the analyzer from dosing too large increments, the maximum volume increment is limited. To avoid unnecessarily long titration times, the minimum volume increment is also limited. The potential difference that the titration system should reach after each increment addition is defined as the target value. These measures ensure fast and accurate titration results in most cases.

An example of a computer-controlled multistation titration system for simultaneous work is shown in Fig. 5.5.

The introduction of electronic semiconductor components at the microprocessor integration level in instrumental analysis also opened up new possibilities for automatic titrators. Examples include coordination of individual steps in the form of sequence control, monitoring of the titration process by means of control functions and evaluation of data via arithmetic operations, working with predefined program systems [194, 195]. Various manufacturers have developed high-performance automatic titrators and improved them over several generations of devices. The degree of automation has been extended, reliability increased, ease of use optimized and methodological innovations achieved with advanced software.

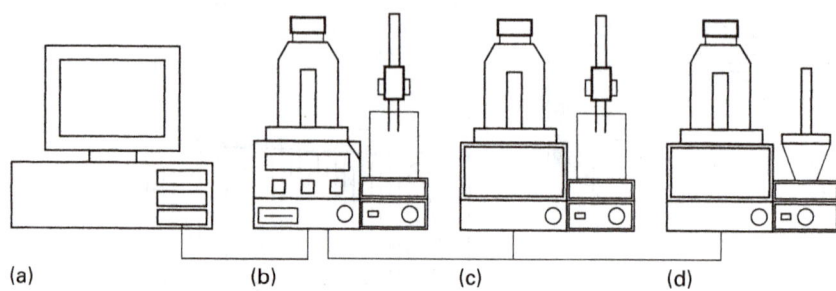

Fig. 5.5: Computer-controlled titration system with three workstations: (a) computer; (b) determination of free fatty acids, reagent NaOH solution; (c) determination of KOH content, reagent HCl solution; and (d) determination of water according to Karl Fischer, KF reagent.

6 Overview of the history of volumetric analysis

The beginnings of titrimetry can be traced back to the middle of the eighteenth century. They date back to the age in which increasing attention was paid to the quantitative relationships in chemical reactions. This was the epoch that followed the period of a more qualitative approach to chemical processes, as was prevalent among the followers of the phlogiston theory (around 1700 to 1780), and whose beginning was characterized by the name **Antoine Laurent Lavoisier** (1743–1794). The discovery of the principle of the conservation of matter, Dalton's atomic hypothesis and the discovery of stoichiometric laws paved the way for quantitative analytical chemistry. However, quantitative analyses had also been carried out long before this; we need only mention the cupellation, which dates back to the pre-Christian era, i.e. the metal and ore testing by dry means, as well as the investigations in the seventeenth century on quantitative examination by wet means [167].

Volumetric analysis developed in close connection with the industrialization of chemical production, which was particularly influenced by the emerging textile industry (bleaching and dyeing agents). Fast and reliable methods were needed to assess the quality of sulfuric acid, soda, hypochlorite and chlorinated lime. This necessity is the real reason for the invention of titrimetry [167]. At first, no absolute determinations were carried out; one was content with the result of whether a substance was good or not. It was judged to be good if a weighed portion of the substance to be analyzed, when added to a solution of known content that triggers a visible reaction, consumed a certain, empirically determined volume of the solution by the time this reaction was completed. Later, the evaluation was carried out by calculation. The first practitioners who developed such methods are not known.

The first documented titration method with an analytical objective was probably the determination of acidity in vinegar 168], which **Claude Joseph Geoffroy** (1683–1752) described in a paper presented to the French Academy of Sciences in 1729: ". . . I took 2 drachms from each variety, weighed them exactly and added small portions of finely powdered and dried potash to them in rounded flasks until the fermentation stopped. Now the acidity of the vinegar was consumed and the liquid became salty. I can tell the strength of the vinegar from how much potash was needed to swallow the acid." His "standard solution" was solid potash, and the indicator was the gas development. In 1747, **Louis Guillaume Le Monnier** (1717–1799) examined the carbonate content of mineral waters by evaporating the samples, adding sulfuric acid drop by drop to a certain proportion of the residue and counting the number of drops until the effervescence was complete. A color indicator was used for the first time by **Gabriel François Venel** (1723–1775) in 1750 when he analyzed the red coloration of violet extract when sulfuric acid was added. He concluded from his experiments that the mineral water contained no alkali and that the acid was only used to color the indicator red, as he had achieved the same result with snow water. In 1756, **Francis Home** (1720–1813) in

https://doi.org/10.1515/9783111350127-006

Edinburgh used volume measurement for the first time in the examination of potash when he added nitric acid by the teaspoon to the weighed sample and observed the foaming. His book *Experiments of Bleaching* from 1756 also describes the first precipitation titration: Home determined the hardness of the water by repeatedly adding a few drops of alkali metal carbonate in soft water to the sample of hard water, each time waiting for the milky turbidity to settle and the absence of turbidity indicating the end of the titration. In 1767, **William Lewis** in England was the first to determine the absolute content of potash by comparing the results for commercial products with those for pure potassium carbonate. He used hydrochloric acid as a standard solution, the consumption of which he determined by weighing, and blotting paper soaked in litmus solution. **Vittorio Amadeo Gioanetti** (1729–1815) described a water analysis in 1779 in which, after evaporation, the carbonate in the residue was determined using acetic acid, which he had previously adjusted against soda. For his salt-petre factory in Dijon in 1782, **Louis Bernard Guyton de Morveau** (1737–1816) developed a titrimetric method for determining the hydrochloric acid and nitric acid in the mother liquor. Alkali metal carbonate solution was added to it until paper strips with tincture of curcuma and pernambuco changed color; the consumption was determined by weighing. In a second sample, Guyton de Morveau determined the hydrochloric acid solely by precipitation titration with lead nitrate solution. He calculated the nitric acid content from the difference between the two weight titrations. In another work on the volumetric determination of carbon dioxide in water by adding lime water until the initial turbidity disappears, the first mention of the original form of a burette was described by Guyton de Morveau as a "gaso-mètre" a cylindrical glass tube covered with paper on which a graduation was drawn. In 1784, **Richard Kirwan** (1735–1812) used the first time a standard solution of potassium hexacyanoferrate(II) for the determination of iron. The Finnish chemist **Johan Gadolin** (1760–1852), who discovered the element yttrium, developed a similar method independently of Kirwan in 1788. The first redox titration was carried out by **François Antoine Henri Descroizilles** (1751–1825) by determining the relative strength of chlorine-containing bleaching solutions for the textile industry by adding portions of sulfuric acid indigo solution of known content from a pipette until the blue coloration no longer disappeared. The use of this method spread very quickly. It was described by various authors from 1789 onwards before Descroizilles himself published it in 1795. His essay also contains a description of the first titration devices, a burette – which Descroizilles called a "Berthollimètre," a simple graduated cylinder – and two pipettes which he used to determine chlorine. **George Fordyce** (1736–1802) in London was the first to use standard solutions of alkali metal hydroxides instead of carbonates for the titration of acids in 1792; he used violet extract as an indicator. **Joseph Black** (1728–1799) was the first to discover the interfering effect of dissolved carbon dioxide when determining the alkali content of water by titration with sulfuric acid and observed the consumption of the indicator litmus in the standard solution (indicator error); he also carried out the first back-titration (1803). Until the beginning of the nineteenth century, volumetric determinations were carried out almost exclusively in the form of weight titrations. Determining the used standard solution by volume only became widespread after Descroizilles presented his "alkalimeter" in a paper in 1806. which was used to assess potash by titration with sulfuric acid. The alkalimeter, a graduated glass tube that was emptied by pouring, is also a prototype of the burette. Descroizilles described a volumetric flask for the first time in 1809.

Charles Bartholdi (died 1849) in Colmar made further contributions to precipitation titration: in 1792 he published a method for determining the sulfate content of madder extract with lime solution, and in 1798 he reported on the titration of sulfate with barium acetate solution and of chloride with silver nitrate solution, in each case until precipitate formation was absent in the water analysis after dissolution of the evaporation residue. This is the first time that an argentometric titration was used.

But it was **Joseph Louis Gay-Lussac** (1778–1850), whose name is known for his discovery of the laws governing the influence of temperature on gas volumes and the volume ratios of gases in chemical reactions, systematically developed the existing practical approaches to volumetric analysis into a usable scientific method. Additionally, he created new titrimetric procedures in order to avoid tedious gravimetric determinations and save time in analytical work. He is therefore often regarded as the actual founder of volumetric analysis. In 1824, he published a paper on the determination of chlorinated lime of Descroizilles with indigo solution, the concentration of which he adjusted with a solution prepared by introducing 1 liter of chlorine into milk of lime; the chlorine was released from hydrochloric acid by a calculated portion of manganese dioxide. The devices used for the titration and their handling are described in detail. The names pipette (petite measure) and burette appear for the first time. In his work "Essai des potasses du commerce," which appeared in 1828, he deals with acidimetry and alkalimetry and uses the French word "titre" both in the sense of content and thus quality designation for the potash product under investigation (titre ponderal) – as **Claude Louis Berthollet** (1748–1822) already used it in a book in 1804 – as well as in the sense of the consumption of a solution of a certain concentration to determine the content (titre alcalimetrique). Around the middle of the nineteenth century, it then gave the name titrimetric method to the determination methods based on the measurement of the standard solution consumed. Gay-Lussac adjusted his sulfuric acid solution diluting concentrated sulfuric acid so that 50 ml of it just neutralized the solution of a weighed sample of pure potash; he used litmus tincture as an indicator. If he used the same weight for a commercial product, he could immediately indicate the percentage of potash by mass when reading the used solution from his 50 ml burette. In 1832, Gay-Lussac worked out his now-famous precipitation titration for the determination of silver with sodium chloride solution. One year later, Liebig obtained a German translation of the publication [99]. Gay-Lussac's method replaced the centuries-old cupellation method which until then had been used for the quantitative analysis of coin metals for their silver content by dry methods. The saline solutions were prepared in such a way that 100 g or 100 ml indicated just 1 g of silver. Gay-Lussac never filled the pipettes by mouth aspiration but constructed a complicated filling device for this purpose. In 1835,

Gay-Lussac introduced arsenic acid as a standard solution for the titration of hypochlorite. He recognized the end point by the decolorization of the indigo solution, of which he added only a few drops. He thus used the dye as a redox indicator for the first time but not as a standard solution like Descroizilles. Potassium hexacyanoferrate(II) and mercury(I) nitrate solutions were also used by him for the titration of hypochlorite.

Despite Gay-Lussac's carefully elaborated and scientifically sound titration methods, which delivered good results, it took another quarter of a century for volumetric analysis to gain general acceptance and assume a position equal to gravimetry in quantitative analysis. The generalization of titrimetry was hampered by the use of empirical standard solutions whose concentrations were determined by practical considerations and were not chemically based; each method also required different measuring instruments. In the following two decades, numerous new methods of volumetric analysis were developed, but only some of them became established in practice. Only the most important discoveries are listed below.

At the same time, iodometry and permanganometry appeared in research. **Alphonse Du Pasquier** (1793–1848) used an alcoholic iodine solution for the first time in 1840 for the titration of hydrogen sulfide with starch as an indicator. This was followed in 1843 by the determination of sulfurous acid and thiosulfate by **Mathurin Joseph Fordos** (1816–1878) and **Amadée Gélis** (1815–1882) and in 1845 the determination of iron by **Adolphe Duflos** (1802–1889), who titrated the iodine, precipitated during the reaction of iron(III) with potassium iodide, with tin(II) chloride solution, which he had prepared on an equivalent basis. In 1846, **François Gaultier de Claubry** (1792–1878) determined tin iodometrically. **Robert Wilhelm Bunsen** (1811–1899) generalized the method in 1853 by reducing numerous oxidizing substances with hydrochloric acid, introducing the resulting chlorine into potassium iodide solution and titrating the released iodine with sulfurous acid as a standard solution.

Frédéric Margueritte introduced the potassium permanganate solution in 1846 (chameleon solution) for the determination of iron and thus founded permanganometry. **Antoine Brutus Bussy** (1794–1883) used it in 1847 for the titration of arsenious acid, **Théophile Jules Pelouze** (1807–1864) in the same year for the determination of nitrate by back titration of iron(II). It was used in the same way in 1849 by **Karl-Heinrich Schwarz** (1823–1890) for the determination of chromate. **Hempel** titrated oxalic acid for the first time in 1853 permanganometrically.

This period also saw the development of other methods of volumetric analysis: **Berthet** 1846 used a potassium iodate solution for the titration of iodide. In the same year, **Étienne Ossian Henry** (1798–1873) published a paper in which he described the precipitation titrimetric determination of potassium with sodium perchlorate solution in alcohol and described a new type of burette, a glass tube with a copper stopcock at the lower end, which, however, did not prevail over Gay-Lussac's "pouring glass." In 1847, **Thomas Clark** (1801–1867) published the determination of water hardness with a soap standard solution and **Eugène Melchior Péligot** (1811–1890) published the determination of ammonia after absorption in acid solution and back titration of the acid with calcium hydroxide solution. **Frederick Penny** (1816–1869) and **Jacob Schabus** (1825–1867) first used a potassium dichromate solution to titrate iron in 1850 and 1851, respectively. They applied it in this manner, whereby the end point was determined by spotting with potassium hexacyanoferrate(II). The oldest complexation titration goes back to **Justus von Liebig** (1803–1873), who in 1851 titrated alkaline cyanide solution with silver nitrate solution until the resulting precipitate no longer dissolved when shaken [100]. The chloride determination reported by Liebig in 1853 is based on the formation of slightly dissociated mercury(II) chloride with a standard solution of mercury(II) nitrate in which he added urea and indicated the end point by the opalescence that occurred.

Although there were numerous publications on titrimetric methods for the most important sub-areas of volumetric analysis (oxidation and reduction titrations, neutralization and precipitation titrations) around the middle of the nineteenth century, the method only really became widespread after the publication of the first summarizing books on titrimetry. In 1850, Schwarz published *Praktische Anleitung zu Maaßanalysen (Titrir-Methode)* (*Practical instructions for volumetric analyses – titration methods*, **second edition 1853**), in which the word "Maaßanalyse (*sic*)" (volumetric analysis) was used for the first time, and in addition to empirical solutions, solutions on an equivalent basis were also mentioned and described by Schwarz as "rational solutions". In his book, Schwarz also replaced the standard solution of sulfurous acid specified by Bunsen for iodine titration, which was unstable and difficult to prepare, with sodium thiosulfate solution. In 1855, **Friedrich Mohr** (1806–1879) wrote *Lehrbuch der chemisch-analytischen Titrirmethode* (*Textbook of the chemical-analytical titration method*) [102], which went through many editions (the last in 1914, edited by **Alexander Classen**). In his two-volume work, Mohr systematically compiled the previously known methods, which he had largely improved on the basis of his own experiments, and added many new ones. He rejected the standard solutions of arbitrary content that were still in common use at the time and strongly advocated the use of the solutions commonly used today, which contained "the small atomic weight expressed in grams or one tenth of an atom of active substance" per liter [102], solutions that were later referred to as standard solutions. Mohr thus made a significant contribution to the simplification and clarity of volumetric analysis. The idea for these solutions probably originated in England, as the measurement system used there did not allow the mass fraction of the substance in question to be expressed as a percentage from the weight and volume of the standard solution used, either directly or by multiplying it by a power of 10 [167]. Mohr stated, without giving further details, that a man named **Griffin** had already used such solutions before him. The idea was first documented in 1844 by **Andrew Ure** (1778–1857); however, the use of these solutions only became common practice after the publication of Mohr's book [167]. His classic work, which he constantly expanded and revised, became the model for many later textbooks on volumetric analysis. Mohr was a very critical and controversial man, rich in knowledge and ideas. A number of terms are still associated with his name today: Mohr's salt, Mohr's chloride determination, Mohr's stopcock burette, Mohr's balance. The cork borer and the graduated pipette are his inventions, and he improved the cooler later named after Liebig. In 1837, he had already written about the principle of the conservation of energy – probably without fully realizing its significance – which was then formulated in 1842 by **Robert Julius Mayer** (1814–1878).

Over the next few decades, many researchers continued to work in the field of volumetric analysis, improving the known methods and discovering numerous valuable innovations. Details of the development must be omitted here; only a few examples are mentioned: In 1858/59, **L. Péan de Saint Gilles** (1832–1863) described the permanganometric titration of iodide, nitrite and organic compounds. **Friedrich Christian Kessler** (1824–1896) found in 1863 that the titration of iron(II) with permanganate – that works interference-free in a sulfuric acid solution – only in the presence of manganese(II) salts gives correct results in a hydrochloric acid solution. In 1855, Kessler had already suggested using tin(II) chloride instead of zinc for the determination of iron(II) with dichromate for the reduction of iron(III) and removing the excess tin(II) with mercury(II) chloride, but this was forgotten. In 1881, **Julius Clemens Zimmermann** again observed the influence of manganese(II) and in 1889 **C. Reinhardt** gave the process its final form by combining it with Kessler's suggestions. The reduction of iron was adopted in 1889 by **Harry Clair Jones** (1856–1916), who carried out the reduction of iron on a zinc dust column. **P. W. Shimer** improved the reducer in 1899 by amalgamation. The titration of manganese(II) in hot solution with permanganate, now named after Volhard and Wolff, was mentioned in 1862 by **A. Guyard** for the first time. In 1879, **Jacob Volhard** (1834–1910) recognized that divalent metal ions must be present, and **N. Wolff** finally found in 1884 that correct results could first be achieved with the addition of a zinc oxide slurry. Volhard also extended the range of titrimetric precipitation analyses in the 1870s – by introducing the potassium thiocyanate standard solution and ammonium iron(III) sulfate as an indicator – for chloride and silver determination and for the titration of mercury. **Lajos Winkler** (1863–1939) determined the oxygen dissolved in water in 1888 via the oxidation of manganese(II) in an alkaline solution by iodometric titration. **R. T. Thomson** found in 1893 that boric acid can be titrated as a weak acid in the presence of glycerol. As early as 1861, **Th. Lange** used a cerium(IV) sulfate solution for the first time for the determination of iron(II), which, despite its advantages over the potassium permanganate solution, was not yet able to establish itself due to the lack of suitable redox indicators, a shortcoming that also hindered its further development in other areas of volumetric analysis.

For a long time, only extracts of plant dyes were available as indicators (detailed description in [167]). Toward the end of the nineteenth century, the first synthetic dyes appeared, which were suitable as acid–base indicators: phenolphthalein was recommended in 1877 by **E. Luck**, Tropäolin 00 in 1878 by **M. Müller** and methyl orange by **Georg Lunge**. The range of these indicators expanded in the following years, and in 1893 **H. Tromssdorff** already named 14 synthetic acid-base indicators in his book on chemical-technical examination methods. They were superior to natural dyes because their color change was more clearly visible and because it was possible to distinguish whether it occurred in the acidic or alkaline range. In 1908, methyl red by **E. Rupp** and **R. Loose** was added, and in 1915 **William Mansfield Clark** and **Herbert August Lubs** introduced the sulfophthaleines, which became very important for acid-base titrations due to their diverse color changes.

The first theory about the color change was put forward in 1894 by **Wilhelm Ostwald** (1853–1932) [169]. According to this theory, indicators are weak acids or bases which, as undissociated compounds, have a different color than their ions. The application of the law of mass action to the dissociation equilibrium (protolysis equilibrium) allowed a simple mathematical treatment of the indicator turnover. The fact that this occurs for the individual indicators at different pH values could be explained by the different values of the acid or base constants of the indicators. However, various objections were soon raised against the explanation of the color change solely from the dissociation process. For example, the change in various indicators occurs too slowly for an ionic reaction. In addition, Ostwald's theory did not correspond to the empirical fact that constitutively invariable inorganic and organic acids, bases and salts exhibit the same color in the dissociated and undissociated state. The ion theory was therefore subsequently expanded and deepened. In 1906/1908, **Arthur Rudolf Hantzsch** (1857–1935) provided a significant addition, which was based on the chromophore theory introduced in 1876 by **Otto Nikolaus Witt** (1853–1915), the first attempt to find a connection between chemical constitution and color [170]. Hantzsch explained the color change of the indicators with a structural change of the indicator molecule (pseudoacid or base) during the transition to an ionogenic form capable of dissociation ("aciform" or "basoform", respectively), whereby a quinoid ring is formed. The quinoid group was already recognized in 1882 by **H. E. Armstrong** as an important chromophore. Ostwald's basic idea that the indicators should be regarded as weak acids or bases was retained, but the cause of the color change was seen in a constitutional change in the indicator molecule that runs parallel to the dissociation. Since the intramolecular rearrangement of the pseudoform into the ionogenic form can occur slowly as a molecular reaction, the observation that some indicator changes are time reactions provided an explanation. In the following decades, numerous researchers contributed to the further development of chemical color theory, so that the narrowly defined concept of chromophore could no longer be adhered to. The advances in knowledge about chemical bonding, the concept of mesomerism and the excitability of π electrons in double bonds during light absorption led to new views, the theoretical justification of which is given with the help of quantum theory and wave mechanics (see p. 93 f.).

Around four decades after the introduction of synthetic acid-base indicators, the diphenylamine as the first usable redox indicator was discovered by **Josef Knop** and used for the dichromatometric determination of iron and in 1923/1924 also for other redox titrations. Already in 1835, Gay-Lussac titrated hypochlorite with arsenious acid acid and recognized the end point in the decolorization of indigo, but the method lacked the necessary precision; indigo is not a reversible indicator. As no better indicators were available, the end point was subsequently recognized by spotting. In 1840, **Walter Crum** (1796–1867) was the first to use potassium hexacyanoferrate(III) as a spot indicator for the determination of hypochlorite with iron(II) sulfate solution. For the titration of hypochlorite with arsenous acid (in 1852), **Penot** applied iodine-starch solution as spot indicator. Attempts to introduce other substances as redox indicators were unsuccessful in the following decades. After diphenylamine, Knop and **O. Kubelková-Knopová** (in 1929 and later) (see also p. 178 f.) and other researchers proposed numerous other redox indicators. The term "redox indicator" comes from **Leonor Michaelis** (1875–1940), who also wrote the first monograph on redox titrations [171]. Fluorescein was already known in 1876 by **F. Krüger** as a fluorescence indicator for acid-base titrations. However, this type of indicator only came into use after 1910 when **H. Lehmann** recommended observation in UV light. Fluorescein was also the first substance that **Kasimir Fajans** and **Odd Hassel** introduced in 1923 as an adsorption indicator for argentometric chloride determination (see p. 145 f.).

Of the numerous papers published in the first half of the twentieth century in the field of volumetric analysis and which advanced its development, two more are worth mentioning because of their particular importance: the iodometric method for water determination which was proposed by **Karl Fischer** in 1935 [172], and the introduction of aminopolycarboxylic acids for complexation titrations by **Gerold Schwarzenbach** since 1945 [128] (see p. 222 ff.).

The development of physical chemistry and its analytical methods has had a considerable influence on volumetric analysis, with electrical methods, in particular, being used to indicate the titration end point. Potentiometry, for example, originated from work carried out at the turn of the twentieth century. **Robert Behrend** (1856–1926) carried out the first potentiometric titration in 1893 and recorded titration curves for the precipitation of mercury(I) and silver halides [173]. **Wilhelm Böttger** (1871–1949) first described the potentiometric titration of acids and bases using a hydrogen electrode in 1897 [174] and in 1900 **F. Crotogino** used a platinum electrode for the potentiometric indication of a redox titration during the oxidation of halide ions with permanganate [175]. **Paul Dutoit** (1873–1944), together with **G. Weisse** proposed the use of a polarized platinum electrode in 1911. After the First World War, the development of potentiometric titration continued apace. Many renowned chemists were involved in its development and introduced numerous innovations. In 1909, for example, **Fritz Haber** (1868–1934) and **Zygmunt Klemensiewicz** (1886–1963) introduced the glass electrode as an indicator electrode [176]. The first monograph on potentiometric titrations was written in 1923 by **Erich Müller** (1870–1948) [177], who made great efforts to work out the fundamentals and further develop the electrometric methods and convincingly demonstrated their versatile application possibilities.

Conductometry dates back to the work of **Friedrich Kohlrausch** (1840–1910) and numerous other researchers. The method was first used for volumetric analysis in 1903 by **Friedrich Wilhelm Küster** (1861–1917) and **M. Grüters** [178] and in 1905–1909 by **A. Thiel** and in 1910 by **P. Dutoit**. In analytical practice, conductivity titration initially received little attention because the measurement method was too inconvenient and insensitive. First with improved measurement technology, the conductometric titrations became established and developed into a method that could be used successfully for analytical problem solving. The first monograph was written by **Isaac Mauritz Kolthoff** [179]. The development was significantly influenced by the work of **Gerhart Jander** (1892–1961) [180], **H. T. S. Britton** [173] and **G. Jones** promoted. The use of high-frequency alternating currents to measure the change in conductivity during a titration was first described in the mid-1940s by **G. G. Blake** [182], **F. W. Jensen** and **A. L. Parrack** [183] as well as **J. Foreman** and **D. J. Crips** [184]. High-frequency titration was particularly promoted in Germany by the work of **Kurt Cruse** [147].

Voltametry, the measurement of the potential difference between two metal electrodes polarized with the aid of an external current, is a titrimetric indication method and goes back to investigations by **H. H. Willard** and **F. Fenwick** in 1922, who worked with two platinum electrodes, where one of which was polarized [185]. The measurement of the polarographic diffusion limit current in the course of a titration, the amperometry was first recommended in 1927 by **Jaroslav Heyrovsky** [186]. This indication method was later developed by **Vladimir Majer** who called it polarometry [187]. The use of two electrodes that are polarized by an externally applied potential difference is older than this method with one polarized electrode. As early as 1897, **E. Salomon** described this technique for the titration of chloride with silver nitrate [188]. In 1905, **Walter Nernst** (1864–1941) and **E. S. Merriam** used two polarized hydrogen electrodes to indicate acid-base titrations [189]. The method was then forgotten. It was not used again until 1926 by **C. W. Foulk** and **A. T. Bawden** who used it for the titration of iodine with thiosulfate. Because of the sudden drop in current at the equivalence point, the authors gave it the name dead-stop titration [190]. However, the method only became widespread two decades later with the indication of water determination by Karl Fischer titration for which it was first used in 1943 by **G. Wernimont** and **F. J. Hopkinson** in 1943 [191]. Further methodological advances include titrations in non-aqueous systems [192] and phase titrations as well as titrations with photochemically generated reagents. Two-phase titrations (distribution and extractive titrations) are used, for example, to determine cationic detergents as well as amines and quaternary ammonium compounds in pharmaceutical analysis.

Throughout the entire period of methodological development, the equipment used for volumetric analysis remained essentially the same. In 1936, **Schellbach** proposed a stripe named after him for better readability of burettes. However, the development made great progress after 1955, when the piston burette designed by **K. Schlotterbeck** and **J. Städtler** came onto the market. This was followed by the motorized burette and, in 1958, the potentiograph with a recording automatic titrator and the end point titrator. The use of now-improved indicator electrodes for electrometric end point determination in combination with highly developed electronics led to instrumental volumetric analysis using microprocessor-controlled titrators in the 1960s [192].

In industrial routine laboratories today, volumetric analysis is often used for quality control of inorganic and organic substances. In recent years, more and more areas of application for volumetric analysis have been developed in organic, biochemical and pharmaceutical analysis. Advances in automation and measurement technology have also led to increased use of microanalysis. Today, the focus in the practical application of volumetric methods is on physicochemical indications and automation. The large number of papers indexed in reference journals (*Chemical Abstracts, Analytical Abstracts*) demonstrates the importance of volumetric analysis today [16].

7 Selected didactic approaches

Tell me and I will forget it. Show me and I might keep it. Let me do it, and I will be able to do it.
(Confucius, * 551 BC, † 479 BC, Chinese philosopher)

Performing chemical analyses in the laboratory is an essential component of scientific education. In the beginning, the participants learn how to handle the materials and equipment as well as how to carry out and evaluate the experiments according to a given set of instructions. With sufficient expertise, the participants can be confronted with problem-oriented tasks that help them to understand and deepen the previously learned theory as well as to gain new insights.

Below are some suggestions for methodological and didactic applications of the textbook content in practical courses.

A. Selection and use of measuring devices in the preparation of solutions

Didactic objective: To learn the correct choice of measuring instruments and how to use them correctly when preparing solutions

Task: Describe the production of the solutions below! Indicate which glassware you will use for this and why!

The following solutions are to be prepared:
a) 1 l of a 0.1 mol/l EDTA standard solution from EDTA disodium salt (see Section 3.4.4);
b) 100 ml of a 0.01 mol/l EDTA standard solution from a 0.1 mol/l EDTA standard solution:

Determine the expanded measurement uncertainty ($P = 95\%$) of the EDTA solution produced if the measuring instruments used belong to accuracy class A or accuracy class B. If the uncertainty of the 0.1 mol/l EDTA solution is unknown, assume that its relative value is 0.05%. Compare the results and discuss them.

c) 100 ml of an approximately 2 mol/l sulfuric acid solution from approximately 96 wt% (~18 mol/l) sulfuric acid solution.

B. Influence of some interfering effects in titrations
a) Interference of a permanganometric titration by chloride ions or
b) Influence of carbon dioxide on a potentiometric acid-base titration of phosphoric acid

Didactic objective: Recognize the faults of a determination method and minimize these faults

https://doi.org/10.1515/9783111350127-007

Task definition:

a) Prepare a 0.25 mol/l iron(II) sulfate solution. Then prepare Erlenmeyer flasks with the following solutions:

   i)   10.0 ml of the prepared iron(II) sulfate solution and 10 ml of a 1:4 diluted sulfuric acid (~4.5 mol/l) make up to approximately 200 ml with demineralized water;

   ii) 10.0 ml of the prepared iron(II) sulfate solution, 20 ml of a 0.1 M hydrochloric acid and 10 ml of a 1:4 diluted sulfuric acid (~4.5 mol/l) make up to approximately 200 ml with demineralized water;

   iii) 20 ml of a 0.1 M hydrochloric acid and 10 ml of a 1:4 diluted sulfuric acid (~4.5 mol/l) make up to approximately 200 ml with demineralized water;

   iv) 10.0 ml of the prepared iron(II) sulfate solution, 20 ml of a 0.1 M hydrochloric acid and 10 ml of the Reinhardt-Zimmermann solution make up to approximately 200 ml with demineralized water.

   The contents of the Erlenmeyer flask are heated to approximately 80 °C and titrated with a KMnO$_4$ solution, $c(\frac{1}{5}$ KMnO$_4) = 0.1$ mol/l until a weak pink coloration is obtained (see Section 3.3.2).

   Discuss the influence of chloride ions on the permanganometric titration based on the results.

b) Carry out the titration of phosphoric acid (see Section 4.4.6) with a heated and an unheated Coca-Cola® sample solution. Compare the results of the two titrations and discuss them.

C. Influence of the pH value on a complexometric determination

Didactic objective: Finding optimal conditions for a determination method

Task: Dilute 20.0 ml of a 0.01 mol/l magnesium chloride solution in each of four Erlenmeyer flasks to approximately 100 ml with demineralized water and then bring to the following pH values:
i) pH ≈ 2.5, ii) pH ≈ 7, iii) pH ≈ 10 and iv) pH ≈ 14.

Add an indicator suitable for this pH value to the respective Erlenmeyer flask and titrate the solutions with a 0.01 mol/l EDTA solution until the corresponding color change occurs. Compare and discuss your results.

D. Determination of an analyte using two independent titration methods

Didactic objective: Learning the correct use of measurement methods and critical assessment of the measurement results.

Task: Prepare a solution containing iron(III) with $c(Fe^{3+}) = 0.1$ mol/l. Examine this solution using two independent methods – for example permanganometrically (after reduction to Fe^{2+}) and complexometrically! Prepare 20.0 ml of the prepared iron(III) solution for the respective titration and titrate it accordingly (see Section 3.3.2 or 3.4.4).

Carry out the respective determinations at least three times! Calculate the corresponding uncertainties! Discuss the advantages and disadvantages of the methods when comparing the results.

E. Application of the method to a real sample
Didactic objective: Development of a quantitative analysis of real samples based on the specified measurement methods and critical evaluation of the result.

Task: An effervescent tablet containing magnesium or calcium with a known content (package information) is to be dissolved and titrated, e.g. complexometrically (see Section 3.4.4).

To do this, first calculate the mass of the effervescent tablet that is to be placed in the solution to enable determination without further dilution or enrichment steps. The following factors should be taken into account:
–   concentration of the standard solution and
–   available devices.
–   required amount of sample solution for a statistically sufficient number of titrations.
–   solubility of the effervescent tablet (usually up to approximately 4 g in 1 l of water with approximately 25 ml 69 wt% HNO_3).

Compare your measured values with the package information and discuss them.

Note: Such a task is very complex and requires appropriate guidance depending on the knowledge of the participants.

Appendix

Content data for common laboratory solutions.

Solution	Concentration c in mol/l	Mass fraction w in %	Density ρ in g/ml
Hydrochloric acid fuming	12	37	1.19
Concentrated hydrochloric acid	10.2	32	1.16
Concentrated hydrochloric acid	7.7	25	1.12
Diluted hydrochloric acid	2	7	1.033
Nitric acid fuming	24	100	1.51
Concentrated nitric acid	14.4	65	1.40
Diluted nitric acid	2	12	1.065
Concentrated sulfuric acid	18	96	1.84
Diluted sulfuric acid	1	9.3	1.061
Concentrated perchloric acid	11.6	70	1.67
Diluted perchloric acid	2	18	1.114
Glacial acetic acid	17.6	100	1.057
Diluted acetic acid	2	12	1.015
Concentrated hydrofluoric acid	22	40	1.126
Diluted hydrofluoric acid	2	4	1.012
Concentrated phosphoric acid	14.8	85	1.71
Diluted phosphoric acid	1	9.3	1.050
Concentrated sodium hydroxide	14.3	40	1.43
Diluted sodium hydroxide	2	7.4	1.081
Concentrated ammonia	16.5	32	0.88
Concentrated ammonia	13.4	25	0.91
Diluted ammonia	2	3.5	0.983
Sodium carbonate	1	9.7	1.099
Hydrogen peroxide	9.8	30	1.112
Hydrogen peroxide	1	3	1.010

https://doi.org/10.1515/9783111350127-008

Chemical elements

In the table, the proton number (atomic number) and the relative atomic mass A_r are listed for each element. The numerical values correspond to the data of the International Union of Pure and Applied Chemistry (IUPAC) as of 2005 [196]. The relative atomic mass of an element is related to $\frac{1}{12}$ of the atomic mass of the carbon nuclide ^{12}C; the term is applied to the natural isotopic composition of nuclide mixtures (concerning mixed elements). As a ratio of two quantities of the same type, the relative atomic mass has the unit 1. The numerical values of A_r are identical to the numerical values for the atomic masses measured in the atomic mass unit u, the 12th part of the mass of an atom of the nuclide ^{12}C (1 u = (1.6605402 ± 0.0000010) · 10^{-27} kg) [197], and equal to the numerical value of the molar mass M in g/mol.

Elements marked with an asterisk (*) do not have stable nuclides.

Name	Icon	Proton number	Relative atomic mass A_r
Actinium*	Ac	89	
Aluminum	Al	13	26.9815386 ± 8
Americium*	Am	95	
Antimony	Sb	51	121.760 ± 1^g
Argon	Ar	18	39.948 ± $1^{g, r}$
Arsenic	As	33	74.92160 ± 2
Astatine*	At	85	
Barium	Ba	56	137.327 ± 7
Berkelium*	Bk	97	
Beryllium	Be	4	9.012182 ± 3
Bismuth	Bi	83	208.98040 ± 1
Bohrium*	Bh	107	
Boron	B	5	10.811 ± $7^{m, r, g}$
Bromine	Br	35	79.904 ± 1
Cadmium	Cd	48	112.411 ± 8^g
Cesium	Cs	55	132.9054519 ± 2
Calcium	Ca	20	40.078 ± 4^g
Californium*	Cf	98	
Carbon	C	6	12.0107 ± $8^{g, r}$
Cerium	Ce	58	140.116 ± 1^g
Chlorine	Cl	17	35.453 ± $2^{g, m, r}$
Chrome	Cr	24	51.9961 ± 6
Cobalt	Co	27	58.933195 ± 5
Copper	Cu	29	63.546 ± 3^r
Curium*	Cm	96	
Darmstadtium*	Ds	110	
Dubnium*	Db	105	
Dysprosium	Dy	66	162.500 ± 1^g
Einsteinium*	Es	99	
Erbium	Er	68	167.259 ± 3^g

(continued)

Name	Icon	Proton number	Relative atomic mass A_r
Europium	Eu	63	151.964 ± 1^g
Fermium*	Fm	100	
Fluorine	F	9	18.9984032 ± 5
Francium*	Fr	87	
Gadolinium	Gd	64	157.25 ± 3^g
Gallium	Ga	31	69.723 ± 1
Germanium	Ge	32	72.64 ± 1
Gold	Au	79	196.966569 ± 4
Hafnium	Hf	72	178.49 ± 2
Hassium*	Hs	108	
Helium	He	2	$4.002602 \pm 2^{g,\,r}$
Holmium	Ho	67	164.93032 ± 2
Hydrogen	H	1	$1.00794 \pm 7^{g,\,m,\,r}$
Indium	In	49	114.818 ± 3
Iodine	I	53	126.90447 ± 3
Iridium	Ir	77	192.217 ± 3
Iron	Fe	26	55.845 ± 2
Krypton	Kr	36	$83.798 \pm 2^{g,\,m}$
Lanthanum	La	57	138.9047 ± 7^g
Lawrencium*	Lr	103	
Lead	Pb	82	$207.2 \pm 1^{g,\,r}$
Lithium	Li	3	$6.941 \pm 2^{g,\,m,\,r}$
Lutetium	Lu	71	174.967 ± 1^g
Magnesium	Mg	12	24.3050 ± 6
Manganese	Mn	25	54.938045 ± 5
Meitnerium*	Mt	109	
Mendelevium*	Md	101	
Mercury	Hg	80	200.59 ± 3
Molybdenum	Mo	42	95.94 ± 1^g
Neodymium	Nd	60	144.242 ± 3^g
Neon	Ne	10	$20.1797 \pm 6^{g,\,m}$
Neptunium*	Np	93	
Nickel	Ni	28	58.6934 ± 2
Niobium	Nb	41	92.90638 ± 2
Nitrogen	N	7	$14.0067 \pm 2^{g,\,r}$
Nobelium*	No	102	
Osmium	Os	76	190.23 ± 3^g
Oxygen	O	8	$15.9994 \pm 3^{g,\,r}$
Palladium	Pd	46	106.42 ± 1^g
Phosphorus	P	15	30.973762 ± 2
Platinum	Pt	78	195.084 ± 9
Plutonium*	Pu	94	
Polonium*	Po	84	
Potassium	K	19	39.0983 ± 1
Praseodymium	Pr	59	140.90765 ± 2
Promethium	Pm	61	

(continued)

Name	Icon	Proton number	Relative atomic mass A_r
Protactinium*	Pa	91	231.03588 ± 2^Z
Radium*	Ra	88	
Radon*	Rn	86	
Rhenium	Re	75	186.207 ± 1
Rhodium	Rh	45	102.90550 ± 2
Roentgenium*	Rg	111	
Rubidium	Rb	37	85.4678 ± 3^g
Ruthenium	Ru	44	101.07 ± 2^g
Rutherfordium*	Rf	104	
Samarium	Sm	62	150.36 ± 3^g
Scandium	Sc	21	44.955912 ± 6
Seaborgium*	Sg	106	
Selenium	Se	34	78.96 ± 3^r
Silicon	Si	14	28.0855 ± 3^r
Silver	Ag	47	107.8682 ± 2^g
Sodium	Na	11	22.989776928 ± 2
Strontium	Sr	38	$87.62 \pm 1^{g,\ r}$
Sulfur	S	16	$32.065 \pm 5^{g,\ r}$
Tantalum	Ta	73	180.9488 ± 2
Technetium*	Tc	43	
Tellurium	Te	52	127.60 ± 3^g
Terbium	Tb	65	158.92535 ± 2
Thallium	Tl	81	204.3833 ± 2
Thorium*	Th	90	$232.03806 \pm 2^{g,\ Z}$
Thulium	Tm	69	168.93421 ± 2
Tin	Sn	50	118.710 ± 7^g
Titanium	Ti	22	47.867 ± 1
Tungsten	W	74	183.84 ± 1
Uranium*	U	92	$238.02891 \pm 1^{g,\ m,\ Z}$
Vanadium	V	23	50.9415 ± 1
Xenon	Xe	54	$131.293 \pm 6^{g,\ m}$
Ytterbium	Yb	70	173.04 ± 3^g
Yttrium	Y	39	88.90585 ± 2
Zinc	Zn	30	65.409 ± 4
Zirconium	Zr	40	91.224 ± 2^g

[g]Geologically unusual samples are known in which the element has an isotopic composition outside the limits for normal material. The difference between the atomic mass of the element in such samples and that given in the table can considerably exceed the stated uncertainty.

[m]Modified (altered) isotopic compositions can be found in commercially available material because it has been subjected to an undisclosed or unknown isotope separation.

[r]Variations in isotopic composition in normal terrestrial material prevent more accurate values than those given. The table values should be applicable to all normal materials.

[Z]Elements without stable nuclides. However, long-lived nuclides of these elements occur in characteristic compositions, so a relative atomic mass can certainly be specified.

Bibliography

The compilation contains textbooks, monographs and original works without any claim to completeness. The works of a general nature are arranged by subject and preceded by the citations referred to in the individual chapters and sections. Under each heading, you will find references to the works in the general literature section that are relevant for further information.

General literature

Analytical chemistry

Manuals, tables and reference works

[1]   Fresenius, W., Jander, G. (Eds.), Handbuch der Analytischen Chemie, Springer-Verlag, Berlin/
        Heidelberg/New York from 1940.
[2]   Meyers, R. A. (Ed.), Encyclopedia of Analytical Chemistry, Wiley, London/New York 2011.
[3]   Günzler, H., Williams, A. (Eds.), Handbook of Analytical Techniques, Wiley-VCH, Weinheim 2001.
[4]   D'Ans, J., Lax, E., Taschenbuch für Chemiker und Physiker, Vol. 1, Springer-Verlag, Berlin/
        Heidelberg/New York: Physikalisch-chemische Daten, Lechner, M. D. (Ed.), 4th ed. 1992 Vol. 3:
        Elemente, anorganische Verbindungen und Materialien, Minerale, Blachnik, R. (Ed.), 4th rev.
        ed. 2012.
[5]   Küster, F. W., Thiel, A., Rechentafeln für die Chemische Analytik, 107th edition, edited by Ruland,
        A., Ruland, U., Verlag Walter de Gruyter, Berlin/New York 2011.
[5a]   Rauscher, K., Voigt, J., Wilke, I., Chemische Tabellen und Rechentafeln für die analytische Praxis,
        11th corr. ed., continued by Friebe, R., Verlag Harri Deutsch, Thun/Frankfurt am Main 2000.
[6]   Haynes, W. M. (Ed.). CRC Handbook of Chemistry and Physics, 96th edition, CRC Press, Boca
        Raton, FL 2015.
[7]   DIN Deutsches Institut für Normung e. V. (Ed.). DIN-Taschenbuch 22, Einheiten und Begriffe für
        physikalische Größen, 9th edition, Beuth Verlag, Berlin April 2009.
[8]   Liebscher, W., Fluck, E., Die systematische Nomenklatur der anorganischen Chemie, Springer-
        Verlag, Berlin/Heidelberg/New York 1999. Liebscher, W./GDCh (Ed.), Nomenclature of Inorganic
        Chemistry, Wiley-VCH, Weinheim 2009. IUPAC/Homann, K.-H. (Ed.), Quantities, Units and Symbols
        in Physical Chemistry, Wiley-VCH, Weinheim 1996.
[8a]   Pharmacopoeia Europaea. (European Pharmacopoeia) 10.0-10.1, basic edition 2020 incl. 1st
        supplement, Deutscher Apotheker Verlag, Stuttgart 2020.
[8b]   Hager's Handbook of Pharmaceutical Practice, Springer Verlag, Berlin 1998.
[8c]   Bracher, F., Heisig, P., Langguth, P., Mutschler, E., Schirmeister, T., Scriba, G. K. E., Stahl-Biskup, E.,
        Troschütz, R., Arzneibuch-Kommentar, Wissenschaftliche Erläuterungen zum Arzneibuch, 1st ed.
        incl. 66th act. lfg. 2020, Wissenschaftliche Verlagsgesellschaft, Stuttgart.
[8d]   Roth, H. J., Blaschke, G., Pharmazeutische Analytik, 2nd edition, Thieme Verlag, Stuttgart 1981.
[8e]   Surmann, P., Quantitative Analysis of Drug Substances and Drug Preparations, Wissenschaftliche
        Verlagsgesellschaft, Stuttgart 1987.
[8f]   Böhme, H., Hartke, K., Deutsches Arzneibuch. 7th edition 1968. commentary, Wissenschaftliche
        Verlagsgesellschaft Stuttgart; Govi-Verlag Frankfurt, 1973.
[8g]   Kaiser, H., Pharmazeutisches Taschenbuch, Wissenschaftliche Verlagsgesellschaft, Stuttgart 1968.

https://doi.org/10.1515/9783111350127-009

Theoretically oriented works

[9] Seel, F., Grundlagen der analytischen Chemie, 7th edition, Verlag Chemie, Weinheim 1979.

[10] Fluck, E., Becke-Goehring, M., Einführung in die Theorie der quantitativen Analyse, 7th edition, Steinkopff Verlag, Darmstadt 2013.

[11] Kunze, U. R., Schwedt, G., Grundlagen der qualitativen und quantitativen Analyse, 6th updated and supplemented edition, Wiley-VCH, Weinheim 2009.

[12] Latscha, H. P., Klein, H. A., Linti, G., Analytische Chemie (Chemie-Basiswissen III), 4th fully revised edition, Springer Verlag, Berlin/Heidelberg/New York 2004.

[13] Danzer, K., Than, E., Molch, D., Analytik. Systematischer Überblick, 2nd edition, Wissenschaftliche Verlagsgesellschaft, Stuttgart 1998.

[14] Doerffel, K., Geyer, R., Müller, H. (Eds.). Analytikum, 9th revised edition, Wiley-VCH, Weinheim 1994.

[15] Kolditz, L. (Ed.). Anorganikum, Part 2, 13th revised edition, Wiley-VCH, Weinheim 1993.

[16] Schwedt, G., Analytical Chemistry. Grundlagen, Methoden und Praxis, 2nd fully revised and expanded edition, Wiley-VCH, Weinheim 2008.

[17] Skoog, D. A., Fundamentals of Analytical Chemistry, 9th edition, Thompson Learning, London 2013.

[18] Harris, D. C., Quantitative Chemical Analysis, 9th edition, W. H. Freeman, New York 2015; Harris, D. C., Lehrbuch der Quantitativen Analyse, German translation of the 8th edition, Springer Verlag, Berlin/Heidelberg/New York 2014.

[19] Fritz, J. S., Schenk, G. H., Lüderwald, I., Gros, L., Quantitative Analytical Chemistry, Fundamentals, Methods, Experiments, Vieweg & Sohn, Braunschweig/Wiesbaden 1989.

[20] Otto, M., Analytical Chemistry, 4th revised and amended edition, Wiley-VCH, Weinheim 2011.

[21] Kellner, R., Mermet, J.-M., Otto, M., Valcárcel, M., Widmer, H. M. (Eds.). Analytical Chemistry, 2nd edition, Wiley-VCH, Weinheim 2004.

[22] Skoog, D. A., Leary, J. J., Instrumentelle Analytik. Grundlagen, Geräte, Anwendungen, Springer Verlag, Berlin/Heidelberg/New York 1996.

[23a] Cammann, K. (Ed.), Instrumentelle Analytische Chemie. Verfahren, Anwendungen, Qualitätssicherung, Spektrum Akademischer Verlag, Heidelberg/Berlin, 2000.

[23b] Gernand, W., Sommer, M.-J., Steckenreuter, K., Wieland, G. (Eds.), The ABC of Titration, Merck KGaA, Darmstadt.

Practically oriented works

[24] Jander, G., Blasius, E., Anorganische Chemie I – Einführung & Qualitative Analyse, 19th edition, completely revised by Schweda, E., Hirzel Verlag, Stuttgart 2022.

[24a] Jander, G., Blasius, E., Anorganische Chemie II – Quantitative Analyse & Präparate, 18th ed., completely revised by Schweda, E., Hirzel Verlag, Stuttgart 2022.

[24b] Jander, G., Blasius, E., Einführung in das anorganisch-chemische Praktikum (einschließlich der quantitativen Analyse), 13th edition, revised by Strähle, J., Schweda, E., Hirzel Verlag, Stuttgart 1990.

[24c] Metrohm Application Bulletin No. 233/4e, Detergents.

[25] Müller, G.-O., Lehr- und Übungsbuch der anorganisch-analytischen Chemie, Vol. 3, Quantitativ-anorganisches Praktikum, 7th completely revised edition, Verlag Harri Deutsch, Thun/Frankfurt am Main 1992.

[26] Lux, H., Fichtner, W., Praktikum der quantitativen anorganischen Analyse, 9. neubearb. edition, Springer Verlag, Berlin/Heidelberg/New York 1992.

[27] Jander, G. (Ed.), Neuere maßanalytische Methoden, 4th edition. (Die chemische Analyse, 33rd vol.), Ferdinand Enke Verlag, Stuttgart 1956.

[28]   Poethke, W., Kupferschmid, W., Praktikum der Maßanalyse, 3rd revised edition, Verlag Harri
         Deutsch, Thun and Frankfurt am Main 1987.
[29]   Gübitz, T., Haubold, G., Stoll, C., Analytisches Praktikum: Quantitative Analyse, 2nd edition, Wiley-
         VCH, Weinheim 1993.
[30]   Mendham, J., Denney, R. C., Barnes, J. D., Thomas, M. J. K. (Eds.), Vogel's Textbook of Quantitative
         Inorganic Analysis (including Elementary Instrumental Analysis), 6th edition, Pearson
         Education, 2009.
[30a]   Martens-Menzel, R., Physikalische Chemie in der Analytik, 2nd edition, Vieweg+Teubner,
         Wiesbaden 2011.

Inorganic chemistry

[31]   Blaschette, A., Chemie, A., Allgemeine Chemie, Bd. 1: Atome, Moleküle, Kristalle, 2nd edition, Bd.
         2: Chemische Reaktionen, 3rd ed. 1993, Aula-Verlag, Wiesbaden 1993.
[32]   Christen, H. R., Meyer, G., Grundlagen der allgemeinen und anorganischen Chemie, Moritz
         Diesterweg/Otto Salle. Verlag, Frankfurt am Main and Verlag Sauerländer, Aarau 1997.
[33a]   Latscha, H. P., Klein, H. A., Anorganische Chemie (Chemie-Basiswissen I), 10th edition, Springer
         Verlag, Berlin/Heidelberg/New York 2011.
[33b]   Binnewies, M., Jäckel, M., Willner, H., Rayner-Canham, G., Allgemeine und Anorganische Chemie,
         3rd edition, Spektrum Akademischer Verlag, Heidelberg/Berlin, 2016.
[34]   Riedel, E., Janiak, C., Inorganic Chemistry, 9th edition, Walter de Gruyter, Berlin/New York 2015.
[35]   Holleman, A. F., Wiberg, E., Wiberg, N., Lehrbuch der Anorganischen Chemie, 103. umgearb.
         u. verb. edition, Walter de Gruyter, Berlin/New York 2016.

Physical chemistry

[36]   Atkins, P. W., De paula, J., Physical Chemistry, 5th edition, Wiley-VCH, Weinheim 2013.
[37]   Czeslik, C., Seemann, H., Winter, R., Basiswissen Physikalische Chemie, 4th edition,
         Vieweg+Teubner, Wiesbaden 2010.
[38]   Wedler, G., Lehrbuch der Physikalischen Chemie, 6th fully revised edition, Wiley-VCH,
         Weinheim 2012.
[39]   Hamann, C. H., Vielstich, W., Elektrochemie, 5th edition, Wiley-VCH, Weinheim 2005.

Text-related literature

1 Introduction and basic concepts
[3, 9, 13, 14, 16, 18, 23b, 24, 30]

2 Practical basics of volumetric analysis

[40]   Bureau International des Poids et Mesures. Le Système International d'Unités (SI), 6th edition,
         1991. Taylor, B. N., Guide for the Use of the International System of Units (SI), National Institute of
         Standards and Technology (NIST), Gaithersburg 1995. Physikalisch-Technische Bundesanstalt
         (Ed.), The SI base units. Definition, Development, Realization, Braunschweig/Berlin 1997.
         Physikalisch-Technische Bundesanstalt (Ed.), Leitfaden für den Gebrauch des Internationalen
         Einheitensystems, Braunschweig/Berlin 1998.

[41]   Act on Units of Measurement of 22. 2. 1985, Federal Law Gazette 1985, Part I, 408. Implementing ordinance to the law on units of measurement of 13. 12. 1985 Amendment ordinance of 22. 3. 1991. Sacklowski, A., Einheitenlexikon: Entstehung, Anwendung, Erläuterung von Gesetz und Normen, neubearb. von Drath, P., DIN Deutsches Institut für Normung (Hrsg.), Beuth-Verlag, Berlin/Köln 1986.

[42]   DIN 1301-1, Units – Part 1: Unit names, unit symbols, October 2010. DIN 1301–2, Units – Part 2: Submultiples and multiples for general use, February 1978.

[43]   Haeder, W., Gärtner, E., Die gesetzlichen Einheiten in der Technik, 5th edition, Beuth-Verlag, Berlin/Cologne 1980. Bender, D., Pippig, E., Einheiten, Maßsysteme, SI, 5th edition, Akademie Verlag, Berlin 1986.

[44]   Springer, G. (Ed.), Größen, Formelzeichen, Einheiten in Naturwissenschaften und Technik, Verlag Europa-Lehrmittel, Nourney, Vollmer GmbH & Co, Haan-Gruiten, 1991 Fischer, R., Vogelsang, K., Größen und Einheiten in Physik und Technik, 6th completely revised and expanded edition, Verlag Technik GmbH, Berlin/Munich 1993.

[45]   Aylward, G. H., Findlay, T. J. V., Data Collection Chemistry in SI Units, 3rd revised and revised., Wiley-VCH, Weinheim 1999.

2.1 Devices for volume measurement

[17, 18, 19, 29, 30]

[46]   Scholze, H., Glas – Natur, Struktur und Eigenschaften, 3rd edition, Springer Verlag, Berlin/ Heidelberg/New York 1988. DIN ISO 3585, Borosilicate glass 3.3 – Properties, October 1999; identical to ISO 3585, July 1998.

[47]   DIN 12600, Laboratory volumetric instruments; conformity testing and certification, April 1990.

[48]   DIN EN ISO 1042, Laboratory glassware – One-mark volumetric flasks, August 1999.

[49]   DIN 12242-1, Laboratory glassware – Interchangeable conical ground joints, dimensions, tolerances, July 1980.

[50]   DIN 12252, Laboratory glassware; interchangeable ground stoppers, April 1979.

[51]   DIN EN ISO 4788, Laboratory glassware – Graduated measuring cylinders, August 2005.

[52]   DIN EN ISO 384. Laboratory glass and plastics ware - Principles of design and construction of volumetric instruments (ISO 384:2015), May 2016.

[53]   DIN 12681, Plastics laboratory ware – Graduated measuring cylinders, March 1998.

[54]   DIN EN ISO 648, Laboratory glassware – Single-volume pipettes, January 2009.

[55]   Kühne, W. H., Überlegungen zur Volumenmessung (1970), Information on volume measurement (2006), Schriften der Firma Brand GmbH, Wertheim.

[56]   DIN EN ISO 835, Laboratory glassware – Graduated pipettes, July 2007.

[57]   DIN 12621, Laboratory glassware – Colour marking of pipettes; arrangements, dimensions, characterizing colours, July 1971.

[58]   German Social Accident Insurance. Expert Committee Chemistry, Safe Working in Laboratories - Principles and Guidelines, DGUV Information 213-850, previously BGI/GUV - I-850-0, version 10/ 2011, edition 2, reprint with editorial changes 03/2015, Jedermann-Verlag Heidelberg.

[59]   Schirm, P., Jahns, A., Dosiersysteme im Labor – Technologie und Anwendung, 3. vollst. überarb. und erweitert. Edition, by Ewald, K., Verlag moderne Industrie, Landsberg 2005.

[60]   DIN EN ISO 8655, Piston-operated volumetric apparatus. Part 1: Definitions, general requirements and recommendations for use (ISO 8655-1:2002), Corrigendum 1:2009-07; Part 2: Piston-operated pipettes (ISO 8655-2:2002), Corrigendum 1:2009-07; Part 3: Piston burettes (ISO 8655-3:2002), Corrigendum 1:2009-07; Part 4: Dilutors (ISO 8655-4:2002), Corrigendum 1:2009-07; Part 5:

Dispensers (ISO 8655-5:2002), Corrigendum 1:2009-07; Part 6: Gravimetric test methods (ISO 8655-6:2002), Corrigendum 1:2009-07; Part 7: Non-gravimetric test methods (ISO 8655-7:2005), Corrigendum 1:2009–07.

[61] DIN EN ISO 385, Laboratory glassware – Burettes, July 2005.

[62a] Friedman, H. B., LaMer, V. K., Ind. Engng. Chem., Anal. Ed. 2(54), (1930).

[62b] Hahn, F. L., Z. Anal. Chem. 167(104), (1959). Rice, T. D., Anal. Chim. Acta 97, 213 (1976).

[62c] Szebelledy, L., Clauder, O., Z Anal. Chem. 105(24), (1936).

[62d] Rellstab, W., GIT Fachz. Lab. 13, 1053 (1969).

[62e] Kratochvil, B., Maitra, C., Internat. Lab. 4(24), (1983).

[62f] Saur, D., Spahn, E., GIT Fachz. Lab. 38(934), (1994). Spahn, E., Saur, D., CLB-Chem. in Lab. u. Biotech. 47, 6 (1996). Spahn, E., Galvanotechnik 88, 3 (1997).

[63] Kromidas, S. (Ed.), Qualität im analytischen Labor, Verlag Chemie, Weinheim 1995. Kromidas, S., Validation in analytics, Wiley-VCH, Weinheim 2011.

[64] DIN EN ISO 4787, Laboratory glass and plastic ware – Volumetric instruments – Methods for testing of capacity and for use (ISO 4787: 2010, corrected version 2010-06-15), May 2011.

[65] Doerffel, K., Statistik in der analytischen Chemie, 5th revised edition, Wiley-VCH, Weinheim 1990. Gottwald, W., Statistics for Users, Wiley-VCH, Weinheim 2000.

2.2 Solutions for volumetric analysis

[17, 19, 24, 25, 29, 30]

[66] Seel, F., Valenztheoretische Begriffe, Angew. Chem. 66, 581 (1954).

[67] DIN 32625, Quantities and units in chemistry – quantities of substances and quantities derived therefrom, terms and definitions, December 1989, withdrawn 2004.

[68] DIN 32629, Portion of substance; concept, characterization, November 1988, withdrawn 2004.

[69] Weninger, J., Stoffportion, Stoffmenge und Teilchenmenge, Diesterweg Salle, Frankfurt am Main/ Berlin/Munich, 1970.

[70] Cordes, J. F., Größen- und Einheitensysteme; SI-Einheiten, in: Bock, R., Fresenius, W., Günzler, H., Huber, W., Tölg, G. (Eds.), Analytiker-Taschenbuch, Vol. 2, Springer-Verlag, Berlin/Heidelberg/ New York 1981.

[70a] Physikalisch-Technische Bundesanstalt, 2019, web link:

[71] DIN 1310, Composition of (gaseous, liquid and solid) mixtures; concepts, symbols, February 1984.

[72] Kullbach, W., Mengenberechnungen in der Chemie, Verlag Chemie, Weinheim 1980.

[73] Brinkmann, H., Rechnen mit Größen in der Chemie, Verlag Moritz Diesterweg/Otto Salle Verlag, Frankfurt am Main and Verlag Sauerländer, Aarau 1980.

2.3 Statement of the analysis result

[73a] Brinkmann, B., International Dictionary of Metrology. Basic and general terms and associated designations (VIM) – German-English version ISO/IEC Guide 99:2007, Beuth Verlag GmbH, 2012.

[73b] JCGM 104:2009, Evaluation of measurement data – An introduction to the "Guide to the expression of uncertainty in measurement" and related documents, German edition, 2011-03-30.

[73c] Krystek, M., Calculation of measurement uncertainty, 3rd edition, Beuth Verlag GmbH, 2020.

[73d] Ellison, S. L. R., Williams, A. (Eds.), Eurachem/CITAC Guide CG4: Eurachem/CITAC, Quantifying uncertainty in analytical measurement, 3rd edition, Eurachem, 2012.

[73e] DIN ISO 5725-1, Accuracy (trueness and precision) of measurement procedures and results, Part 1: General principles and terminology. November 1997.

[73f] ISO 16269-4, Statistical interpretation of data – Part 4: Detection and treatment of outliers, October 2010.

[73g]   Grubbs, F. E., Beck, G., Technometrics. 14, 847 (1972).

[73h]   DIN 1333, Presentation of numerical data, February 1992.

3 Volumetric analysis with chemical end point determination

3.1 Acid-base titrations

[8a-12, 15, 17–19, 24–26, 28–30]

[74]   Bliefert, C., pH-value calculations, Verlag Chemie, Weinheim 1978.

[75]   Kielland, J., J. Amer. Chem. Soc. 59, 1675 (1937).

[76]   Sörensen, S. P. L., Biochem. Z. 21, 131 (1909).

[77]   Merck company, buffer substances, buffer solutions, buffer titrisols, Darmstadt.

[78]   DIN 19260, pH measurement – General terms and definitions, October 2012.

[79]   DIN 19266, pH measurement – Reference buffer solutions for the calibration of pH measuring equipment, May 2015.

[80]   Stegemann, K., Kienbaum, F., Verwendung von Farb- und Fluoreszenzindikatoren bei der Acidi- und Alkalimetrie (Titrationsfehler), in: [27].

[81]   Rast, K., Indikatoren und Reagenspapiere, in: Müller, E. (Ed.), Methoden der organischen Chemie (Houben-Weyl), 4th edition, Vol. III/2, Georg Thieme Verlag, Stuttgart 1962.

[82]   Bishop, E. (Ed.), Indicators, Pergamon Press, Oxford/New York 1972.

[83]   Schmidt, V., Mayer, W. D., Indikatoren und ihre Eigenschaften, in: Bock, R., Fresenius, W., Günzler, H., Huber, W., Tölg, G., (eds.), Analytiker-Taschenbuch, Vol. 2 (1981) and Vol. 3 (1983), Springer-Verlag, Berlin/Heidelberg/New York.

[84]   Company Riedel-de Haën, indicators, indicator and reagent papers, Seelze/Hanover.

[85]   Demuth, R., Kober, F., Grundlagen der Spektroskopie, Verlag Moritz Diesterweg/Otto Salle Verlag, Frankfurt am Main/Berlin/Munich and Verlag Sauerländer, Aarau/Frankfurt am Main/ Salzburg. 1977.

[86]   Hesse, M., Meier, H., Zeeh, B., Spektroskopische Methoden in der organischen Chemie, 6th edition, Georg Thieme Verlag, Stuttgart/New York 2002.

[87]   Williams, D. H., Fleming, I., Strukturaufklärung in der organischen Chemie. An introduction to spectroscopic methods, 6th edition, transl. and ed. by Zeller, K. P., Wiley-VCH, Weinheim 1991.

[88]   Staab, H. A., Einführung in die thoretische organische Chemie, 3rd reprint of the 4th edition, Verlag Chemie, Weinheim 1975.

[89]   Kolthoff, I. M., Die Maßanalyse, 2nd edition, vol. I (1930), vol. II (1931), Springer-Verlag, Berlin.

[90]   Kolthoff, I. M., Säure-Base-Indikatoren, Springer-Verlag, Berlin 1932.

[91]   Sörensen, S. P. L., Andersen, A. C., Z. Anal. Chem. 44, 156 (1905).

[92]   Incze, G., Z. Anal. Chem. 56, 177 (1917).

[93]   Küster, F. W., Z. anorg. allg. Chem. 13(134), (1897).

[94]   Sörensen, S. P. L., Biochem. Z. 21(186), (1909).

[95]   Winkler, C., Praktische Übungen in der Maßanalyse, 5th edition, edited by Brunck, O., Verlag Felix, Leipzig 1920.

[96]   Kjeldahl, J., Z. Anal. Chem. 22(366), (1883).

[96a]   Görlitzer, K., PharmuZ, 1976 /5. 145 (1976).

[97]   Schäfer, H., Sieverts, A., Z. Anal. Chem. 121, 170 (1941).

[98]   Blasius, E., The use of ion exchangers in volumetric analysis, in: [27].

[98a]   Eger, K., Troschütz, R., Roth, H. J., Arzneistoffanalytik, 4th edition, Wissenschaftliche Verlagsgesellschaft, Stuttgart 1999.

3.2 Precipitation titrations

[8a-12, 15, 17–19, 24–26, 28–30]

[99]   Gay-Lussac, J. L., Vollständiger Unterricht über das Verfahren Silber auf nassem Weg zu probieren, Friedrich Vieweg, Braunschweig 1833.

[100]   Liebig, J., Ann. Chem. Pharm. 77(102), (1851).

[101]   Volhard, J., J. prakt. Chem. 117(217), (1874).

[102]   Mohr, F., Lehrbuch der chemisch-analytischen Titrirmethode, Friedrich Vieweg und Sohn, Braunschweig 1855.

[103]   Fajans, K., Hassel, O., Z. Elektrochem. angew. phys. Chem. 29(495), (1923). 221. Fajans, K., Wolff, H., Z. anorg. allg. Chem. 137, 221 (1924).

[104]   Fajans, K., Adsorption indicators for precipitation titrations, in: [27.

[105]   Caldwell, J. R., Moyer, H. V., Ind. Eng. Chem., Anal. Edit. 7(38), (1935).

[106]   Alary, J., Bourbon, P., Escrient, C., Vandaele, J., Fresenius, Z., Anal. Chem. 322, 777 (1985).

3.3 Oxidation and reduction titrations

[9–12, 15, 17–19, 24–26, 28–30]

[107]   Pauling, L., Die Natur der chemischen Bindung, 2nd reprint of the 3rd edition, Verlag Chemie, Weinheim 1968. 1976.

[108]   Jones, C., Chem. News. 60(163), (1889).

[109]   Stegemann, K., Metals as reducing agents (metal reducers), in: [27].

[110]   Zimmermann, C., Ber. dt. chem. Ges. 14, 779 (1881).

[111]   Laitinen, H. A., Chemical Analysis, McGraw-Hill, New York 1960.

[112]   Kolthoff, I. M., Stenger, V. A., Volumetric Analysis, Vol. 1, 2nd reprint of 2nd ed. (1947), vol. 2, 2nd ed. (1947), vol. 3, 2nd ed. (1957), Interscience Publishers, New York.

[113]   Reinhardt, C., Chemiker-Ztg, 13, 323 (1889).

[113a] Handbuch für das Eisenhüttenlaboratorium (Volume 2), VDEh, Düsseldorf, 1966.

[113b] DIN EN ISO 8467, Water quality – Determination of permanganate index, May 1995.

[114]   Furman, N. H., Cerium(IV) solutions as oxidizing agents for volumetric analysis, in: [27].

[115]   Brennecke, E., Oxidation-Reduction-Indicators, neubearb. von Blasius, E., in: [27].

[116]   Blasius, E., Wittwer, G., in: [115].

[117]   Cramer, F., Einschlussverbindungen, Springer-Verlag, Berlin/Göttingen/Heidelberg 1954.

[118]   Saenger, W., Naturwissenschaften. 71(31), (1984).

[118a] Stevens, J. W., Ind. Eng. Chem. Anal. Ed. 10(5), 269 (1938).

[118b] Silva, C. R., Simoni, J. A., Collins, C. H., Volpe, P. L. O., J. Chem. Educ. 76(10), 1421 (1999).

[118c] Suntornsuk, L., Gritsanapun, W., Nilkamhank, S., Paochom, A., J. Pharm. Biomed. Anal. 28(5), 849–855 (2002).

[119]   de Haën, E., Ann. Chem. Pharm. 91(237), (1854).

[120]   Bruhns, G., Chemiker-Ztg. 42(301), (1918).

[121]   Bastius, H., Fresenius Z. Anal. Chem. 250, 169 (1970).

[121a] Hütter, L. A., Wasser und Wasseruntersuchung, Salle+Sauerländer, Frankfurt/M, 1994.

[121b] DIN 38408–21, German standard methods for the examination of water, waste water and sludge; gaseous constituents (group G); determination of dissolved oxygen in water, Winkler iodometric method (G 21), May 1984

[121c] Procter Smith, H., Chem. News. 90(237), (1904).

[121d] Mukerjee, B. C., Analyst. 52, 689 (1927).

[121e] Stamm, H., Angew. Chem. 47, 791 (1934).

[121f] Stamm, H., Angew. Chem. 48, 150 (1935).

[121g]   Stamm, H., Die Reduktion von Permanganat zu Manganat als Grundlage eines maßanalytischen Verfahrens, in: [27].

[122]   Stegemann, K., Stegemann, E., Iodatometry, P., Iodatometry, Periodatometry, Bromatometry and Bromometry. Application for iodometric methods with special consideration of chloramine (sodium salt of p-toluenesulphone chloramide), in: [27].

[123]   Berka, A., Vulterin, J., Zýka, J., Maßanalytische Oxidations- und Reduktionsmethoden, Akademische Verlagsgesellschaft Geest und Portig, Leipzig 1964.

[124]   Stamm, H., Die Reduktion von Permanganat zu Manganat als Grundlage eines maßanalytischen Verfahrens, in: [27].

[125]   Jerschkewitz, H. G., Rienäcker, G., Titrations with chromium(II) and titanium(III) salt solutions, in: [27].

3.4 Complex formation titrations

[9–12, 15, 17–19, 20, 24–26, 30]

[126]   Fick, R., Ulrich, H., I. G. Farbenindustrie AG, D. R. P. 638071 dated 9. 11. 1936.

[127]   Ender, W., Fette u. Seifen 45, 144 (1938).

[128]   Schwarzenbach, G., Schweiz. Chemiker-Ztg. Techn.-Ind. 28, 181, 377 (1945).

[129]   Schwarzenbach, G., Helv. Chim. Acta. 29, 1338 (1946).

[130]   Schwarzenbach, G., Flaschka, H., Die komplexometrische Titration, 5th edition, Ferdinand Enke Verlag, Stuttgart 1965.

[131]   Schwarzenbach, G., Schneider, W., Komplexometrische Titrationsmethoden, in: [27].

[132]   Biedermann, W., Schwarzenbach, G., Chimia 2, 1, (1948).

[133]   Schwarzenbach, G., Biedermann, W., Helv. Chim. Acta . 31, 678 (1948).

[134]   Körbl, J., Přibil, R., Emr, A., Collection Czech. Chem. Commun. 22(961), (1957).

[135]   Přibil, R., Komplexometrie, Vol. I–IV, translated by Emr, A., VEB Deutscher Verlag für Grundstoffindustrie, Leipzig 1962–1966.

[136]   Kober, F., Grundlagen der Komplexchemie, Verlag Moritz Diesterweg/Otto Salle Verlag, Frankfurt am Main/Berlin/Munich and Verlag Sauerländer, Aarau/Frankfurt am Main/Salzburg 1979).

[137]   Umland, F., Janssen, A., Thierig, D., Wünsch, G., Theorie und Praktische Anwendung von Komplexbildern, Akademische Verlagsgesellschaft, Frankfurt am Main 1971.

[138]   Flaschka, H., Püschel, R., Z. Anal. Chem. 143, 330 (1954).

[139]   Merck, C. E., Complexometric Determination Methods with Titriplex, Darmstadt.

[140]   Company Riedel-de Haën, Idranal reagents for complexometry, Seelze/Hanover.

[140a] Pearson, R. G., J. Am. Chem. Soc. 85, 3533 (1963).

4 Volumetric analysis with physical end point determination

[2, 10–12, 14–21, 30, 39]

[141]   Kraft, G., Fischer, J., Indikation von Titrationen, Walter de Gruyter, Berlin/New York 1972.

[142]   Kraft, G., Elektrochemische Analyseverfahren, in: Kienitz, H., Bock, R., Fresenius, W., Huber, W., Tölg, G. (Eds.), Analytiker-Taschenbuch Bd. 1, Springer-Verlag, Berlin/Heidelberg/New York 1980.

[143]   Schumacher, E., Umland, F., Neue Titrationen mit elektrochemischer Enpunktsanzeige, in: Bock, R., Fresenius, W., Günzler, H., Huber, W., Tölg, G. (Eds.), Analytiker-Taschenbuch Bd, Vol. 2, Springer-Verlag, Berlin/Heidelberg/New York 1981.

[144]   Oehme, F., Richter, W., Instrumentelle Titrationstechnik, Dr. Alfred Hüthig Verlag, Heidelberg 1983.

[145]   Kortüm, G., Kolorimetrie,, Photometrie und Spektrometrie, 4th edition, Springer-Verlag, Berlin 1962.

[146]   Abrahamczik, E., Methoden der organischen Chemie, in: Houben-Weyl, M. E. (Ed.), Potentiometrische und konduktometrische Titrationen, 4th edition, Vol. III/2, Georg Thieme Verlag, Stuttgart 1962.

[147]   Cruse, K., Huber, R., Hochfrequenztitration, Verlag Chemie, Weinheim 1957.

[148]   Eisenman, G., Adv. Anal. Chem. Instrum. 4, 215 (1965).

[149]   Cammann, K., Galster, H., Das Arbeiten mit ionenselektiven Elektroden, 3rd edition, Springer-Verlag, Berlin/Heidelberg/New York 1996.

[149a] Murphy, J. J., Chem. Educ. 60(5), 420 (1983).

[150]   Scholz, E., Karl-Fischer-Titration, Methoden zur Wasserbestimmung, Springer-Verlag, Berlin/Heidelberg/New York 1984. Karl-Fischer-Titration.

[150a] Bruttel, P., Schlink, R., Wasserbestimmung durch Karl-Fischer-Titration, Metrohm Monogr. (2003).

[151]   Mitchell, J. Jr, Smith, D. M., Aquametry, Part III (The Karl Fischer Reagent), 2nd edition, John Wiley & Sons, New York 1980.

[152]   Eberius, E., Wasserbestimmungen mit Karl-Fischer-Lösung, Verlag Chemie, Weinheim 1958.

[153]   Abresch, K., Claasen, I., Die coulometrische Analyse, Verlag Chemie, Weinheim 1961.

[154]   Karsten, P., Kies, H. J., Bergshoeff, G., Chem. Weekblad. 48(734), (1952).

[155]   Gran, G., Analyst. 77, 661 (1952).

[156]   Johansson, A., Svensk Papperstidn. 50(no.11 B), 124 (1947).

[157]   Jungnickel, J. L., Peters, E. D., Anal. Chem. 27, 450 (1955).

[158a] Scholz, E., Fresenius, Z. Anal. Chem. 312(462), (1982).

[158h] Schöffski, K., Die Wasserbestimmung mit Karl-Fischer-Titration, Chem. unserer. Zeit. 34(3), 170 (2000).

5 Instrumental volumetric analysis

[159]   Oehme, F., Fresenius Z. Anal. Chem. 222(244), (1966).

[160]   Philips, J. P., Automatic Titrators, Academic Press, London 1954.

[161]   Squirrel, D. C. M., Automatic Methods in Volumetric Analysis, Hilger Watts, London 1964.

[162]   Tubbs, C. F., Anal. Chem. 26, 1670 (1954).

[163]   Ebel, S., Fresenius Z, Anal. Chem. 245(108), (1969).

[164]   Ebel, S., Parzefall, W., Experimentelle Einführung in die Potentiometrie, Verlag Chemie, Weinheim 1975.

[165]   Hahn, F. L., pH und potentiometrische Titrierungen, Akademische Verlagsgesellschaft, Frankfurt am Main 1964.

[166]   Ebel, S., Seuring, A., Angew. Chem. 89, 129 (1977).

6 Overview of the history of volumetric analysis

[167]   Szabadváry, F., Geschichte der analytischen Chemie, Friedr. Vieweg und Sohn, Braunschweig 1966. Szabadváry, F., Robinson, A., The History of Analytical Chemistry in: Wilson + Wilson's Comprehensive Analytical Chemistry, Vol. X, Elsevier Scientific Publishing Comp., Amsterdam/Oxford/New York 1980.

[168]   Rancke-Madsen, E., Development of Titrimetry Analysis till 1806, Copenhagen 1958.

[169]   Ostwald, W., Die wissenschaftlichen Grundlagen der analytischen Chemie, Verlag Wilhelm Engelmann, Leipzig 1894.

[170]   Hantzsch, A. R., Ber. dt. chem. Ges. 39(1084), (1906). 40, 3071 (1907), 41, 1187 (1908).

[171]   Michaelis, L., Oxidations- und Reduktionspotentiale, 2nd edition, Springer-Verlag, Berlin 1926 1933.

[172]   Fischer, K., Angew. Chem. 48(394) (1935).

[173]   Behrend, R., Z. Physik. Chem. 11(466), (1893).

[174]   Böttger, W., Z. Physik. Chem. 24, 253 (1897).

[175]   Crotogino, F., Z. Anorg. Chem. 24, 225 (1900).

[176]   Haber, F., Klemensiewicz, Z., Z. Physik. Chem. 67(385), (1909).

[177]   Müller, E., Die elektrometrische (potentiometrische) Maßanalyse, 7th edition, Verlag Theodor Steinkopff, Dresden/Leipzig 1944.

[178]   Küster, F. W., Grüters, M., Z. Anorg. Chem. 35(454), (1903).

[179]   Kolthoff, I. M., Konduktometrische Titrationen, Verlag Theodor Steinkopff, Dresden/Leipzig, 1923.

[180]   Jander, G., Manegold, E., Z. Anorg. Chem. 134, 283 (1924) Jander, G., Pfundt, O., Z. Anorg. Chem. 153, 219 (1926) Jander, G., Pfundt, O., Die konduktometrische Maßanalyse u. a. Anwendungen der Leitfähigkeitsmessungen auf chemischem Gebiet unter besonderer Berücksichtigung der visuellen Methode, Ferdinand Enke Berlag, Stuttgart 1945 Jander, G., Pfundt, O., Die Leitfähigkeitstitration, in: Böttger, W. (ed.), Physikalische Methoden der analytischen Chemie, Part 2, 2nd ed., Akademische Verlagsgesellschaft, Leipzig 1949.

[181]   Britton, H. T. S., Conductometric Analysis, Chapman & Hall, London 1934.

[182]   Blake, G. G., J. Sci. Instrum. 22, 174 (1945).

[183]   Jensen, F. W., Parrack, A. L., Ind. Eng. Chem. Anal. Ed. 18, 595 (1946).

[184]   Foreman, J., Crips, D. J., Trans. Faraday Soc. 42A(186), (1946).

[185]   Willard, H. H., Fenwick, F., J. Am. Chem. Soc. 44(2504), 2516 (1922).

[186]   Heyrovsky, J., Bull. Soc. Chim., Françe. 41, 1224 (1927).

[187]   Majer, V., Z. Elektrochem. angew. phys. Chem. 42(120), 123 (1936).

[188]   Salomon, E., Z. Phys. Chem. 24(55), (1897).

[189]   Nernst, W., Merriam, E. S., Z. phys. Chem. 53(235), (1905).

[190]   Foulk, C. W., Bawden, A. T., J. Am. Chem. Soc. 48, 2045 (1926).

[191]   Wernimont, G., Hopkinson, F. J., Ind. Eng. Chem. Anal. Ed. 15, 272 (1943).

[192]   Huber, W., Titration in nichtwässrigen Lösemitteln, Akademische Verlagsgesellschaft, Frankfurt am Main 1964. Gyenes, I., Titrations in nichtwässrigen Medien, Ferdinand Enke Verlag, Stuttgart 1970. Fritz, J. S., Acid-Base Titrations in Nonaqueous Solvents, Allyn and Bacon, Boston 1973. Safarik, L., Stransky, Z., Titrimetric Analysis in Organic Solvents, in: Comprehensive Analytical Chemistry, Vol. XXII, Elsevier, Amsterdam 1986.

[193]   Geil, J. V., Reger, H., Richter, W., Schäfer, J., Laborpraxis 5, 1032 (1981) 6, 24 (1982).

[194]   Efferenn, K., Reger, H., Schäfer, J., Laborpraxis. 23(12), 24 (1999).

[195]   Dettenrieder, A., Reger, H., Schiefke, GIT Fachz. Lab. 44, 906 (2000).

Appendix

[5, 7, 8]

[196]   International Union of Pure and Applied Chemistry, Inorganic Chemistry Division, Atomic weights of the elements 2005, Pure Appl. Chem. 78, 2051 (2006).

[197]   Codata Bulletin No. 63, Codata Secretariat, Paris November 1986.

Subject index

https://doi.org/10.1515/9783111350127-010

Persons index

https://doi.org/10.1515/9783111350127-011

www.ingramcontent.com/pod-product-compliance
Lightning Source LLC
Chambersburg PA
CBHW080704220326
41598CB00033B/5302